G. K. CHESTERTON

G. K. CHESTERTON

A Half Century of Views

EDITED BY
D. J. CONLON

Oxford   New York
OXFORD UNIVERSITY PRESS
1987

Oxford University Press, Walton Street, Oxford OX2 6DP
Oxford New York Toronto
Delhi Bombay Calcutta Madras Karachi
Petaling Jaya Singapore Hong Kong Tokyo
Nairobi Dar es Salaam Cape Town
Melbourne Auckland
and associated companies in
Beirut Berlin Ibadan Nicosia

Oxford is a trade mark of Oxford University Press

Introduction and selection © D. J. Conlon 1987

All rights reserved. No part of this publication may be reproduced, stored in a retrieval system, or transmitted, in any form or by any means, electronic, mechanical, photocopying, recording, or otherwise, without the prior permission of Oxford University Press

This book is sold subject to the condition that it shall not, by way of trade or otherwise, be lent, re-sold, hired out or otherwise circulated without the publisher's prior consent in any form of binding or cover other than that in which it is published and without a similar condition including this condition being imposed on the subsequent purchaser

British Library Cataloguing in Publication Data
G.K. Chesterton: a half century of views.
1. Chesterton, G.K.—Criticism and interpretation
I. Conlon, D.J.
828'.91209   PR4453.C4Z/
ISBN 0-19-212260-6

Library of Congress Cataloging in Publication Data
G.K. Chesterton: a half century of views.
1. Chesterton, G. K. (Gilbert Keith), 1874-1936—
Criticism and interpretation. I. Conlon, D. J.
PR4453.C4Z64625 1987   828'.91209   86-12458
ISBN 0-19-212260-6

Set by Hope Services
Printed in Great Britain by
Butler & Tanner Ltd.
Frome, Somerset

FOREWORD

A Composite Essay from G. K. Chesterton's
Autobiography

ART may be long, but schools of art are short and very fleeting; and there have been five or six since I attended an art school. Mine was the time of Impressionism; and nobody dared to dream there could be such a thing as Post-Impressionism or Post-Post-Impressionism. The very latest thing was to keep abreast of Whistler, and take him by the white forelock, as if he were Time himself. Since then that conspicuous white forelock has rather faded into a harmony of white and grey, and what was once so young has in its turn grown hoary. But I think there was a spiritual significance in Impressionism, in connection with this age as the age of scepticism. I mean that it illustrated scepticism in the sense of subjectivism. Its principle was that if all that could be seen of a cow was a white line and a purple shadow, we should only render the line and the shadow; in a sense we should only believe in the line and the shadow, rather than in the cow. In one sense the Impressionist sceptic contradicted the poet who said he had never seen a purple cow. He tended rather to say that he had only seen a purple cow; or rather that he had not seen the cow but only the purple. Whatever may be the merits of this as a method of art, there is obviously something highly subjective and sceptical about it as a method of thought. It naturally lends itself to the metaphysical suggestion that things only exist as we perceive them, or that things do not exist at all. The philosophy of Impressionism is necessarily close to the philosophy of Illusion. And this atmosphere also tended to contribute, however indirectly, to a certain mood of unreality and sterile isolation that settled at this time upon me; and I think upon many others.

What surprises me in looking back on youth, and even on boyhood, is the extreme rapidity with which it can think its way back to fundamental things; and even to the denial of fundamental things. At a very early age I had thought my way back to thought itself. It is a very dreadful thing to do; for it may lead to thinking that there is nothing but thought. At this time I did not very clearly distinguish between dreaming and waking; not only as a mood, but as a metaphysical doubt, I felt as if everything might be a dream. It was as if I had myself projected the universe from within, with all its trees and stars; and that is so near to the notion of being God that it is manifestly even nearer to going mad. Yet I was not mad, in any medical or physical sense; I was simply carrying the scepticism of my time as far as it would go. And I soon found it would go a great deal further than most of the sceptics went. While dull atheists came and explained to me that there was nothing but matter, I listened with a sort of calm horror of detachment, suspecting that there was nothing but mind. I have always felt that there was something thin and third-rate about materialists and materialism ever since. The atheist told me so pompously that he did not believe there was any God; and there were moments when I did not even believe there was any atheist.

And as with mental, so with moral extremes. There is something truly menacing in the thought of how quickly I could imagine the maddest, when I had never committed the mildest crime. Something may have been due to the atmosphere of the Decadents, and their perpetual hints of the luxurious horrors of paganism; but I am not disposed to dwell much on that defence; I suspect I manufactured most of my morbidities for myself. But anyhow, it is true that there was a time when I had reached that condition of moral anarchy within, in which a man says, in the words of Wilde, that 'Atys with the blood-stained knife were better than the thing I am.' I have never indeed felt the faintest temptation to the particular madness of Wilde; but I could at this time imagine the worst and wildest disproportions and distortions of more normal passion; the point is that the whole mood was overpowered and oppressed with a sort of congestion of imagination. As Bunyan, in his morbid period, described himself as prompted to utter blasphemies, I had an overpowering impulse to record or draw horrible ideas and images; plunging deeper and deeper as in a blind spiritual suicide. I dug quite low enough to discover the devil; and even in some dim way to recognize the devil. At least I never, even in this first vague and sceptical stage, indulged very much in the current arguments about the relativity of evil or the unreality of sin. Perhaps, when I eventually emerged as a sort of a theorist, and was described as an Optimist, it was because I was one of the few people in that world of diabolism who really believed in devils.

In truth, the story of what was called my Optimism was rather odd. When I had been for some time in these, the darkest depths of the contemporary pessimism, I had a strong inward impulse to revolt; to dislodge this incubus or throw off this nightmare. But as I was still thinking the thing out by myself, with little help from philosophy and no real help from religion, I invented a rudimentary and makeshift mystical theory of my own. It was substantially this: that even mere existence, reduced to its most primary limits, was extraordinary enough to be exciting. Anything was magnificent as compared with nothing. Even if the very daylight were a dream, it was a daydream; it was not a nightmare. The mere fact that one could wave one's arms and legs about (or those dubious external objects in the landscape which were called one's arms and legs) showed that it had not the mere paralysis of a nightmare. Or if it was a nightmare, it was an enjoyable nightmare. What I meant, whether or no I managed to say it, was this; that no man knows how much he is an optimist, even when he calls himself a pessimist, because he has not really measured the depths of his debt to whatever created him and enabled him to call himself anything. At the back of our brains, so to speak, there was a forgotten blaze or burst of astonishment at our own existence. The object of the artistic and spiritual life was to dig for this submerged sunrise of wonder; so that a man sitting in a chair might suddenly understand that he was actually alive, and be happy. When I did begin to write, I was full of a new and fiery resolution to write against the Decadents and the Pessimists who ruled the culture of the age.

I began by being what the pessimists called an optimist; I have ended by being what the optimists would very probably call a pessimist. And I have never in fact been either, and I have never really changed at all. I began by defending vermilion pillar-boxes and Victorian omnibuses although they were ugly. I have ended by denouncing modern advertisements or American films even when they are beautiful. The thing that I was trying to say then is the same thing that I am trying to say now; and even the deepest revolution of religion has only confirmed me in the desire to say it. For indeed, I never saw the two sides of this single truth stated together anywhere, until I happened to open the Penny Catechism and read the words, 'The two sins against Hope are presumption and despair.'

Long before this it was apparent that the centre of gravity in my existence had shifted from what we will (for the sake of courtesy) call Art to what we will (for the sake of courtesy) call Literature. The agent in this change of intention was, in the first instance, my friend Ernest Hodder Williams, afterwards the head of the well-known publishing firm. He was attending University College while I was attending, or not attending, to the art instructions of the Slade School. Hodder Williams and I often talked about literature, and he conceived a fixed notion that I could write; a delusion which he retained to the day of his death. In consequence of this, and in connection with my art studies, he gave me some books on art to review for the *Bookman*, the famous organ of his firm and family. I need not say that, having entirely failed to learn how to draw or paint, I tossed off easily enough some criticisms of the weaker points of Rubens or the misdirected talents of Tintoretto. I had discovered the easiest of all professions; which I have pursued ever since.

When I look back on these things, and indeed on my life generally, the thing that strikes me most is my extraordinary luck; but it is against all the proper principles that even any such measure of good fortune should have come to the Idle Apprentice. In the case of my association with Hodder Williams, it was against all reason that so unbusinesslike a person should have so businesslike a friend. In the case of the choice of a trade, it was outrageously unjust that a man should succeed in becoming a journalist merely by failing to become an artist. I say a trade and not a profession; for the only thing I can say for myself, in connection with both trades, is that I was never pompous about them. If I have had a profession, at least I have never been a professor. But in another sense there was about these first stages an element of luck, and even of accident. I mean that my mind remained very much abstracted and almost stunned; and these opportunities were merely things that happened to me, almost like calamities. To say that I was not ambitious makes it sound far too like a virtue, when it really was a not very disgraceful defect; it was that curious blindness of youth which we can observe in others and yet never explain in ourselves. But, above all, I mention it here also because it was connected with the continuity of that unresolved riddle in the mind, which I mentioned at the beginning of this chapter. The essential reason was that my eyes were turned inwards rather than outwards; giving my moral personality, I should imagine, a very unattractive squint. I was still oppressed with the metaphysical nightmare of negations about mind and matter, with the morbid imagery of evil, with the burden of my own mysterious brain and body; but by this time I was in revolt against them; and trying to construct a healthier conception of cosmic life, even if it were one that should err on the side of health. I even called myself an optimist, because I was so horribly near to being a pessimist. It is the only excuse I can offer. All this part of the process was afterwards thrown up in the very formless form of a piece of fiction called *The Man Who Was Thursday*. The title attracted some attention at the time; and there were many journalistic jokes about it. Some, referring to my supposed festive views, affected to mistake it for 'The Man Who Was Thirsty'. Others naturally supposed that Man Thursday was the black brother of Man Friday. Others again, with more penetration, treated it as a mere title out of topsyturvydom; as if it had been 'The Woman Who was Half-past Eight', or 'The Cow Who Was Tomorrow Evening'. But what interests me about it was this; that hardly anybody who looked at the title ever seems to have looked at the sub-title; which was 'A Nightmare', and the answer to a good many critical questions.

I pause upon the point here, because it is of some importance to the understanding of that time. I have often been asked what I meant by the monstrous pantomime ogre who was called Sunday in that story; and some have suggested, and in one sense not untruly, that he was meant for a blasphemous version of the Creator. But the point is that the whole story is a nightmare of things, not as they are, but as they seemed to the young half-pessimist of the Nineties; and the ogre who appears brutal but is also cryptically benevolent is not so much God, in the sense of religion or irreligion, but rather Nature as it appears to the pantheist, whose pantheism is struggling out of pessimism. So far as the story had any sense in it, it was meant to begin with the picture of the world at its worst and to work towards the suggestion that the picture was not so black as it was already painted. I explained that the whole thing was thrown out in the nihilism of the Nineties in the dedicatory lines which I wrote to my friend Bentley, who had been through the same period and problems; asking rhetorically: 'Who shall understand but you?' In reply to which a book reviewer very sensibly remarked that if nobody understood the book except Mr Bentley, it seemed unreasonable to ask other people to read it.

Amid all this scattered thinking, sometimes not unfairly to be called scatter-brained thinking, I began to piece together the fragments of the old religious scheme; mainly by the various gaps that denoted its disappearance. And the more I saw of real human nature, the more I came to suspect that it was really rather bad for all these people that it had disappeared. Many of them held, and still hold, very noble and necessary truths in the social and secular area. But even these it seemed to me they held less firmly than they might have done, if there had been anything like a fundamental principle of morals and metaphysics to support them. Men who believed ardently in altruism were yet troubled by the necessity of believing with even more religious reverence in Darwinism, and even in the deductions from Darwinism about a ruthless struggle as the rule of life. Men who naturally accepted the moral equality of mankind yet did so, in a manner, shrinkingly, under the gigantic shadow of the Superman of Nietzsche and Shaw. Their hearts were in the right place; but their heads were emphatically in the wrong place, being generally poked or plunged into vast volumes of materialism and scepticism, crabbed, barren, servile and without any light of liberty or of hope. I began to examine more exactly the general Christian theology which many execrated and few examined. I soon found that it did in fact correspond to many of these experiences of life; that even its paradoxes corresponded to the paradoxes of life.

It was about this time that I had published some studies on contemporary writers such as Kipling and Shaw and Wells; and feeling that each of them erred through an ultimate or religious error, I gave the book the title of *Heretics*. It was reviewed by Mr G. S. Street, the very delightful essayist, who casually used the expression that he was not going to bother about his theology until I had really stated mine. With all the solemnity of youth, I accepted this as a challenge; and wrote an outline of my own reasons for believing that the Christian theory, as summarized in the Apostles' Creed, would be found to be a better criticism of life than any of those that I had criticized. I called it *Orthodoxy*, but even at the time I was very much dissatisfied with the title. It sounded a thinnish sort of thing to be defending through thick and thin. Even then I fancy I had a dim foreshadowing that I should have to find some better name for it before I died.

But there did remain one rather vague virtue about the title, from my point of view; that it was provocative. And it is an exact test of that extraordinary modern society that it really was provocative. I had begun to discover that, in all that welter of inconsistent and incompatible heresies, the one and only really unpardonable heresy was orthodoxy. And through this experience I learned two very interesting things, which serve to divide all this part of my life into two distinct periods. Very nearly everybody, in the ordinary literary and journalistic world, began by taking it for granted that my faith in the Christian creed was a pose or a paradox. The more cynical supposed that it was only a stunt. The more generous and loyal warmly maintained that it was only a joke. It was not until long afterwards that the full horror of the truth burst upon them; the disgraceful truth that I really thought the thing was true. Critics were almost entirely complimentary to what they were pleased to call my brilliant paradoxes; *until* they discovered that I really meant what I said. Since then they have been more combative; and I do not blame them.

I first made this discovery at a dinner party given by the staff of the *Clarion*, the important and popular Socialist paper of the period. I remember that there was, sitting next to me at this dinner, one of those very refined and rather academic gentlemen from Cambridge who seemed to form so considerable a section of the rugged stalwarts of Labour. There was a cloud on his brow, as if he were beginning to be puzzled about something; and he said suddenly, with abrupt civility, 'Excuse my asking, Mr Chesterton, of course I shall quite understand if you prefer not to answer, and I shan't think any the worse of it, you know, even if it's true. But I suppose I'm right in thinking you don't really *believe* in those things you're defending?' I informed him with adamantine gravity that I did most definitely believe in those things I was defending. His cold and refined face did not move a visible muscle; and yet I knew in some fashion it had completely altered. 'Oh, you *do*,' he said, 'I beg your pardon. Thank you. That's all I wanted to know.' And he went on eating his (probably vegetarian) meal. But I was sure that for the rest of the evening, despite his calm, he felt as if he were sitting next to a fabulous griffin.

I have never understood why a solid argument is any less solid because you make the illustrations as entertaining as you can. If you say that two sheep added to two sheep make four sheep, your audience will accept it patiently—like sheep. But if you say it of two monkeys, or two kangaroos, or two sea-green griffins, people will refuse to believe that two and two make four. They seem to imagine that you must have made up the arithmetic, just as you have made up the illustration of the arithmetic. And though they would actually know that what you say is sense, if they thought about it sensibly, they cannot believe that anything decorated by an incidental joke can be sensible. Perhaps it explains why so many successful men are so dull—or why so many dull men are successful.

Apart from vanity or mock modesty (which healthy people always use as jokes) my real judgement of my own work is that I have spoilt a number of jolly good ideas in my time. There is a reason for this; and it is really rather a piece of autobiography than of literary criticism. I think *The Napoleon of Notting Hill* was a book very well worth writing; but I am not sure that it was ever written. I think that a harlequinade like *The Flying Inn* was an extremely promising subject, but I very strongly doubt whether I kept the promise. I am almost tempted to say that it is still a very promising subject—for somebody else. I think the story called *The Ball and the Cross* had quite a good plot, about two men perpetually prevented by the police from fighting a duel about the collision of blasphemy and worship, or what all respectable people would call, 'a mere difference about religion'. I believe that the suggestion that the modern world is organised in relation to the most obvious and urgent of all questions, not so much to answer it wrongly, as to prevent it being answered at all, is a social suggestion that really has a great deal in it; but I am more doubtful about whether I got a great deal out of it, even in comparison with what could be got out of it. Considered as stories, in the sense of anecdotes, these things seem to me to have been more or less fresh and personal, but considered as novels, they were not only not as good as a real novelist would have made them, but they were not as good as I might have made them myself, if I had really even been trying to be a real novelist. And among many more abject reasons for not being able to be a novelist, is the fact that I always have been and presumably always shall be a journalist.

But it was not the superficial or silly or jolly part of me that made me a journalist. On the contrary, it is such part as I have in what is serious or even solemn. A taste for mere fun might have led me to a public-house, but hardly to a publishing-house. And if it led me to a publishing-house, for the publishing of mere nonsense-rhymes or fairy tales, it could never thus have led me to my deplorable course of endless articles and letters in the newspapers. In short, I could not be a novelist; because I really like to see ideas or notions wrestling naked, as it were, and not dressed up in a masquerade as men and women. But I could be a journalist because I could not help being a controversialist. I do not even know if this would be called mock modesty or vanity, in the modern scale of values; but I do know that it is neither.

The profound problem of how I ever managed to fall on my feet in Fleet Street is a mystery; at least it is still a mystery to me. It used to be said by critics that falling on my feet was only a preliminary to standing on my head. But in fact Fleet Street, not to mention my head, was a rather sea-sick and earthquaky sort of thing to stand on. On the whole, I think I owe my success (as the millionaires say) to having listened respectfully and rather bashfully to the very best advice, given by all the best journalists who had achieved the best sort of success in journalism; and then going away and doing the exact opposite. For what they all told me was that the secret of success in journalism was to study the particular journal and write what was suitable to it. And, partly by accident and ignorance and partly through the real rabid certainties of youth, I cannot remember that I ever wrote any article that was at all suitable to any paper.

On the contrary, I think I became a sort of comic success by contrast. I have a notion that the real advice I could give to a young journalist, now that I am myself an old journalist, is simply this: to write an article for the *Sporting Times* and another for the *Church Times*, and put them into the wrong envelopes. Then, if the articles were accepted and were reasonably intelligent, all the sporting men would go about saying to each other, 'Great mistake to suppose there isn't a good case for us; really brainy fellows say so'; and all the clergymen would go about saying to each other, 'Rattling good writing on some of our religious papers; very witty fellow.' This is perhaps a little faint and fantastic as a theory; but it is the only theory upon which I can explain my own undeserved survival in the journalistic squabble on the old Fleet Street. I wrote on a Nonconformist organ like the old *Daily News* and told them all about French cafés and Catholic cathedrals; and they loved it, because they had never heard of them before. I wrote on a robust Labour organ like the old *Clarion* and defended medieval theology and all the things readers had never heard of; and their readers did not mind me a bit. What is really the matter, with almost every paper, is that it is much too full of things suitable to the paper. But in these later days of the solidification of journalism, like everything else, into trusts and monopolies, there seems to be even less likelihood of anyone repeating my rare and reckless and unscrupulous manœuvre; of anyone waking up to find himelf famous as the only funny man on the *Methodist Monthly*; or the only serious man on *Cocktail Comics*. Anyhow, all will agree that I was an accident in Fleet Street. Some will say a fatal accident, such as is proclaimed on the placards of Fleet Street.

My name achieved a certain notoriety as that of a writer of these murderous short stories, commonly called detective stories; certain publishers and magazines have come to count on me for such trifles; and are still kind enough, from time to time, to write to me ordering a new batch of corpses; generally in consignments of eight at a time.

Any who have come upon traces of this industry may possibly know that a large number of my little crime stories were concerned with a person called Father Brown; a Catholic priest whose external simplicity and internal subtlety formed something near enough to a character for the purposes of this sketchy sort of story-telling. And certain questions have arisen, especially questions about the identity or accuracy of the type, which have not been without an effect on more important things.

As I have said, I have never taken my novels or short stories very seriously, or imagined that I had any particular status in anything so serious as a novel. But I can claim at the same time that it was novel enough to be novel, in the sense of not being historical or biographical; and that even one of my short stories was original enough to do without originals. The notion that a character in a novel must be 'meant' for somebody or 'taken from' somebody is founded on a misunderstanding of the nature of narrative fancy, and especially of such slight fancies as mine. Nevertheless, it has been generally said that Father Brown had an original in real life; and in one particular and rather personal sense, it is true.

When a writer invents a character for the purposes of fiction, especially of light or fanciful fiction, he fits him out with all sorts of features meant to be effective in that setting and against that background. He may have taken, and probably has taken, a hint from a human being. But he will not hesitate to alter the human being, especially in externals, because he is not thinking of a portrait but a picture. In Father Brown, it was the chief feature to be featureless. The point of him was to appear pointless; and one might say that his conspicuous quality was not being conspicuous. His commonplace exterior was meant to contrast with his unsuspected vigilance and intelligence; and that being so, of course, I made his appearance shabby and shapeless, his face round and expressionless, his manners clumsy, and so on. At the same time I did take some of his inner intellectual qualities from my friend, Father John O'Connor of Bradford, who has not, as a matter of fact, any of these external qualities. He is not shabby, but rather neat; he is not clumsy, but very delicate and dexterous; he not only is but looks amusing and amused. My Father Brown was deliberately described as a Suffolk dumpling from East Anglia. That, and the rest of his description, was a deliberate disguise for the purpose of detective fiction. In short, I permitted myself the grave liberty of taking my friend and knocking him about; beating his hat and umbrella shapeless, untidying his clothes, punching his intelligent countenance into a condition of pudding-faced fatuity, and generally disguising Father O'Connor as Father Brown. The disguise, as I have said, was a deliberate piece of fiction, meant to bring out or accentuate the contrast that was the point of the comedy. There is also in the conception, as in nearly everything I have ever written, a good deal of inconsistency and inaccuracy on minor points; not the least of such flaws being the general suggestion that Father Brown had nothing in particular to do, except to hang about in any household where there was likely to be a murder.

Now why do I offer here this handful of scrappy topics, types, metaphors all totally disconnected? Because I am finishing a story; rounding off what has been to me at least a romance, and very much of a mystery-story. A man does not grow old without being bothered; but I have grown old without being bored. Existence is still a strange thing to me; and as a stranger I give it welcome.

This rude and primitive religion of gratitude did not save me from ingratitude; from sin which is perhaps most horrible to me because it is ingratitude. My morbidities were mental as well as moral; and sounded the most appalling depths of fundamental scepticism and solipsism. I found that the Church had gone before me and established her adamantine foundations; that she had affirmed the actuality of external things; so that even madmen might hear her voice; and by a revelation in their very brain begin to believe their eyes.

Finally, I tried, however imperfectly, to serve justice; I saw our industrial civilization as rooted in injustice, long before it became so common a comment as it is today. Anybody who cares to turn up the files of the great newspapers, even those supposed to be Radical newspapers, and see what they said about the Great Strikes, and compare it with what my friends and I said at the same date, can easily test whether this is a boast or a brute fact. From the very beginning my instinct about justice, about liberty and equality was somewhat different from that current in our age; and from all the tendencies towards concentration and generalization. It was my instinct to defend liberty in small nations and poor families; that is, to defend the rights of man as including the rights of property; especially the property of the poor. I did not really understand what I meant by Liberty, until I heard it called by the new name of Human Dignity.

G. K. C.

1936

CONTENTS

Introduction — xxiii
A Chronology of Gilbert Keith Chesterton — xxvii

H. MARSHALL MCLUHAN
G. K. Chesterton: A Practical Mystic (1936) — 1

HERBERT PALMER
G. K. Chesterton and His School (1938) — 10

MAURICE EVANS
Background and Influences on Chesterton (1938) — 18

FRANK SWINNERTON
A True Edwardian (1938) — 26

HILAIRE BELLOC
On the Place of Gilbert Chesterton in English Letters (1940) — 39

RONALD KNOX
G. K. Chesterton: The Man and his Work (1941) — 46

THEODORE MAYNARD
Chesterton (1943) — 50

L. A. G. STRONG
Chesterton's Drinking Songs (1943) — 53

IVOR BROWN
A Multiple Man (1944) — 57

GRAHAM GREENE
G. K. Chesterton (1944) — 59

PATRICK BRAYBROOKE
Chesterton and Charles Dickens (1945) — 60

JAMES STEPHENS
The 'Period Talent' of G. K. Chesterton (1946) — 64

C. S. LEWIS
On Stephens on Chesterton (1946) — 69

EVELYN WAUGH
   The Man Who Was Thursday (1947) ... 72
H. MARSHALL MCLUHAN
   Where Chesterton Comes In (1948) ... 75
HUGH KENNER
   The Word and the World (1948) ... 78
HUGH KINGSMILL
   G. K. Chesterton (1948) ... 98
GEORGE ORWELL
   Great is Diana of the Ephesians (1948) ... 102
ANONYMOUS (from the Times Literary Supplement)
   Chesterton as Essayist: The Mirror of a Silver Age (1950) ... 104
KENNETH M. HAMILTON
   G. K. Chesterton and George Orwell: A Contrast in Prophecy (1951) ... 111
MICHAEL ASQUITH
   G. K. Chesterton: Prophet and Jester (1952) ... 118
DOROTHY L. SAYERS
   Chesterton's The Surprise (1952) ... 123
ALFRED NOYES
   The Centrality of Chesterton (1953) ... 126
RONALD KNOX
   Chesterton's Father Brown (1954) ... 133
SIR ARTHUR BRYANT
   Chesterton (1955) ... 139
HESKETH PEARSON
   G. K. Chesterton (1956) ... 142
D. B. WYNDHAM LEWIS
   Diamonds of the Gayest (1956) ... 145
LANCE SIEVEKING
   'Mr Tame Lion.' Reminiscences of G. K. Chesterton (1957) ... 153

CONTENTS

JOHN RAYMOND
Jeekaycee (1957) — 157

WILFRID SHEED
On Chesterton (1958) — 162

R. G. G. PRICE
A Check-up on Chesterton's Detective (1958) — 172

BERNARD LEVIN
Pantomime Horse (1958) — 177

BERNARD BERGONZI
Chesterton and/or Belloc (1959) — 180

GARRY WILLS
Rhyme and Reason (1961) — 188

HESKETH PEARSON
Gilbert Keith Chesterton (1962) — 202

NEVILLE BRAYBROOKE
The Poet of Fleet Street (1962) — 215

JOHN WAIN
Manalive, *a Good Bad Book* (1962) — 219

CHRISTOPHER HOLLIS
Chesterton's Paradoxes (1963) — 224

MALCOLM MUGGERIDGE
GKC (1963) — 225

CHRISTOPHER HOLLIS
G. K. Chesterton (1966) — 229

ROBERT HAMILTON
The Rationalist from Fairyland (1967) — 232

MICHAEL MASON
Chesterbelloc (1968) — 241

KINGSLEY AMIS
An Unreal Policeman (1968) — 247

DENIS BROGAN
The Chester-Belloc's Better Half (1969) — 248

CONTENTS

ANTHONY BURGESS
  The Level of Eternity (1961) — 251

JOHN GROSS
  A Man of Letters (1969) — 255

W. H. AUDEN
  Chesterton's Non-Fictional Prose (1970) — 262

KINGSLEY AMIS
  The Poet and the Lunatics (1971) — 269

KINGSLEY AMIS
  Four Fluent Fellows (1974) — 273

IAN BOYD
  Chesterton and Distributism (1974) — 283

LEO A. HETZLER
  Chesterton's Writings in his Teenage Years (1974) — 291

R. C. CHURCHILL
  The Man Who Was Sunday (1974) — 300

KATHERINE WHITEHORN
  The Return of G. K. Chesterton (1974) — 305

RICHARD INGRAMS
  The Mystic beneath the Sombrero (1974) — 310

BERNARD LEVIN
  The Case for Chesterton (1974) — 314

BENNY GREEN
  Father of Father Brown (1974) — 317

W. H. AUDEN
  The Gift of Wonder (1974) — 318

V. S. PRITCHETT
  Secret Terrors (1974) — 324

DAVID LODGE
  Dual Vision: Chesterton as a Novelist (1974) — 326

GARRY WILLS
  The Man Who Was Thursday (1975) — 335

CONTENTS

BENNY GREEN
   Defender of the Faith (1977)   343

JOHN CAREY
   For Beer and Liberty (1978)   345

JOHN ATKINS
   Styles in Treachery (1984)   347

P. J. KAVANAGH
   Chesterton Reappraised (1984)   347

A. N. WILSON
   Glimpses of Chesterton (1985)   364

ALLAN MASSIE
   The Master Writer beneath the Card (1985)   365

ROY HATTERSLEY
   Dragon-maker (1985)   366

Acknowledgements   369

INTRODUCTION

CHESTERTON was at least consistent in his own assessment of his literary achievement, for we find him writing in 1905 in much the same vein he used in 1936:

I have no feeling for immortality. I don't care for anything except to be in the present stress of life as it is. I would rather live now and die, from an artistic point of view, than keep aloof and write things that will remain in the world hundreds of years after my death. What I say is subject to some modification. It so happens that I couldn't be immortal: but if I could, I shouldn't want to be.

Public Opinion, 29 September 1905

Since his death in 1936 many have been willing to take him at this own evaluation and to remember GKC, the front legs of the Chesterbelloc, the fat man in the cloak and brigand's hat forever stopping for a pork pie and a beer while he scribbled yet another poem or article on his cuff or on the back of a sugar packet. Everyone seemed to know a story about him, but although there was some real foundation for most of the anecdotes, they did sound unlikely once Chesterton himself had gone. Even the wide scope of his genius proved inconvenient and puzzling—was he really a superb cut-and-thrust debater, a fine humorist and wit, a poet, an essayist, a novelist, a dramatist, a writer of detective stories featuring Father Brown, one of the century's most incisive critics, a Christian apologist, a commentator on political affairs, the founder and pillar of the Distributist movement and party (which soon faded without him), the editor of the *New Witness* and *G.K.'s Weekly*, a weekly columnist on the *Illustrated London News*, a popular broadcaster on the BBC, and the first president of the Detection Club? Well, yes he was, but, over years when copyright difficulties seem to have kept many of his books out of print, it was easy to dismiss his work as ephemeral journalism. Between 1892, when 'The Song of Labour' appeared in *The Speaker*, and 1936 he had written 105 books and an untold number of essays, prefaces, poems and short stories which are still uncollected; such a vast output was bound to be uneven in quality, although it was invariably better than was suggested by his worst denigrator, himself: 'My real judgement of my own work is that I have spoilt a number of jolly good ideas in my time.'

There are those who would agree that those jolly good ideas had been spoilt because he was what he called a 'jolly journalist'. However, although most people know what Chesterton meant by being jolly, it is usually forgotten that journalism has changed a good deal since he began his career in 1900. He was always fiercely independent, demanding the full freedom of a freelance to say whatever he wished to say on whatever topic happened to appeal to him, and if that brought him into conflict with the official policy of any paper or periodical to which he was contributing, then so much the better. Unfortunately, as far as his own interests were concerned, it was often so much the worse for them, for large sections of the Press were to be closed to his work, including Cadbury's *Daily News* when in praising wine he penned the lines:

> Cocoa is a cad and coward,
> Cocoa is a vulgar beast.

Chesterton had contributed a weekly article to the *Daily News* from 1901 until 1913, when he moved to the *Daily Herald*. He also contributed 'Our Note-book' to the *Illustrated London News* from 1905 until his death. But after a severe illness in 1914 and more especially following his brother's death in 1918 his main journalistic function was the editorial chair of the *New Witness* and its successor *G.K.'s Weekly*, very aptly defined in recent years by John Sullivan as *Private Eye avant le coup*.

How did this West Londoner, a scion of a well-known estate agency, manage to produce his tremendous flood of books? Most probably by not being the incorrigible optimist everybody took him for. Indeed, by being born in Kensington in the shadow of the water-tower on Campden Hill, he was not even the cockney George Orwell thought him to be. What is certain is that he learnt his debating skills at St Paul's School, sharpened them in discussions at the Fabian Society, the Pharos Club and the Christian Social Union, and then practised them on an ever-widening social circle which included Yeats and Belloc and Shaw. His training at the Slade (his own accounts of his non-training are belied by his skill with chalk and pencil) gave him an excellent eye for detail which, with Ernest Hodder Williams's encouragement, he was able to bring to the printed page. What is often overlooked is the early drafting in the 1890s of chapters of novels such as *The Napoleon of Notting Hill* and *Manalive*, together with short stories published in the Slade School magazine *The Quarto*. He tells us that at the time he was tormented by doubts amounting to solipsism, but he still produced, in 'A Picture of Tuesday' from *The Quarto*, an outline of one of the main themes to be found in *The Man Who Was Thursday* years later in 1908. Like many other novelists and dramatists he was almost always to expand or combine short stories to produce longer narratives and dramas; unfortunately, later in his career editorial responsibilities did not leave him the time to iron out the seams, and so all his novels, with the exception of the second half of *The Return of Don Quixote*, date from before 1914. Strangely enough, he did continue to write his often overlooked plays well into the 1930s, while the essays, poems, and short stories, were, as usual, produced at the drop of a hat, and in the closing years of his life he achieved great success as a broadcaster, the sole surviving BBC recording even discounting the often-repeated assertion that this mountain of a man had a mouse of a voice.

What emerges very clearly from critical comment is that GKC's very fecundity in every known literary form and medium has enabled each critic to pick his own Chesterton and to draw a portrait which is frequently at violent odds with that sketched by someone else. We meet him as both poet and versifier, apologist and apostate, stylist and scribbler, Zionist and anti-Semite, the embodiment of almost any pair of opposites it is possible to imagine. His enemies were few and, like Lloyd George and the Isaacs brothers, usually political in their background. His opponents in other spheres are usually at pains to praise him as a man of integrity and a lovable companion to be clearly distinguished from the undesirable company they felt he kept. Even his friend Belloc encountered a problem when he came, in 1940, to write the book for which he had contracted at Chesterton's funeral in 1936: he seems to have been unfamiliar with most of his friend's work, and those books he appears to have read were those calculated to support Belloc's picture of a saintly man whose main achievement had been to embody the spirit of resurgent Catholicism. In other words Belloc, despite the title he chose, did not know Chesterton as a man of letters but only as a polemicist. A similar point of view gained popularity in certain Catholic circles which tended to imply that any attempt to assess GKC's standing as a literary or political figure detracted from his image as a defender of the faith he had adopted in 1922, and McLuhan, Kenner, and Wyndham Lewis among others have been influenced by this. Luckily, those who first encountered him through his fiction and essays redress any imbalance by underlining the quality of his writing. Of course, there were purple patches, there were the good bad books (a description of Chesterton's own invention!), and there was the overdone alliteration, but it is incredibly difficult to find a consensus as to what forms the dross and what the gold. What causes the disagreements is most probably Chesterton's ability to appreciate the obvious and to take an almost palpable joy in revealing what most other people have misssed; the medieval buffoon may have talked a good deal of sense amid the nonsense, but its import was not always welcomed or even understood, and so it was with the latter-day jester GKC, who proclaimed that the best present anyone could expect to find in their stocking was a leg. Such a statement of the fact that it was good to be alive was sometimes misunderstood, and few notice that his so-called optimism is very pessimistic indeed, his novels and stories quite often taking us into a nightmare world where multi-schizoid personalities battle with each other before coming to terms and resolving themselves into one healthy person. In suchlike fashion, the multi-faceted GKC needs to be refocused so that we can see his real worth fifty years after his death.

Popular interest in Chesterton was rekindled in the 1950s by a well-received film in which Alec Guinness took the role of Father Brown, and again in the early 1970s when Kenneth More appeared in the same role in a television series; unhappily, More's untimely illness and death prevented the continuation of that success.

Chesterton was certainly never dull and, as Malcolm Muggeridge points out, he has more often been proved right in his judgements than most of his contemporaries. He proposed to free men from the demands of a consumer age which would use them as cogs in the wheels of mass production by maintaining that the individual was more important than any system, capitalist, socialist or communist. Any flaw lay in his inability to give practical expression to his ideas which were usually sound but involved turning the world upside-down. The topsy-turviness of his proposals hid the inherent sanity on which they were based. In the present age, however, his insistence on owning one's own property as a hedge against inflation to help maintain some small measure of independence, and his ideas on worker participation in the control and management of industry no longer seem as silly as they once did. Nor does the proposition, in *The Man Who Was Thursday*, that the head of Scotland Yard may be the chairman of the anarchist-terrorist brotherhood now seem quite so stupid as a plot for a novel.

Following the fiftieth anniversary of Chesterton's death, this collection presents the views of a variety of critics from Britain and North America in the hope that they will give an impression of his impact over the last half century. He himself once said (*pace* T. S. Eliot) that he intended to go with a bang not a whimper, and, at the very least, a man critics compare to Franz Kafka, consider to have something in common with Henry Miller, and whose novel *Manalive* is thought to be a chaste precursor of Pinter's play, *The Lovers*, must be a very interesting author indeed. Unless the critics are wrong, but that is a question the reader must decide.

D. J. C.

1986

A CHRONOLOGY OF
GILBERT KEITH CHESTERTON
1874–1936

1874 (29 May) Born at 32 Sheffield Terrace, Campden Hill, London W8

1881 The Chesterton family moves to 11 Warwick Gardens, London W14

1887 Attends St Paul's School

1892 Publishes 'The Song of Labour' in *The Speaker*

1892–5 Attends the Slade School of Art and University College

1895–1901 Employed as a reader by the publishers Redway, and later by Fisher Unwin

1900 *Greybeards at Play* and *The Wild Knight*. First meeting with Hilaire Belloc. Member of the Pharos Club and the Christian Social Union

1901 Marries Frances Blogg and moves to Overstrand Mansions, Battersea, London SW11. Begins his association with the *Daily News*. *The Defendant*

1902 *Twelve Types*

1903 *Robert Browning*

1904 *G.F. Watts* and *The Napoleon of Notting Hill*

1905 *The Club of Queer Trades* and *Heretics*. Begins to write 'Our Note-book' in the *Illustrated London News*. First half of *The Ball and the Cross* in *Commonwealth*

1906 *Charles Dickens*

1908 *The Man Who Was Thursday*, *All Things Considered* and *Orthodoxy*

1909 Moves to Overroads, Beaconsfield. *George Bernard Shaw* and *Tremendous Trifles*

1910 *The Ball and the Cross*, *What's Wrong with the World*, *Alarms and Discursions*, and *William Blake*

1911 Begins to contribute to the *Eye-Witness*, later the *New Witness*. *Appreciations and Criticisms of the Works of Charles Dickens*, *The Innocence of Father Brown*, and *The Ballad of the White Horse*

1912 *Manalive* and *A Miscellany of Men*

1913 *The Victorian Age in Literature* and *Magic*, a play. Helps to reveal the Marconi scandal and is embittered by its corruption

1914 *The Flying Inn*, *The Wisdom of Father Brown* and *The Barbarism of Barlin*. First half of *The Return of Don Quixote*. Taken seriously ill and looses consciousness for three months

1915 Recovers from his illness, which was a kind of dropsy. *Letters to an Old Garibaldian*, *Poems*, *The Crimes of England*

1916 (October) Replaces his brother as editor of the *New Witness*

1917 *A Short History of England*
1918 Visits Ireland
1919 *Irish Impressions.* Visits Palestine
1920 Undertakes a lecture tour in the United States. *The Superstition of Divorce, The Uses of Diversity,* and *The New Jerusalem*
1922 *Eugenics and other Evils, What I Saw in America,* and *The Man Who Knew Too Much.* Received into the Catholic Church
1923 *Fancies versus Fads* and *St Francis of Assisi*
1925 (21 March) Editor of *G.K.'s Weekly* until his death. *The Supersititions of the Sceptic, Tales of the Long Bow, The Everlasting Man,* and *William Cobbett*
1926 *The Incredulity of Father Brown, The Outline of Sanity,* and *The Queen of Seven Swords.* Start of the Distributist League
1927 Visits Poland. *The Catholic Church and Conversion, The Return of Don Quixote, Collected Poems, The Secret of Father Brown, The Judgement of Dr Johnson, Robert Louis Stevenson*
1928 *Generally Speaking.* A broadcast debate with Bernard Shaw published as *Do We Agree?*
1929 Visits Rome. *The Poet and the Lunatics* and *The Thing*
1930 *Four Faultless Felons, The Resurrection of Rome* and *Come to Think of It*
1930–1 Undertakes a second lecture tour of the United States
1931 *All is Grist*
1932 *Chaucer, Sidelights on New London and Newer York,* and *Christendom in Dublin.* Begins to broadcast regularly on the BBC
1933 *All I Survey* and *St Thomas Aquinas*
1934 *Avowals and Denials.* Visits Rome and Sicily
1935 *The Scandal of Father Brown* and *The Well and the Shallows*
1936 *As I Was Saying.* (14 June) Dies at Beaconsfield. *Autobiography*
1937 *The Paradoxes of Mr Pond*

H. MARSHALL McLUHAN
G. K. Chesterton: A Practical Mystic

WHEN it is seen that there are two principal sides to everything, a practical and a mystical, both exciting yet fruitful, then the meaning and effect of Chesterton can become clear even to those who delight to repeat that he stands on his head. That tireless vigilance in examining current fashion and fatalism, which has characterized him for more than thirty years, clearly depends upon his loyalty to a great vision: 'His creed of wonder was Christian by this absolute test, that he felt it continually slipping from himself as much as from others.' Therefore the prosaic is invariably the false appearance of things to fatigued intellect and jaded spirit. This is the basis of Chesterton's inspiriting opposition to the spread of officialdom and bureaucracy. The cynical social legislation of today, undertaken in supposed accord with unyielding economic circumstance, is often light-headed because it is not lighthearted. And Chesterton is a revolutionary, not because he finds everything equally detestable, but because he fears lest certain infinitely valuable things, such as the family and personal liberty, should perish. That Beauty transformed the Beast only because she loved it while it was yet a beast, is timelessly significant.

It is necessary to define the sense in which Chesterton is a mystic, before the relation of this to the practical side can be judged. He once wrote: 'Real mystics don't hide mysteries, they reveal them. They set a thing up in broad daylight, and when you've seen it, it is still a mystery. But the mystagogues hide a thing in darkness and secrecy; and when you find it, it's a platitude.' The mysteries revealed by Chesterton are the daily miracles of sense and consciousness. His ecstasy and gratitude are for what has been given to all men. He rejoices at 'the green hair' on the hills, or 'the smell of Sunday morning', and

> Those rolling mirrors made alive in me,
> Terrible crystal more incredible
> Than all the things they see.

Here it is possible only to speak of the sacramental sense of the life of earth and sea and sky, of tillage and growth, and of food and wine, which informs his work. For to him existence has a value utterly inexpressible, and absolutely superior to any arguments for optimism or pessimism. What truce could such a great lover of life make with agnostic humanitarianism and world-weary eugenists? To those 'who have snarled through the ages' he hurls this reply:

> Know that in this grotesque old masque
> Too loud we cannot sing,
> Or dance too wild or speak too wide
> To praise a hidden thing.

It is the spirit of Christendom, as it was the spirit of R. L. Stevenson; but it is not the spirit of the age. In this sense Chesterton is reactionary. If universal nature made shipwreck tomorrow, yet would

> One wild form reel on the last rocking cliff,
> And shout: 'The daisy has a ring of red.'

Chesterton has stepped beyond the frontiers of poetry, to what Maritain in speaking of Rimbaud calls, 'The Eucharistic passion which he finds in the heart of life.'

As a comment on one of his characters, he wrote: 'He had somehow made a giant stride from babyhood to manhood, and missed that crisis in youth when most of us grow old.' Chesterton himself is full of that child-like surprise and enjoyment which a sophisticated age supposes to be able to exist only in children. And it is to this more than ordinary awareness and freshness of perception that we may attribute his extraordinarily strong sense of fact. We can apply to him what he wrote of a young poet in one of his stories: 'What are to most men impressions or half impressions, were to him incidents ... The slope of a hill or the corner of a house checked him like a challenge. He wrestled with it seriously, until he could put something like a name to his nameless fancy.' This profound humility in the face of reality is the very condition of honest art and all philosophy, and it explains Chesterton's imaginative sympathy with popular legends and proverbs; just as it is the reason for his energetic revival of tradition in so far as it dignifies and illuminates any present activity. In short, he is original in the only possible sense, because he considers everything in relation to its origins. It is because he is concerned to maintain our endangered institutions that he earnestly seeks to re-establish agriculture and small property, the only basis of any free culture.

But most of all does his strong sense of fact account for the recurrence of seeming paradoxes in his writings:

The more plain and satisfying our state appears, the more we may know we are living in an unreal world. For the real world is not satisfying. The more clear become the colours and facts of Anglo-Saxon superiority, the more surely we may know we are in a dream. For the real world is not clear or plain. The real world is full of bracing bewilderments and brutal surprises. Comfort is the blessing and the curse of the English ... For there is but an inch of difference between the cushioned chamber and the padded cell.

All profound truth, philosophical and spiritual, makes game with appearances, yet without really contradicting common sense. That is why Chesterton accuses the Victorians of believing in real paradoxes, such as expecting all men to have the same morals when they had different religions, or supposing that it was practical to be illogical. A little attention shows how he consciously causes a clash between appearances in order to attract attention to a real truth transcending such a conflict. There is no hint or hue of meaning amidst the dizziest crags of thought that is safe from his swift, darting, pursuit. We return safely and lucidly from the exhilarating chase of an idea to its logical conclusion. Such a world, rigid with thought and brilliant with colour, is the very antithesis of the pale-pink lullaby-land of popular science. It is the difference between a cathedral window and blank infinity. That is why modern life, thoughtless and unpoised, has degenerated from a dance into a race, and history is regarded as a toboggan slide. But Chesterton has exposed the Christless cynicism of the supposedly iron laws of economics, and has shown that history is a road that must often be reconsidered and even retraced. For, if Progress implies a goal, it does not imply that all roads lead to it inevitably. And today, when the goal of Progress is no longer clear, the word is simply an excuse for procrastination.

It is scarcely necessary now, when philosophy and art have been revitalized by the study of medieval achievements, to explain that Chesterton does not want 'to go back to the middle ages', and never did. 'There is none rides back to pick up a glove or a feather.' But the merest reference to anything prior to the Reformation starts a clockwork process in the mind of the nineteenth-century journalists who still write most of our papers: 'Mr Chesterton is a medievalist; and he is therefore quite justified (from his own benighted standpoint) in indulging as he does in the sport of tearing out the teeth of Jews, burning hundreds of human beings alive, and perpetually seeking for the Philosopher's Stone.' Without these automatic and irresponsible reactions to anything resembling serious thought, there could not be that vast and increasing mountain of printed paper which indicates that progress is proceeding. For, as Stevenson noted, man lives not by bread alone, but by catchwords also. It all began with Luther's anathemas against Reason, and Descartes's expressed contempt for Aristotle and Aquinas.

The conspiracy, hatched at that time, to ignore history, which in practice meant the middle ages, had not been generally found out when Chesterton began to write. Certainly he knew there must have been something right about centuries whose architectural remains were admired by every class of mind. He was absolutely certain that people who were capable of an intensely significant use of colours, whose dress was as many-hued as the walls and windows of their churches, could not be as black as they were painted. So much he had in common with Ruskin, Rossetti, and Morris; but much deeper was his interest in the origins of that magnificent and complex culture. His was the intellectual interest of Newman rather than the aesthetic interest of Burne-Jones. 'If we want the flowers of chivalry, we must go back to the roots of chivalry, Theology.' And since it is always the world of ideas that determines the climate of sensation and opinion, the Troubadours, and Dante, and the Metaphysicals are today throbbing again with vital interest for us. Similarly the moral atmosphere of the Victorian time was prepared by Locke and Bentham, and, though inchoate, had stirred the anger of Keats:

> Fools! make me whole again that weighty pearl
> The Queen of Egypt melted, and I'll say
> That ye may love in spite of beaver hats.

Now Chesterton has never written a book to praise the Middle Ages, but he has written books to praise and explain the life of St Francis and the thought of Aquinas. For, as he explains, 'real conviction and real charity are very much nearer than men suppose'. It is plain that he is literally a radical, because he goes to the roots of things. And it is for that reason that he is very hopeful for this generation, which has been forced back to roots and origins. Thus because it is sceptical even about scepticism, 'the sophisticated youth, who has seen through the sophistical old men, may even yet see something worth seeing'.

Although Chesterton has never entertained any desire to restore the Middle Ages, he shows that certain timeless principles were then understood which have since been foolishly forgotten. Though they were not the right place, they were the right turning, and subsequent history has in a deep sense been an ignoble retreat from their difficult and untried ideas. It was a rout which distorted and diminished the Renaissance, and nullified its proud promises to us. Chesterton's concentration on history has been a splendid effort to rescue a civilization weakened by capitalism from the logical conclusion of capitalism, which is, either the servile or the communistic State:

> The highest use of the imagination is to learn from what never happened. It

G. K. CHESTERTON: A PRACTICAL MYSTIC

is to gather the rich treasures of truth stored up as much in what never was as in what was and will be. We are accused of praising and even idealising retrograde and barbaric things. What we praise is the progress which was for those retrograde things prevented, and the civilization that those barbarians were not allowed to reach. We do not merely praise what the Middle Ages possessed. It would be far truer to say that we praise ... what they were never allowed to possess. But they had in them the potentiality of the possession; and it was *that* that was lost in the evil hour when all other possessions became a matter of scramble and pillage. The principle of the guild was a sound principle; and it was the principle and not only the practice that was trampled under foot...

This tragic theme has found memorable expression in such poems as 'The Secret People'.

> Smile at us, pay us, pass us; but do not quite forget;
> For we are the people of England, that have never spoken yet.

For, equally with Dickens, Chesterton is the champion of the English poor. In equal degree he is a hater of class and privilege, of cant and bureaucracy. But he is a great demagogue who has been shouted down by Publicity, that 'voice loud enough to drown any remarks made by the public'. He is the mouthpiece of the poor who cannot hear him. He is their memory and their poet, the cherisher of their traditions stamped out by misery, and the singer of virtues they have almost forgotten. And his sympathies are a proof of his splendid lineage:

> I saw great Cobbett riding,
> The horseman of the shires;
> And his face was red with judgement
> And a light of Luddite fires.

'Cobbett was defeated because the English people was defeated. After the frame-breaking riots, men as men were beaten ... Ireland did not get home rule because England didn't.' This instinctive espousal of the cause of the oppressed and of the harassed and pitiable things is nowhere better expressed than in 'The Donkey', which needs not quoting. There also will be found that inexpressible and mystic certainty, that the big battalions will one day be confounded by that weak and scattered remnant which has survived a hundred defeats. Yet Chesterton does not fight with any hope of success, but rather for the continuance of a cause and to witness to a truth:

> In a world of flying loves and fading lusts
> It is something to be sure of a desire.

It is probably his sympathy with the humble yet joyous traditions of

England, together with his strong but sensitive feeling for her unrivalled roads and hills and villages, that has contributed to make him subtly and unmistakably English. He is not least English in his deep regard for morals, and a fondness for pertinently pointing morals even about morals: 'We have grown to associate morality in a book with a kind of optimism and prettiness; according to us, a moral book is about moral people. But the old idea was almost exactly the opposite; a moral book was about immoral people.' We have but to recall Hogarth and Fielding.

But in nothing is the peculiar quality of the autochthonous Englishman seen more clearly than in Chesterton's patriotism. To people unacquainted with the profound patriotism of Sir Thomas More, Johnson, Cobbett and Dickens, it may be puzzling. For it is anything but an uncritical allegiance to the acts of professional politicians, which Johnson rightly designated as the last refuge of a scoundrel. Nor is it that still more pernicious metaphysical and spiritual perversion that has disfigured Prussia since Frederick. For he heartily hates what England often seems trying to become, and this produces 'a sort of fierce doubt or double-mindedness which cannot exist in vague and homogeneous Englishmen'. In his book *The Crimes of England*, he wrote:

> I have passed the great part of my life in criticising and condemning the existing institutions of my country: I think it is infinitely the most patriotic thing a man can do.

The high standard he sets for his own country makes him quick and generous in his recognition of the rights and virtues of others. And the French, American, and Irish peoples have no more discriminating admirer and interpreter than Chesterton. But like Socrates, he is a gadfly stinging his own countrymen into awareness of the crime of complacency:

> We have made our public schools the strongest walls against a whisper of the honour of England ... What have we done and where have we wandered, we that have produced sages that could have spoken with Socrates and poets who could walk with Dante that we should talk as if we had never done anything more intelligent than found colonies and kick niggers?

Chesterton opposed jingoism at the beginning of the century for the same reason that he opposes pacifism today. For both are craven moods that seek the exhilaration or security to be found in mass sentiment and unrealistic unanimity.

He believes in the ordinary man, and in no other sort of man. But unlike another very great journalist, Daniel Defoe, Chesterton has never failed to pay the ordinary man the compliment of reasoning with him. That is why he is such a poor salesman. For he does not wish to persuade without first convincing. His appeal is always to reason. In fact, that he turns a few cart-wheels out of sheer good spirits by way of enlivening his pages has annoyed a certain type of person, and is sure to puzzle a lazy or a fatigued mind. Such, for instance, is his unparalleled power of making verbal coincidences really coincide. But the irreducibly simple language and the appeal to common experience are unmistakably directed to the ordinary person. T. S. Eliot, writing to the Three Thousand people who comprehend, according to his own computation, the total culture of the West, describes the poet's mind as 'being a more finely perfected medium in which special and very varied feelings are at liberty to enter into new combinations'. He says no more than Chesterton before him, who wrote that 'nobody believes that an ordinary civilized stockbroker can really produce, out of his own inside, noises which denote all the mysteries of memory and all the agonies of desire'. The fact is that Chesterton has always that additional human energy and intellectual power which constitute humour.

That purely esoteric use of his powers which might have gone to the creation of fine art was made impossible by the appeal of a turbid and chaotic time to his great democratic sympathies. The extraordinary extent and variety of his writings and discussions proportioned to the desperate need for direction and unity in an age that has 'smothered man in men.' For external complexity he produced an insane simplification of thought, preying upon personal variety and spontaneous social expression:

> We have hands that fashion and heads that know,
> But our hearts we lost—how long ago!

What Chesterton has written of the power of St Thomas to fix even passing things as they pass, and to scorch details under the magnifying lens of his attention, is strikingly true of himself. His is the power to focus a vast range of material into narrow compass; and his books though very numerous are extremely condensed. They might even be considered as projections of his mastery of epigram and sententious phrase. What had seemed a dull and formless expanse of history is made to shine with contemporary significance, and contemporary details are made to bristle with meaning. It is a great labour of synthesis and reconstruction in which Chesterton has been engaged. He has fixed his attention on the present and the past, because he is concerned lest our future steps be blindly mistaken. And a strong and growing group of like-minded writers indicates the impression he has made upon his generation. For there are many people who no longer regard Herbert Spencer as a philosopher, or think Chesterton a medieval buffoon.

Although his thunders of laughter may beguile some readers, there is no more serious master of debate and controversy. When his exuberant fancy may decorate an argument as a gargoyle embellishes a buttress, the buttress is there; and just as the buttress is there, so is the lofty edifice which it supports. There is a perfect continuity between his ideas even when they are most subtle; for he has seen modern problems from the beginning in all their complexity and connections, in other words, with wisdom. Therefore he has no nostrum and no novelty, no panacea and no private aim; nor is he deceived by his own metaphors. And nothing is more characteristic of him than scrupulous care in the definition of terms.

Chesterton has commanding vigour of expression, and appreciates the genius of the English language which is full of combative and explosive energy, especially found in short epithets: 'A young man grows up in a world that often seems to him to be intolerably old. He grows up among proverbs and precepts that appear to be quite stiff and senseless. He seems to be stuffed with stale things; to be given the stones of death instead of the bread of life; to be fed on the dust of the dead past; to live in a town of tombs.' This passage may also be regarded as an instance of what some hold to be Chesterton's vice of alliteration. But he is doing something quite different from a Swinburne lulling the mind by alliterating woolly, caterpillar words. His energetic hatchet-like phrases hew out sharply defined images that are like a silhouette or a wood-cut. And these are of a piece with the rigorous clarity of his thought.

The artist in Chesterton has been far from subdued by the philosopher and controversialist. His poetry and stories are as important as they are popular. It is because they are popular that very little need be said of them just now. But this part of his work is not easily appreciated in isolation from the more abstract portion. For instance, Chesterton regards the soul of a story as 'the ordeal of a free man' ... 'There is no such thing as a Hegelian story, or a monist story, or a determinist story ... Every short story does truly begin with the creation and end with the last judgement.' The detective story won his praise from the start, because 'it is the earliest and only form of popular literature in which is expressed some sense of the poetry of modern life'. Such stories are 'as rough and refreshing as the ballads of

Robin Hood'. They are based upon the poetry of fact which Chesterton has expressed so well. 'The romance of the police force is thus the whole romance of man. It is based on the fact that morality is the most dark and daring of conspiracies.' That is why the inimitable Father Brown is a psychologist rather than a sleuth; and the culprit he exposes is shown to be a sinner rather than a mere criminal.

The stories of Chesterton are as colourful melodrama as can be imagined or desired. Like the old melodrama, they display this world as a battlefield and restore the colours to life. They are proof that 'the finding and fighting of positive evil is the beginning of all fun and even of all farce'. For in *The Napoleon of Notting Hill* and *Manalive* and the others, Chesterton enters faery lands to show that they are not forlorn. They contain a truly brave new world, rather unlike 'the shape of things to come', in which very vivid people are etched upon a background of significantly contrasted colours:

> The broken flowerpot with its red-hot geraniums, the green bulk of Smith and the black bulk of Warner, the blue-spiked railings behind, clutched by the stranger's yellow vulture claws and peered over by his long vulture neck, the silk hat on the gravel and the little cloudlet of smoke floating across the garden as innocently as the puff of a cigarette—all these seemed unnaturally distinct and definite. They existed, like symbols, in an ecstasy of separation.

The most ordinary things become eerily and portentously real. Bodily gestures are stiff with spiritual significance, as in the old pageantry. And the deeps of the subconscious are entered, and monstrous facts from the borderland of the brain impress themselves upon us. As the modern jargon puts it, Chesterton has achieved an objective correlative for his thought and feeling.

The profound joy in Chesterton's poems and stories can be properly appreciated only when the suffocating materialism of pre-war days is remembered:

> A cloud was on the mind of men, and wailing went the weather,

and

> Life was a fly that faded, and death a drone that stung;
> The world was very old indeed when you and I were young.

But, liberated by faith and joy, Chesterton represents a very great increase in sensibility over the world that read the *Idylls of the King*. Any valuable extension of awareness is directly determined by the rediscovery of neglected truths, and there is much the same truth behind 'The Ballad of the White Horse' and 'The Wasteland'. Many of Chesterton's poems have the directness of a shout or a blow, and at times he recaptures the startling simplicity of Chaucer through a combination of sanity and subtlety. Of his inspiriting songs it is unnecessary to speak. They could arouse spirits in a materialist Utopian.

And yet it is no contradiction to say that Chesterton is primarily an intellectual poet. This is too often overlooked even by those who know that his mind is full of nimble and fiery shapes, and that his wit is 'quick, forgetive, apprehensive'. He deserves the praise Eliot affords to Donne for the 'quality of transmuting ideas into sensations, of transforming an observation into a state of mind'. It is necessary only to refer to the great 'Lepanto' or 'The Donkey' as perfect achievement of this kind.

Had Chesterton been merely a quiet intellectual with an ordinary amount of energy, he would certainly have been an artist who was taken seriously by the Three Thousand cultured minds of Europe. The same might be said of Dickens. But in an age of shallow optimism, of crumbling creeds and faltering faith, he has walked securely and wildly, boisterously praising life and heaping benedictions upon decadents. He has become a legend while he yet lives. Nobody could wish him otherwise than as he is.

HERBERT PALMER
G. K. Chesterton and His School

THERE exist schools of poetry whose members hang together so loosely that their influence does no hurt to other fraternities or to outstanding individualists. They form nothing in the nature of an aloof coterie or hostile clique.

G. K. Chesterton at one time seemed to be the centre of such a school or division. Its journals of outlet were *The New Witness* and *G.K.'s Weekly*, and its characteristics were such that one might almost call it the Modern School of the Medieval Elizabethans, which though an unwieldy contradiction in three terms will perhaps partially describe its comely features of quarrelling Romanticism.

The chief members of this division, which began to show itself dimly between the beginning and end of the War were G. K. Chesterton, Hilaire Belloc, Wilfred Rowland Childe, W. R. Titterton, and William Kean Seymour. Its most obvious characteristics were robustiousness, rapture, medievalism, Roman Catholic Christianity, the exaltation of the humble and homely, traditional music, the cult of the tavern, songs of liberty. It was also characterized by Puritanism. Indeed, Chesterton, who pretended to be the great opponent of Puritanism, rode his hobby-horse with a rather grinning mouth and a long tongue in a very bulging cheek. For the essence of Puritanism is the putting of the Christian religion into all the deeds of daily life, speaking about religion in daily speech and writings, the cultivation of justice, of fair play and sincerity of speech, and the exaltation of sex purity in celibate and family life. It is this last which is the plainest badge of the Puritans. It is true that the chief leaders of Puritanism in this country have sprung from the Nonconformist churches, but Nonconformity is not necessarily Puritanism, though the Elizabethan dissenters were called Puritans, and the Puritan movement owes most of its tenets of faith to them. For, today it is probably among the Catholics of Ireland that you will find the greatest power of Puritanism, manifest peculiarly in the rigorous workings of the literary censor (moralist rather than political) in that green and rain-lashed country. So when wilful enthusiasts about the work of D. H. Lawrence (enthusiasts who want to give him special moral backing) claim that he was a great Puritan they do not speak at all correctly about him, for D. H. Lawrence, although the child of Congregationalists (and very lax Congregationalists), was in continual rebellion against the church of his fathers. But when Chesterton wrote:

> But the song of Beauty and Art and Love
> Is simply an utterly stinking song,
> To double you up and drag you down
> And damn your soul alive . . .

he expressed the Puritan attitude in entire completeness towards sensuous and sensual art, the reverse of which D. H. Lawrence may not have quite crystallized in his verse, but which is often flamboyantly and disturbingly manifest in his lurid, though magnificent prose. Where Chesterton showed his severance from Puritanism was in his very non-moral drinking songs, for Puritanism in part means rigid temperance, if not teetotalism, and alcohol as a theme for poetry is taboo. Even his attempt in the ultimate stanza of 'Wine and Water' to effect a sort of compromise (a very droll compromise) does not clear him of a very strong bias:

> But Noah he sinned, and we have sinned; on tipsy feet we trod,
> Till a great big black teetotaller was sent to us for a rod,
> And you can't get wine at a P.S.A., or chapel, or Eisteddfod,
> For the Curse of Water has come again because of the wrath of God.

And water is on the Bishop's board and the Higher Thinker's shrine,
But I don't care where the water goes if it doesn't get into the wine.

All through his poems he sings in praise of the tavern, of jolly roistering, but roistering in which (here is the real Puritan) women and sensual debauch have no place. His drinkers are unattached medievalists, men-at-arms with no mistress, plump-cheeked fighters in the chaste cause of right and liberty, Sir Galahads taking a day off and shouting their exploits over the wine-barrel. He is the apostle of a sort of drunken Puritanism, the poet of Falstaffian Christianity. He almost unites priest and publican, and ignores every movement of the Modern Age save as something to criticize and condemn. Of all English poets of this century he belongs foremost to an earlier age. Paradoxically though, he is a social propagandist—probably more propagandist than any poet of importance ever born. In the years which immediately preceded his death, his poetry became hidden and embedded in a crush of social rhymes, the artist receding more and more into the background. But this is not the characteristic Chesterton, for in him the artist and propagandist often marvellously combine:

> It has not been as the great wind spoke
>   On the great green down that day;
> We have seen, wherever the wide wind spoke,
>   Slavery slaying the English folk:
> The robbers of land we have seen command,
>   The rulers of land obey.
>
> We have seen the gigantic golden worms
>   In the garden of paradise:
> We have seen the great and the wise make terms
> With the peace of snakes and the pride of worms,
> And them that plant make covenant
>   With the locust and the lice.

He is writing of the present England, the people of today, for according to Chesterton, personal liberty has been almost entirely lost under the tyranny of capitalism. And in his strangely communicating, if rather wild and nonsensical poem, 'The Old Song', the revolutionary note is even more passionately intensified:

> I saw the kings of London town,
>   The kings that buy and sell,
> That built it up with penny loaves
>   And penny lies as well:

> And where the streets were paved with gold, the shrivelled paper
>   shone for gold,
> The scorching light of promises that pave the streets of hell.
> For penny loaves will melt away, melt away, melt away,
> Mock the mean that haggled in the grain they did not grow;
> With hungry faces in the gate, a hundred thousand in the gate,
> A thunder-flash on London and the finding of the foe.

And yet he never advocated Communism, nor even Socialism of any very levelling nature. Distribute wealth and goods a little more evenly, and let us be ruled by Englishmen with national and social interests at heart, people who have sprung from the agricultural land of Anglo-Saxondom, and we shall be glorious and happy once again:

> They have given us into the hand of new unhappy lords,
> Lords without anger and honour, who dare not carry their swords.
> They fight by shuffling papers; they have bright dead alien eyes;
> They look at our labour and laughter as a tired man looks at flies.
> And the load of their loveless pity is worse than the ancient wrongs,
> Their doors are shut in the evening; and they know no songs.

A critic who tries to yoke him to any of his English predecessors has a difficult task. Maybe at moments he sounds a little like Lord Macaulay, and at another time a sort of ghostly fusion of Marlowe and Blake, or of Swinburne and Blake; but the resemblance at closest is not very pronounced. He manifestly read all the old ballads written in the English tongue, and he may have read all the few existing epics and fragmentary poems of the Saxon scops. Add to that, if you like, everything else Teutonic, be it Danish, Icelandic, or German, introducing into their midst some very odd or strange foreign matter, such as the employment of paradoxes, and even such a destructive element as French Symbolism (his poem on old age, 'A Second Childhood', sounds uncommonly like a piece of glorified Surrealism), and the Chestertonian framework knottily emerges. I doubt if he ever bothered himself with German poetry (he was no great friend of the Germans), but, turning over the pages of representative German anthologies (one of them a students' drinking-song book), one is continually confronted by reminders, particularly noticeable in the strong racial outlook, and something straightforwardly musical, resonant, and rhetorical in the language—to say nothing of the wine- and beer-imbibing enthusiasm. One is moderately certain that poets like Bürger, Schiller, and Liliencron, and probably also the divine and demoniacal Heine, would have derived pleasure from Chesterton once they had sifted away the paradoxes and the lyrics containing the most extravagant figures of speech. Affinities, too, exist between him and a more recent German poet, Börries von Münchhausen, so eminently Teutonic is the man. There is a royalist strain in him, certainly, for you may sing of liberty and shout 'Down with Tyrants!' as hard as you like, but you cannot be quite so enthusiastic about medievalism without suggesting that you believe in kings more than presidents, soviets, or republics. One can almost believe that in remote times he thrilled the hearts of the thanes guarding King Alfred, and that his hand tintinnabulated on the harp whilst that great monarch at repose was watching his clock-candles and planning another onslaught on the Danes. Yet there is rather more of the modern brass band in his verse than the harp, and considering his verse from the viewpoint of the orchestra, the harp comes midway between the brass instruments and the violins and cellos. His orchestra, if sufficiently complete, is rather too one-sided, and though violins and cellos break through the general blare they are not at all frequent. And yet it seems rather odd to hear a brass band playing this kind of thing:

> You may be tired and tolerant of fancies as they fade,
> But if men doubt the Charter, ye shall call on the Crusade—
> Trumpet and torch and catapult, cannon and bow and blade,
> Because it was My challenge to all the things I made.

Because what have we got to do today with bows and catapults? and though such songs of liberty are given a backward atmosphere, their application is always quite obviously contemporary.

His longest poem, a complete book, and a very long poem, *The Ballad of the White Horse*, takes us right into Saxon times and the press of nobles around King Alfred; and from that and the general tone of his whole output one gets the view that though a medievalist, he is an early medievalist, believing that since the battle of Hastings the English have always been an oppressed and downtrodden people who have yet to come into their own. His best poem, 'The Secret People', in which he briefly surveys the whole of English history since the Norman Conquest, and agreeably fuses propaganda with poetry and historical substance, quite plainly advocates that idea, its concluding lines being among the most memorable things he has written:

> It may be we shall rise the last as Frenchmen rose the first,
> Our wrath come after Russia's wrath and our wrath be the worst.
> It may be we are meant to mark with our riot and our rest
> God's scorn for all men governing. It may be beer is best.
> But we are the people of England; and we have not spoken yet.
> Smile at us, pay us, pass us. But do not quite forget.

His verse is full of preachings, politics, and arresting hymns of hate.

Anger, witty expostulation, impish satire, and hilarious mockery are the currants and raisins in his holly-decked plum-pudding. Green and prickly is the holly, and red are the berries—red for the wounds of Christ and the scarlet robes of an idealized papacy. He is a Christian poet, and therefore being a poet of festivities a prominent Christmas poet:

> There fared a mother driven forth
>   Out of an inn to roam;
> In the place where she was homeless
>   All men are at home.
> The crazy stable close at hand,
> With shaking timber and shifting sand,
> Grew a stronger thing to abide and stand
>   Than the square stones of Rome.

But Chesterton has always been much more popular as a prose writer than as a poet. Probably it is owing to his propagandist vein, for the educated man in the street seems to like his poetry unadulterated, at any rate not infused with any strong sense of social indignation. You may quarrel with capitalism and the Government as much as you like, and be popular, if you do it only in prose; but as soon as you ignore the general aloofness of art to the turmoil of everyday change you are not only apt to spoil your poetry but to lead people into the belief that you are doing so when your hands are exceptionally untarnished. G. K. Chesterton, in spite of his frequent blare and bombast, has been extraordinarily successful in infusing true poetry into his thundering orchestra. God speaking through him, he knows no restraint, but comes at you, marches up the street and round the corner, a rage of music and colour that seeks to hold up the traffic, so that your spirit stands still, and for a moment you stop work and 'down tools'—bemused, feet going, and head awhirl, intoxicated by a rushing rhythm of bacchanal or religious emotion. But poetry that is merely like that does not necessarily contain the enduring line, the wonderful stanza, and it is astonishing that in the verse of Chesterton there is so much of what is really fine, as for instance:

> Ruin is a builder of windows; her legend witnesseth
> Barbara, the saint of gunners, and a stay in sudden death.

Or:

> But all beyond was the wolfish wind
> And the crafty feet of the snow.

Or:

> Tip-toe on all her thousand years and trumpeting to the sun.

Or:

> We have seen the gigantic golden worms
> In the garden of Paradise.

Or:

> There is a game of April Fool that is played behind its door
> Where the fool remains for ever and the April comes no more.

Or:

> There is always a thing forgotten
> When all the world goes well;
> A thing forgotten, as long ago
> When the gods forgot the mistletoe,
> And soundless as an arrow of snow
> The arrow of anguish fell.

Or in a rather more everyday vein:

> The rolling English drunkard made the rolling English road.

Or:

> When you and I went down the lane with ale-mugs in our hands
> The night we went to Glastonbury by way of Goodwin Sands.

Or, triter in speech, but little less memorable:

> The happy men that lose their heads
> They find their heads in heaven.

There is plenty of magic in Chesterton's verse, not exactly the delicate elfish magic of Yeats or Walter de la Mare, a rather flick-in-the-eye magic if you like, but none the less evident. Sometimes, indeed, he achieves it when flying right into the jaws of bombast he steers miraculously clear, or when formulating a paradox he gets beyond the truth of paradox to the creation of the rose that shines upon the lips of truth.

Then there is the Chesterton who is memorable because he says something very droll, even though it be penetratingly satirical:

> Half of two is one.
> Half of four is two,
> But half of four is forty per cent if your name is Montagu:
> For everything else is on the square
> If done by the best quadratics;
> And nothing is low in High Finance
> Or the Higher Mathematics,

winding up with the very brilliant and veracious:

> Where you hide in the cellar and then look down
>   On the poets that live in the attics;
> For the whole of the house is upside down
>   In the Higher Mathematics.

At moments he had a scathing tongue, and did not always avoid the libellous (at any rate not in his prose), though whoever he may mean by Montagu steers clear of that. A more venomous stanza from another poem is fired at the head of an inanimate substance:

> Tea, although an Oriental,
>   Is a gentleman at least;
> Cocoa is a cad and coward,
>   Cocoa is a vulgar beast,
>
> Cocoa is a dull, disloyal,
>   Lying, crawling cad and clown,
> And may very well be grateful
>   To the fool that takes him down.

But Chesterton was a sort of God's fool, the Almighty's chosen jester, who fearlessly took liberties, secure in most instances beneath his cap and bells.

When his first book of poems came out he was linked up with John Davidson, and though his verse is more like that unfortunate poet's than anybody else's, the coupling seems to have been a little displeasing to both of them. A strong Davidson influence was probably present, but it was no more than an influence and for the most part entirely transmuted and transformed. But hear what he himself has said about it in his autobiography:

My little volume of verse was reviewed with warm and almost overwhelming generosity by Mr James Douglas, then almost entirely known as a leading literary critic. Impetuosity as well as generosity was always one of Mr Douglas's most attractive qualities. And he insisted, for some reason, on affirming positively that there was no such person as G. K. Chesterton; that the name was obviously a *nom de plume*; that the work was obviously not that of a novice, but a successful writer; and finally that it could be none other than Mr John Davidson. This naturally brought an indignant denial from Mr John Davidson. That spirited poet very legitimately thanked the Lord that he had never written such nonsense; and I for one very heartily sympathized with him.

As time went on the severance between Chesterton and John Davidson became only too plain, because where Davidson was bitter, Chesterton was passionately indignant or amusingly malicious; where

Davidson was religiously destructive, Chesterton was eminently Christian. Moreover, much in Davidson seems to reveal the influence of Nietzsche—who was Chesterton's stock abhorrence. Chesterton was always on the side of the humble and the downtrodden; but though John Davidson's sympathies also went to the underdog, and though he was in more ways than one the poet of social unrest, admiration for heroic passivity and humility was hardly part of his stock-in-trade. So Chesterton's best-known lyric, 'The Donkey', could not possibly have been written by John Davidson. Moreover, while John Davidson's verse is somewhat formal and syllabic, Chesterton's is infinitely flexible, built upon the ballad, which Davidson did not fully appreciate, in spite of the fact that he called some of his poems 'ballads'. What had John Davidson in his short-lined poems to do with rolling and shifting stress, or sprung rhythm, or anything relative to sprung rhythm? At any rate, not very much.

MAURICE EVANS

Background and Influences on Chesterton

THE works of Chesterton are essentially a product of their age. He began writing in 1901 with a vigorous attack on the spirit of the 1890s, and he continued to attack it ever after. The closing years of the century seem to have left so permanent an impression on his mind that his latest works are still tinged with the colour of his revolt; and we feel that to the end he was refuting forgotten heresies, like the monk, Michael, in his novel, *The Ball and the Cross*. A study of his philosophy must therefore begin with a rough survey of the period and what he thought of it.

His principal accusation is one of pessimism. In *The Defendant* (1901) Chesterton inveighs against a pessimism that had become almost fashionable. 'I have found that every man is disposed to call the green leaf of the tree a little less green than it is.' And at the other end of his life, he looks back with horror on the period. 'I had an overpowering impulse to record or draw horrible ideas and images; plunging deeper and deeper into a blind spiritual suicide' (*Autobiography*). It was to him a time of philosophical anarchy; and he felt the vital need for some form of optimism, even if it were based on the bare minimum of good.

The fundamental cause was pride, the belief that man has the right to criticize the universe, if it is not to his liking. 'What we call a bad civilization', he writes in *The Defendant*, 'is a good civilization not good enough for us.' Man is born into the world by no agency of his own, and yet feels that he can criticize it 'as if he were house hunting, as if he were being shown over a new suite of apartments' (*Orthodoxy*). This is the great spiritual sin; for it not only prevents us from making the best of inevitable and inescapable facts, but limits our existence by making us feel superior to many valuable planes of life. The *Confessions* of George Moore are saturated with this pride, a perpetual scorn of 'the vulgar details of our vulgar age'. And with that young man, exclusiveness leads to boredom.

Evolution is the main philosophical source of pride. It has become erroneously identified with progress and so can easily be used to establish the doctrine of human perfectibility. But evolution leads, at the same time, to moral scepticism, which is one of the chief diseases of the age. Ideals become useless because, as Chesterton points out in *Orthodoxy*, no ideal can be final, if we are perpetually evolving into something higher. The mad Darwinian in *The Ball and the Cross* says: 'Never trust a god you can't improve on.' Consequently morality must fail, because we have no permanent base on which to establish it. What may be immoral today may be moral tomorrow, and everything becomes relative. 'All had grown dizzy with degree and relativity . . . so that there would be very little difference between eating dog and eating darkie, or between eating darkie and eating dago' (*What I saw in America*).

Now even the 1890s did not go as far as cannibalism, but there can be observed a weakening of morality in the interests of evolution. The New Hedonism of Oscar Wilde preaches the sacrifice of accepted morality to the production of a higher type of individual. A man must cultivate a finer sensibility even if it leads him into crime in the search of new experiences. Similarly Nietzsche in Germany, under the evolutionary spell, teaches revolt against all repressive morality in the interests of the superman who is to be evolved. The superman of *Thus spake Zarathustra* exemplifies the survival of the fittest. In man's ruthless struggle for existence the weaker is crushed and the cruel dominant virtues of will destroy the humbler ideals of Christianity. George Moore talks of 'the terrible austere laws of nature which ordain that the weak shall be trampled upon, shall be ground into death and dust, that the strong shall be glorious and sublime'. This denial of accepted morality appears in a milder form in Shaw's golden rule that there is no golden rule, another manifestation of moral scepticism.

But if evolution leads to a conception of the superman, it also produces fatalism. If man has only survived by a struggle, his whole history becomes something harsh and predestined, directed entirely by nature, 'red in tooth and claw'. The works of Hardy illustrate this fatalistic attitude. The characters of *The Dynasts* are puppets in the hands of a callous and impartial divinity; the Mayor of Casterbridge is fated from the beginning, and Tess of the D'Urbervilles is the sport of the Immortals. It is significant, too, that Hardy's novels are frequently tragedies of love in which the animal aspect is greatly emphasized. Marriage in *Jude the Obscure* is, indeed, defined as a licence to make love to someone on certain premises. Through the idea of evolution, sex has become identified with the animal in man, 'the ape and the tiger', and a type of fatalism is postulated from the sway of our lower natures. This too produces moral scepticism, for you cannot blame a man for his actions if he cannot help himself. Evolution, therefore, is one source of the pessimism which Chesterton analyses.

The rationalistic philosophies of the period also end in pessimism. Rationalism implies that human reason is capable of understanding all things and discovering consistency everywhere. In consequence it forces everything into a narrow scheme and limits the rich diversity of the universe, leading logically, though not inevitably, to madness. There is a brilliant chapter on this point in *Orthodoxy*. Imagination accepts contradiction and so is sane; but reason seeks to explain: 'to cross the infinite sea and so make it finite; and the result is physical exhaustion.' The completely logical man is, in fact, the madman. He has an obsession and finds evidence for it in everything he sees. For example, if he has a persecution mania, he will find a conspiratorial significance in 'the ordinary thoughtless actions of a sane man slashing the grass with a stick, kicking his heels'; they will become signals to accomplices, threats of violence or deliberate snubbings, all logically fitted into the framework of the obsession. 'The madman is the man who has lost everything except his reason.' To impose a logical consistency on varied facts in this way is a possible interpretation of those facts, but it is a poor and limited interpretation. It denies all the irrelevant variety of the evidence it incorporates; and so produces a limited experience and happiness.

In the same way rationalistic philosophies like materialism are an interpretation, but a very narrow interpretation, of the facts. It may be possible to argue that 'history has been simply and solely a chain of material causation': but to do so is to deny all the rich diversity of history. 'Materialism', he writes, 'has a sort of insane simplicity ... but if the cosmos of the materialist is a real cosmos, it is not much of a cosmos' (*Orthodoxy*). But materialism, if logically pursued, leads beyond a merely limited view of life, to total scepticism. We can never prove the existence of physical objects corresponding to our sense impressions, for example: we can only believe that they exist. To attempt such a proof leads inevitably to the scepticism of Hume. Chesterton seems to have attempted this proof in his youth and writes of it in the *Autobiography*, 'while dull atheists came and explained to me that there was nothing but matter, I listened with a sort of calm horror of detachment, suspecting that there was nothing but mind'. To be a sceptic in this way is to be imprisoned in one's own mind.

Chesterton also mentions the cult of realism, which insists on stressing the seamy side of life. Zola was extremely influential for a time, while Ibsen, in such plays as *Ghosts*, opened up the new channels of heredity and the horrors it suggests. The genuine Calvinism of Predestination was fortunately dead, but its influence survived in a few clumsy vetoes and taboos still capable of spoiling the major pleasures of life. *The Way of All Flesh* or Sir Edmund Gosse's *Father and Son* show the position of dying Puritanism.

It is against this puritan aspect of Victorianism that the decadents are primarily in revolt. They have all a strong prejudice against religion and common morality, 'the seven deadly virtues'. Dorian Gray has only to call morality middle class to damn it as effectively as the epithet 'bourgeois' nowadays. George Moore tells us how he shook off his belief in Christianity and the 'intellectual savagery' of Catholicism at a very early age with the help of Shelley. 'It is only natural for me', he writes, 'to oppose the routine of daily thought.'

Pater's statement that life should be a work of art is largely the foundation of the aesthetic cult. Man, he says, has only a limited time on earth and must exclude from it all but the finest experiences and choicest sensations. These he must cull with the care of an artist. 'To burn always with this hard gemlike flame, to maintain this ecstasy, is success in life. What we have to do is to be for ever curiously testing new opinions and courting new impressions, never acquiescing in a facile orthodoxy.' The New Hedonism of Wilde takes this as its basis. The aim of life is self-development by experience of the finest and subtlest pleasures: you must never repress an impulse, for that is to deny the ego. Dorian Gray sets out on a deliberate search for sensations 'that would be at once new and delightful and possess that element of strangeness which is so essential to romance'. The hedonist must be eager for every shade of sensation and his criterion must be purely their aesthetic value. 'The aesthete was all receptiveness, like the flea', writes Chesterton at the time, in his book on Shaw. 'His only affair in the world was to feed on its facts and colours like a parasite upon blood. The ego was the all.' This desire for perpetual novelty inevitably drove them into the strange and exotic when the common experiences had been tried out. George Moore loved exotic plants and furnishings in his chambers, while Dorian Gray passed from rarity to rarity. Sometimes he would collect strange perfumes, sometimes wonderful and bizarre weapons or uncommon jewels. Similarly, when the permissible excitements have been tested one goes into crime and perversity. Baudelaire had already set the fashion with his corpses and his 'affreuse juive', and Moore takes a self-conscious delight in frequenting the lowest haunts of Paris. Dorian Gray, once more, lives among prostitutes and perversity; and Oscar Wilde himself illustrates the belief that 'sin is the only colour element in modern life'.

Books such as *Dorian Gray* and George Moore's *Confessions of a Young Man* seem to have filled Chesterton with enormous disgust, and he launches a strong and varied attack on them. Aestheticism is obviously an unsocial creed, and being based on receptivity, an uncreative creed as well. But apart from objective judgements, it is unsatisfactory to those who practise it, he argues, for it is built on faulty psychology. Its aim is to live among perpetual high lights and at unending emotional crises, but this is an impossibility. The fallacy lies in the fact that 'men cannot even enjoy riot when the riot is the rule ... there is no fun in being lawless when lawlessness is the law'. The essence of pleasure is concentration and contrast, for 'when everybody's somebody, then no one's anybody'. The aesthetes themselves realized this and attempted to vary their pleasures as much as possible, alternating sensations of art with sensations of sensuousness or crime. Nevertheless, pleasure was the end of all and the mind became inevitably sated.

In the same way the doctrine of complete self-gratification is bad psychology. Pleasure must be paid for because part of its very intensity comes from resistance or self-control. To gratify every impulse at once destroys this intensity, as the breaking of a dam reduces all water to the same level. The moderns, Chesterton writes in 1901, are suffering in 'the hell of no resistance' and a 'hedonism which is more sick of happiness than an invalid is sick of pain'. True pleasure is in its nature exceptional and must be made distinguished by contrast, by repression, by complete self-committal; otherwise the life of the hedonist will be a long monotony. The decadent is like the butterfly; his round of pleasure 'need be no more amusing than a postman's, since he has no serious spiritual interest in any of his places of call'. This is the real reason for the boredom of the young men, and for Dorian's loss of desire.

Several positive philosophies were tried expressly to overcome this pessimism, but none were found satisfactory. The Church, as we have already indicated, was in very low water, and Butler estimates in *Erewhon* that 90 per cent of the people regarded it with something not far removed from contempt. There is in place of it a certain revival of Paganism, though of a superficial nature. It is not, however, the genuine nature worship which is at the root of the best pagan beliefs; but rather a reaction from Christian humility and mercy. Springing as it does from a general loss of beliefs it tends to exalt the violent animal virtues of strength and will, and it insists on the beauty of the flesh against Victorianism. Swinburne is its prophet in his *Hymn to Proserpine*:

> Thou hast conquered, O pale Galilean, and the world has grown grey from thy breath.

Christianity has driven out the beauty and joy of life with its praise of meekness. George Moore, in the tradition, prefers the 'bold fearless gaze of Venus' to the lowered glance of the virgin, and vindicates the body against Victorian respectability:

> The clean pagan nude, a love of life and beauty, the broad fair breast of a boy, the long flanks, the head thrown back.

Chesterton does not suffer from the prudish aspect of Victorianism, and has much sympathy with Paganism. It is good up to a point, because it insists on the worship of natural forces and the beauty of the world, but it is a dangerous philosophy, because there is nothing to prevent it from running to abuse. 'There is nothing in Paganism to check its own exaggerations', he writes (*The Well and the Shallows*), and describes in *St Francis* the terrible Priapian orgies to which nature worship descended in the decadence of Rome. What begins as love of life and beauty generally ends in the worship of eroticism or force. This is the invariable history of Paganism, and a fact borne out by its revival in the nineteenth century. Nietzsche's doctrine of the will is a worship of force for its own sake, and George Moore is unpleasantly sadistic. He tells us, if we may refer to the *Confessions* once more, that he kept a python in his rooms and took great pleasure in the 'exquisite gourmandize with which it lubricated and swallowed live guinea pigs'. He despised pity, 'that most vile of virtues', and wallows melodramatically in the most bloodthirsty visions—'to see the great gladiators pass, to hear them cry the famous "Ave Caesar", to hold the thumb down, to see the blood flow, to fill the languid hours with the agonies of poisoned slaves'. George Moore seems to have been less dangerous in practice than he would have us believe; but it is significant that the *Confessions* ran through several editions and drew a rather flattering comment of 'audacious' from Pater.

Apart from this danger, Chesterton points out that Paganism is ultimately a sad religion. Staking all on the beauty and joy of this world, it can offer nothing after. 'The Pagan', he writes, 'was in the main happier and happier as he approached the earth, but sadder and sadder as he approached the heavens' (*Orthodoxy*). We are reminded of the strain of melancholy which runs through so much of the Greek drama. Paganism results in what Chesterton calls the *carpe diem* philosophy.

> Gather ye rosebuds while ye may.

But this is not the philosophy of happy men: 'great joy does not gather the rosebuds while it may,' he says in *Orthodoxy*, 'its eyes are fixed on that immortal rose which Dante saw.' Omar Khayyám drinks not out of happiness but because he is very unhappy:

> Drink—for once dead you never shall return.

Neither are the Pagan virtues really helpful ones. Temperance and justice are reasonable virtues, and reason can only convince us of our inevitable failure when difficulties are very great. For happiness you need the mystical virtues, such as hope and faith. 'The only kind of hope that is any use in a battle is a hope that denies arithmetic.' Such is the gist of Chesterton's attitude to Paganism, a position which he defines frequently.

He deals more summarily with Nietzsche. The Nietzschean philosophy is a reaction against materialistic determinism, and has considerable influence on Shaw at the beginning of the present century, besides attracting Oscar Wilde. To Chesterton its teaching consists primarily of revolt and the praise of will, though in fact it contains very much more. In *Orthodoxy* he concentrates on these, attacking them from a variety of angles. Nietzsche, he says, praises revolt for its own sake and denies all limitation, when in practice, to do anything at all is to impose a form of limitation; for every definite action is an exclusion of all other actions; so that the complete denial of limitation is a contradiction in terms. Again, he argues that Nietzsche falsely identifies revolt with the will. All action is a result of will; not to revolt as much as to revolt, and will only becomes distinguished when it produces a distinguished action which need not be a revolt against morality. To praise revolt for its own sake is, indeed, to be paralysed, for if you are in revolt against everything, you cannot revolt in favour of anything. Some of these arguments are perhaps verbal quibblings, but there is truth in the general idea that our revolt must be limited and organized.

Another answer to the pessimism of the time is to be found in the cult of the future which derives from evolution. The air is full of sunny prophecy, and Nietzsche's superman is echoed in Shaw's play and H. G. Wells's giants reared on the food of the gods. The hope of the race depends on eugenics and the present must be sacrificed to the future. Chesterton takes a rather more realistic view of the problem, in *Heretics*. Our knowledge of eugenics is too small to produce any accurate results in the sphere of human relations: you may produce the body but not the mind. Yet even supposing that the science of eugenics has been sufficiently mastered, the question arises, what kind of a superman do we want? And here the difficulty begins, because no two men agree. Wells desires something generally more altruistic; Shaw has a vague conception of a better body and larger brain, while Nietzsche demands more will. No one indeed can be sufficiently isolated from his own age to see what it really needs. 'If you are to breed men as you breed pigs, then you demand a man as much more intelligent than the ordinary man as a man is more intelligent than a pig.' In any case, Chesterton points out that it is doubtful whether the superman is any happier. 'It is to the ordinary man that odd things seem fantastic. To the extraordinary man, they are simply ordinary.'

We must mention also a slight revival of pantheism and oriental philosophy, the danger of which Chesterton seems greatly to have exaggerated. He goes out of his way to refute it in *Heretics* and *Orthodoxy*, and writes his fantastic novel *The Flying Inn* as a warning against its acceptance. Possibly the popularity of Omar Khayyám was responsible for his fears, though that owed its contemporary vogue much more to its pessimism than to its pantheism. Whatever the cause, Chesterton believes that our pessimism may drive us to lose our individualities in a mystical unity with the whole world, and forget our sorrows with ourselves. As a philosophy, pantheism denies the separateness of things, which is to deny love. 'I want to adore the world not as one likes a looking glass, because it is oneself,' he writes in *Orthodoxy*, 'but as one loves a woman because she is entirely different.' This is analogous to Aldous Huxley's argument that there must be a strict division between self and not self for love to exist at all.

In conclusion, something must be said about the optimism of the early 1900s. It is a term which covers many shades of belief from a harassed and hardly won optimism to the cheapest jingoism. Chesterton begins by calling himself an optimist, on the principle that you cannot go on living unless there is some good in the world. His attitude is based on the very minimum of good, thankfulness for the mere fact of life. At the other extreme there is complete satisfaction with the present, and a denial of all evil, in endless vistas of progress. Wells and Kipling both suffer from this to some extent, and Chesterton attacks them for it. It is obvious that such complete optimism will prevent reform as effectively as complete pessimism.

This is roughly Chesterton's analysis of thought at the beginning of the century. It is not a complete nor perhaps a fair analysis, but it is the basis of nearly all his subsequent philosophy. He has seized on to the salient point of pessimism and shown how a variety of philosophies contribute to it.

FRANK SWINNERTON
A True Edwardian

THE true Edwardians, it seems to me, who represent the tumultuous charge of Liberalism passing over the country between 1906 and 1910, were G. K. Chesterton and Hilaire Belloc, with the Victorian William Watson as a sedate and unheeded singer of Liberal songs far from the battlefield. Chesterton and Belloc, both men of genius, were carried to renown by this glorious charge. With their colleague Charles Masterman, they spoke and wrote everywhere, cutting and slashing and waving large banners of purple and gold. But when the charge had spent itself, Belloc and Chesterton were somehow cut off from their fellows. They both worked indefatigably during the Georgian age. But they were not of it. Something happened to their bugles. Instead of sounding triumphantly a further charge, these instruments uttered nothing but the retreat. It was accordingly left to others to lead both Liberalism and literature down steep places into the sea. Chesterton hoped to the end that he was right, and almost with his last breath announced that the true revolution is the counter-revolution; but he lost pace with the times and will become a classic rather than an influence upon our day.

*   *   *   *

A TRUE EDWARDIAN

With Chesterton I once had an anonymous conversation, when he called at J. M. Dent's office after hours with a Dickens preface, and refused to leave it with me because (as a little boy) I could not give him in exchange the agreed remuneration; with Belloc I once, as an adult, shook hands. Yet when I was sixteen, and Belloc and Chesterton, with Wells and Shaw, were all a single radiance in the firmament, I heard Belloc deliver what I thought, and still with less confidence think, must have been the most brilliant speech ever made—to the Fabian Society, on the differences between Collectivism and Co-operativism;—and at the same period I used with a couple of equally devoted friends to tramp round London in order to listen to Chesterton lecturing, as he was ready to do, upon every subject under the sun, from the Dreadful Danger of Liberalism (which was the persistence in the party of hereditary caste) to Puck. So while Belloc and Chesterton did not know me, I knew them, and can testify to their former grandeur in the eyes of the young.

Probably young people now can hardly imagine how these four men, Shaw, Wells, Chesterton, and Belloc, stood out from their fellows and argued among themselves for the enlightenment of extraordinarily mixed audiences. Wells, being no speaker, largely contented himself with printed argument; the others were as happy face to face as they were in column. Sometimes Shaw would debate with Chesterton, and sometimes with Belloc: once, I recall, he debated with Chesterton while Belloc took the chair and rang an infuriating bell: upon another occasion he met Belloc in the large Queen's Hall, when an audience of three thousand people heard these two discuss the question of whether a Democrat who was not also a Socialist could possibly be a Gentleman, and came away with the problem unsolved after two hours of resolute hard hitting. The admiration between these men was not that of the literary nest, but that of the battlefield or the ring; and I think it fair to say that if any one of them wrote or spoke of another it was as an opponent whose nose at all costs must be struck and struck hard. This was especially true of Belloc; for while all felt a tenderness for Chesterton, none—engaged in a battle for self-preservation—had the smallest tenderness for Belloc, and none, certainly, received quarter at his hands.

One reason for the love of Chesterton was that while he fought he sang lays of chivalry and in spite of all his seriousness warred against wickedness rather than a fleshy opponent, while Belloc sang only after the battle, and warred against men as well as ideas, for the love of fighting and the pleasure he took in what might be called the deployment of the intellect. Another was that Chesterton could be distracted by a joke or an absurdity, whether it occurred to his own mind or to the mind of his foe, while Belloc was a master of the divagation or parenthesis, and never lost his place in an argument. But a third, and very important reason, was that it was often supposed that Chesterton might have been convinced if only Belloc had not stiffened him with recalcitrance. As the front legs of a performing horse are supposed to be the leaders (for Front-Legs works the head and ears), so in the monstrous animal conjured up by Shaw's conception of the Chester-belloc Belloc had the teeth and claws, while the warm and jovial heart belonged to Chesterton. The public said 'GBS' and 'GKC': it said neither HB nor HGW. These were a few—by no means all—of the differences between Chesterton and Belloc as they were seen by those who disagreed with both.

* * * * *

Gilbert Keith Chesterton's life corresponded with Belloc's in no particular. He was not born a Catholic; he was a convert. He was not born in France, but in the London district of Kensington, which is as much associated in the popular mind with respectability as Chelsea and Bloomsbury are associated with aesthetic affectation or Tooting with the inexplicable gibes of professional humorists. He was four years younger than Belloc, having been born in 1874. He went to St Paul's School, London, and did not go to a University. Nor did he engage in any military training; but instead attended classes at the Slade School, tried office life without success, and first began writing for the press by reviewing books on art for Robertson Nicoll's monthly magazine, *The Bookman*. From reviewing books on art he passed to reviewing books upon all subjects; and in 1900, when he was 26, he passed from reviewing books to publishing them.

The earliest title-pages to bear his name were those of *The Wild Knight*, a collection of poems which included a poetic drama notable for its brevity, and *Greybeards at Play*, a collection of rhymes inscribed to 'E.C.B.' (long subsequently, of course, the author of *Trent's Last Case*) after this fashion:

> He was, through boyhood's storm and shower,
>   My best, my nearest friend;
> We wore one hat, smoked one cigar,
>   One standing at each end.

These books were followed in the autumn of 1901 by a collection of reprinted essays called *The Defendant*; and from that time Chesterton was what is called an author, in contradistinction to a journalist. But he remained a journalist. Many of his books are made up of material which first appeared in periodicals. Others were written because publishers insisted. At least one such book owes its existence to Chesterton's quixotry. He was invited to write a history of England; and he declined on the ground that he was no historian. Some time later, he agreed with the same publishers to supply a book of essays. Hardly had the agreement been signed before Chesterton discovered that he was bound to give this book of essays to another firm, and in great distress he asked if his newly made contract could be cancelled. In return, he offered to write any book which might be considered an adequate substitute. The publishers, concealing jubilation, sternly recalled their original proposal for a short history of England. Shrieks and groans were heard all the way from Beaconsfield, but the promise was kept. The *Short History of England* was what Chesterton must have called a wild and awful success. It probably has been the most generally read of all his books. But while the credit for it is his he must not be blamed for impudence in essaying history, when the inspiration arose in another's head, and when in fact no man ever went to the writing of a literary work with less confidence. This story will explain one of his publications: the stories of others must be innumerable—as the books themselves are almost innumerable. To his constant lecturing I have already referred. His journalism was never absent from the London press, and it remained vigorous to the end of his life. What profligate expenditure of energy for one who was born a poet!

It was in *The Daily News*, when that paper was edited by A. G. Gardiner, that Chesterton made his earliest reputation. He used to write in its columns upon all manner of books and other pretexts. A single sentence would be enough to set him at work with an antithesis or proposition that brought the stars into Fleet Street and light into many dark places. Gardiner, his former editor, in an affectionate sketch printed in *Prophets, Priests, and Kings*, pictures the young journalist as he was in those days. 'You may track him', said Gardiner, 'by the blotting-pads he decorates with his riotous fancies, and may come up with him in the midst of a group of children, for whom he is drawing hilarious pictures, or to whom he is revealing the wonders of his toy theatre, the chief child of his fancy and invention, or whom he is instructing in the darkly mysterious game of "Guyping," which will fill the day with laughter. "Well," said the aunt to the little boy who had been to tea with Mr Chesterton,—"well, Frank, I suppose you have had a very instructive afternoon?" "I don't know what that means," said

Frank, "but, oh!" with enthusiasm, "you should see Mr Chesterton catch buns with his mouth."'

Charles Masterman, another friend, and one who in old days was as inseparably linked with Belloc and Chesterton as Athos himself was linked with Porthos and the priestly Aramis, once told me how Chesterton used to sit writing his articles in a Fleet Street café, sampling and mixing a terrible conjunction of drinks, while many waiters hovered about him, partly in awe, and partly in case he should leave the restaurant without paying for what he had had. One day—I do not know whether Chesterton was present or absent—the head waiter approached Masterman. 'Your friend,' he whispered, admiringly, 'he very clever man. He sit and laugh. And then he write. And then he laugh at what he write.' It was always essential to Chesterton that he should be amused by what he wrote, and by what he said in public. I have heard him laugh so much at a debate that he gave himself hiccups for the rest of the evening.

In early days he was very nearly as big as he afterwards became, but whereas, towards the end of his life, his much-thinned hair straggled untidily, like a blown wisp of steam, it was then solid and well-brushed. A feature of it was a Whistlerian white plume in the centre; and anybody who sees an old photograph of Chesterton will there find the plume as neat and trim as brush could make it. Presently he grew more like Porthos; then like Doctor Johnson; and at last like the famous portrait by Velasquez of Don Alessandro del Borro. His gigantic aspect became a matter of common reference. The story of the man who gave up his seat to three ladies was associated with him. Finally, a serious illness contracted his figure and forced him to live less strenuously than he had done. He went to Beaconsfield, in Buckinghamshire; and there he remained until his death, blind to the greatness of Disraeli, who took the name of the town as his title when he became a lord, but by no means blind to the forces of tyranny and the mischiefs wrought in the modern world by stupidity and exactitude.

Upon the public platform he swayed his large bulk from side to side; but he did not gesticulate. His speech was prefaced and accompanied by a curious sort of humming, such as one may hear when glee singers give each other the note before starting to sing. He pronounced the word 'I' (without egotism) as if it were 'Ayee', and drawled, not in the highly gentlemanly manner which Americans believe to be the English accent, and which many Englishmen call the Oxford accent, but in a manner peculiar to himself and either attractive or the reverse according to his hearer's taste. As he talked, and as he invented amusing fancies, he punctuated his talk with little breathless grunts or last gasps of laughter, so that he gave the impression—what with the drawl and the breathless grunts—of speaking very slowly indeed. He also gave the impression of speaking without any effort whatever, without raising his voice, or becoming intimidated by his audience or by lack of material to fill the time allotted to him, or feeling anything but sweet charity towards all those—even Jews, politicians, and sophisticates—whom he felt compelled to denounce. To those, accordingly, who care more for character than for opinion, more for talent than for fashion, Chesterton remains one of the great figures of his time. For the rest, and for those easily made impatient by his habit—they call it 'trick'—of antithesis, he was merely an ingenious and snort-provoking creature. He certainly fell far behind the times, and at the hour of his death was regarded as prehistoric; but whether this is a sin or only a misfortune of unbending genius I shall not now attempt to determine.

The explanation of the failure of Belloc and Chesterton to impress younger sceptics is that both were, in a sense, defendants. They attacked the trend of modern society towards mechanization as severely as any other writers whatsoever. But they did so from the standpoint of the Catholic Church. They said that the world was in a very bad way; but they both insisted that it was once—in the Middle Ages—in a very good way. Belloc, in a really masterly short work called *The Servile State*, published in 1912, established the fact that slavery is a familiar condition of European life. It is his argument that slavery was destroyed by Christianity, and that until the end of the Middle Ages it had ceased to exist in the West. He believed, as so did Chesterton, that all the ills of modern England arise from what followed the dissolution of the monasteries by King Henry VIII. If Henry, says Belloc, had done as he intended, and kept in his own hands—the hands of the Crown—the property taken from the clerical body, the country would have had a happy future. Henry was not strong enough to keep his appropriations; he found the rich men of his day too powerful for him, and was forced to hand over the greater part of the spoils. Hence the violent inequality of wealth and power in England out of which grew from the sixteenth century onwards the evils of Capitalism. Hence, in process of time, the inevitability of the servile state, to which all parties by one path or another have led and will lead the people of England.

It is not my business to comment upon the truth or otherwise of this theory; but there can be no question as to the clearness and power of the book in which it is outlined. And there can be no doubt, I think, that possession of such a view of history prevented Belloc and

Chesterton from capturing the imagination of generations increasingly influenced by scientific and mechanical theory and practice. It is one thing to say that the world is wrong—every reformer agrees that the world is wrong—but when, instead of proceeding to say that the world can be set right by something new, a man says that it can only be set right by a return to something old, he is thrown into a defence of the past. And the past, as Chesterton admitted, is no easy subject. 'I can make the future as narrow as myself; the past is obliged to be as broad and turbulent as humanity.' Both Belloc and Chesterton had to rewrite history for purposes of propaganda amid incessant interruptions both from dryasdusts and from ribalds who did not believe a word of what they said. Men who attack the present are always sure of support; men who contrast the present with the delights of an improved future society may be scorned as unpractical idealists, but they cannot be confounded by texts or refutations; men who insist that at some past time an ideal state existed may be challenged by the proven inability of that ideal state to withstand aggression, and they will certainly be floored by extracts from some old charter or pipe roll or antique letter which demolishes the whole structure they have so ingeniously erected.

That is what happened to Belloc and Chesterton. Belloc deliberately, and Chesterton with misgiving, set up a version—in Belloc's case a series of detailed versions—of what happened in England long ago. It was not accepted by Protestants, scientific historians, or sceptics. Belloc went further. He told us in several books how the French Revolution arose, succeeded, and failed (the early chapters of his *Danton* give the clearest exposition of the events preceding the Revolution which I have ever read, but I do not know if they are the truest); he in one book of *Miniatures of French History* told us what must have happened in France at various crucial points from 599 BC to AD 1914. He traced the history of warfare in England. His mind played over the entire history of Europe, and he expressed himself as to that history with a certainty and I imagine a consistency which ought to have satisfied every reader. But he did not satisfy every reader. Every reader could relish the style in which Belloc told his story; but every reader, in spite of the charm and certainty of the narrative, felt that Belloc was a partisan, bent upon proving a case. It is nothing new in historical writing; if the case is the popular case it will be swallowed gladly; but when it conflicts with the case as presented by every Whig and Protestant historian, or with the case as overwhelmingly demonstrated by the ironist Gibbon, it is suspect from the start.

Now Belloc adored Froissart. He owed much to the gargantuan

historical method of Rabelais. He believed that legends and ballads are better authorities than pipe rolls. He used his imagination in describing battles (for which he had a peculiar gift and fondness), political intrigues, and religious and economic influences as to which there are or are not written records. The results have been controversial. I need not dwell upon the horrid pursuit of A. F. Pollard whenever Belloc wrote upon the Tudors; nor upon that description of the campaign of Evesham in 1265 for which he gave as his chief authority the chronicle of Matthew Paris, who died in 1259. All I need say is that in a scientific period a man who sets out to prove upon general philosophical grounds that such and such events must have been caused in such and such a way, and in that way only, must expect to be challenged by all to whom his major premiss seems arbitrary. And that a man who writes history as if it had just happened under his eyes, in a scene as familiar to him as the palm of his hand, and for the glory of the Catholic Church, must expect to be smothered with dust brought directly from the Record Office.

But how delightful Belloc's historical method is! Take the following short passage from a book that is only musingly a historical work at all. Notice the grand introduction, the apparent candour with which two opposed views are presented, the preference for legend declared in the last lines:

It is a debate which will not be decided (for the material of full decision is lacking) whether since the Romans crowded their millions into this Africa, the rainfall has or has not changed. It is certain that they husbanded water upon every side and built great barricades to hold the streams; yet it is certain, also, that their cities stood where no such great groups of men could live today. There are those who believe that under Atlas, towards the desert, a shallow sea spread westward from the Mediterranean and from Syrtis; there are others who believe that the dry water-courses of the Sahara were recently alive with streams, and that the tombs and inscriptions of the waste places, now half buried in the sand, prove a great lake upon whose shores a whole province could cultivate and live. Both hypotheses are doubtful for this reason—that no good legend preserves the record. Changes far less momentous have left whole cycles of ballads and stories behind them.

Would you not rather read history written thus than as it is told by so dead a person as Freeman, learned though Freeman was, or by somebody as accurate as Samuel Rawson Gardiner, or J. B. Bury? I have no doubt of it. But you might feel it wise—if a student—to read a little more deeply?

Chesterton suffered from the same charge of presenting what may be called unduly simplified history. He also had a preference for the legendary. He says:

Probably the rhyme which runs,

> 'When good King Arthur ruled this land
> He was a noble king,
> He stole three pecks of barley meal',

is much nearer the true medieval note than the aristocratic stateliness of Tennyson. But about all these grotesques of the popular fancy there is one last thing to be remembered. It must especially be remembered by those who would dwell exclusively on documents, and take no note of tradition at all. Wild as would be the results of credulity concerning all the old wives' tales, it would not be so wild as the errors that can arise from trusting to written evidence when there is not enough of it. Now the whole written evidence for the first part of our history would go into a small book. A very few details are mentioned, and none are explained. A fact thus standing alone, without the key of contemporary thought, may be very much more misleading than any fable. To know what word an archaic scribe wrote without being sure of what thing he meant, may produce a result that is literally mad. Thus, for instance, it would be unwise to accept literally the tale that St Helena was not only a native of Colchester, but was a daughter of Old King Cole. But it would not be very unwise; not so unwise as some things that are deduced from documents. The natives of Colchester certainly did honour to St Helena, and might have had a king named Cole. According to the more serious story, the saint's father was an inn-keeper; and the only recorded action of Cole is well within the resources of that calling. It would not be nearly so unwise as to deduce from the written word, as some critic of the future may do, that the natives of Colchester were oysters.

After that, and taking into consideration all that the method implies, it is perhaps hardly surprising that there are some—including myself—who believe Belloc's province to be not history, but that exceedingly personal product, the essay, of which he is complete master; and Chesterton's province not history, but poetry, or that beautiful kind of fable which he created for the exploitation of Father Brown, the priest-detective. Belloc at large, sailing, tramping, debating with the reader, singing, observing the country and the people he meets, eating and drinking with gusto, and writing with triumphant relish both of himself and his language (which he handles as a steersman handles a small boat at sea when the wind is freshening); Chesterton taking a common story, ingeniously twisting it, and at last leaving the earth altogether with a style that suggests the upward soaring (never the graceless flap) of a children's kite; both bringing to their art a love of life so uncommon at the present time as to ravish us with a sense of what they mean when the speak of Merrie England and 'that laughter that has slept since the Middle Ages'—do they not thus establish themselves as very important figures in modern literature?

I am sure that most of us would find the Middle Ages an occasion for—it is the 'modern' word—disgust if we were transported back into them. Belloc and Chesterton would not have done so. Belloc would have been happy if he had been born in the days when disputatious scholars went from University to University upon the Continent and argued for their bread. He had the hardiness to sail his own boat across the English Channel in a fog; and a life that gave him wine to drink, brains to test, and a soul to shrive was ever the good life for him. Chesterton was less hardy, but he was one, nevertheless, to whom eating and drinking and good talk could be as well conducted from the floor as from a modern chair. Neither Belloc nor Chesterton had anything of the funereal sleekness of the modern aesthete, the silence, the self-engrossed aloofness, the lack of good fellowship, the pinched assumption that humour is not humour unless it is concerned with sexual symbols. Yet neither Belloc nor Chesterton had the dreary boisterousness of the boon companion. Each was an original.

When Belloc wrote *The Path to Rome* he was following a recognized literary road—the writer of charm going alone upon a trip among strange people, laughing, learning, and then posing a little before the world. Stevenson had done just such a walk through the Cévennes, and his book about the journey had greatly pleased the sentimentalists. Rather earlier in history, Laurence Sterne had also told the story of his travels in France rather better than Stevenson (but as I am going to speak presently of Chesterton I must beware of saying anything ill of one of his idols) and had pleased even those who were not sentimentalists. And so Belloc knew what he was doing when he took this path and set down his account of what happened to him en route. But, miraculously, he avoided sentimentality; there was in him a certain robust and boyish courage and simplicity which, although he could chaff the reader and at times do a little bragging of one sort or another, gave *The Path to Rome* a character of its own. It remains Belloc's best long book, and the one by which most of his admirers would wish him first to be known. It is a tale of dangers run, and fears acknowledged, of hours enlivened by nonsense and accidentally good or evil meals. It is a chronicle of moods, a picture of mountains and forests and small towns, a traveller's tale, an enchanting monologue, and anything else the reader fancies. And it is the early work of a man who could already do anything he wished with his pen.

That, indeed, was one of Belloc's weaknesses—that he was always so much a master of his pen. He was too versatile for the mind of a public that prefers repetition. And furthermore it was sometimes difficult to discover at what point the serious Belloc yielded to the extravagant Belloc; for in the grand manner which he often used an able and humour-filled writer must often check himself sharply lest he guy his own grandeur. Let any reader take Belloc at his most serious, and then turn to his satirical novel, *Emmanuel Burden*: it will be found that *Emmanuel Burden* contains passages, despite its satirical character, which strongly resemble Belloc at his most serious, and barely exceed the sonorousness of his more rhetorical mood. That does not happen in so finished and consummate a work of irony as *The Mercy of Allah*, of course; but *The Mercy of Allah* is extremely mature, whereas *Emmanual Burden*, being experimental, is particularly interesting to the student of Belloc as showing his mind more fitfully playing between anger and pity, and particularly informing as showing the danger to a stylist of the power to burlesque all styles, including his own.

Even so, Belloc is a more persistent and sustained writer than Chesterton, who was best in short fights. To take one single work of his as an example, *The Flying Inn* is a laborious and aimless extravaganza, spoiled by the author's inability to keep upon one plane and by his uncertainty as to the particular plane he is occupying. But the lyrics in it, and single episodes or paragraphs where inspiration caught up the pen and made it caper, are sufficient to make *The Flying Inn* a book known to every man of humour in England. It would seem to have been an effort upon Chesterton's part to write a modern Peacockian novel; but Chesterton's is a rambling wit and not an incisive one, and the resemblance between *The Flying Inn* and the scholarly Peacock's *Melincourt* is in favour of the older book in all except the poetical passages. Here, indeed, Chesterton was on his own ground.

But if we compare a book such as *The Flying Inn* with the volumes of Father Brown stories we can see how admirably the Chestertonian genius is fitted to colour the slightest tale with magic. Whether we observe how a crime can be committed by a postman because nobody sees a postman (or any other quite familiar figure) entering a block of flats; or hear the steps of the tall man who is a guest to the waiters and a waiter to the guests and uses his opportunity to steal all the fish knives and forks of the *Twelve True Fishermen*; or imagine the velocity of a hammer flung from a church tower upon a villain below, we are taken entirely away from the world we live in and into a world of dreams and strangeness. And we are taken, not because the events are strange in themselves, but because the author mesmerizes us with the

A TRUE EDWARDIAN

aid of the genius which he conceals up his sleeve. It is the literary genius, and nothing else, which effects the mesmerism; for the charm of *The Hanmmer of God*, for example, would vanish if the story were told by any other writer. Father Brown mounts with the criminal, a clergyman, to a stone balcony about the tower of a church:

Immediately beneath and about them the lines of the Gothic building plunged outwards into the void with a sickening swiftness akin to suicide. There is that element of Titan energy in the architecture of the Middle Ages that, from whatever aspect it be seen, it always seems to be rushing away, like the strong back of some maddened horse. This church was hewn out of ancient and silent stone, bearded with old fungoids and stained with the nests of birds. And yet, when they saw it from below, it sprang like a fountain at the stars; and when they saw it, as now, from above, it poured like a cataract into a voiceless pit. For those two men on the tower were left alone with the most terrible aspect of the Gothic; the monstrous foreshortening and disproportion, the dizzy perspectives, the glimpses of great things small and small things great; a topsy-turvydom of stone in the mid-air. Details of stone, enormous by their proximity, were relieved against a pattern of fields and farms, pigmy in their distance. A carved bird or beast at a corner seemed like some vast walking or flying dragon wasting the pastures and villages below. The whole atmosphere was dizzy and dangerous, as if men were upheld in air amid the gyrating wings of colossal genii, and the whole of that old church, as tall and rich as a cathedral, seemed to sit upon the sunlit country like a cloud-burst.

In such a scene, how could any man not exchange with Father Brown the glance which confesses to murder?

These two men, Chesterton and Belloc, were essentially writers and dialecticians. They sported with ideas and imaginings for the enjoyment of sport; for as Belloc says:

Human affairs have always in them very strongly and permanently inherent, the character of a sport; the interest (at any rate of males) in the conduct of human life is always largely an interest of seeing that certain rules are kept, and certain points won, according to the rules.

They had greater skill in dialectical writing than any of their contemporaries excepting only Shaw. They had the gift of writing with peculiar simplicity and beauty, and the utmost clearness. But when it comes to what they write, it must be said that Belloc was governed by his passion for propositions; and that Chesterton was governed by his passion for antithesis. Belloc said 'I shall show; I shall prove; I shall establish.' He would show that the French Revolution turned upon and was conditioned by its military history. He would prove that Robespierre, a weak man, did not create the Terror, but resisted it and was unwillingly driven to it by others. He would establish that the dissolution of the monasteries in England in the sixteenth century was the beginning of the industrial revolution and the Capitalist system. But he does not convince us about the French Revolution or about Robespierre or about the dissolution of the monasteries for at least three reasons. The first of these reasons is that we already hold other views (the basis for which he ignores in spite of all clamour) as to the events; the second reason is that his style, being authoritative, is unfitted for persuasion; and the third reason is that despite every ingenuity he is unable in the communications he makes to fulfil the promise he has given. The third reason is the fatal reason.

Chesterton in the same way made propositions; but they were less peremptory and less serious than Belloc's. He did not say 'I shall show.' He said: 'There is one metaphor of which the moderns are very fond; they are always saying, "You can't put the clock back." The simple and obvious answer is "You can" ... There is another proverb, "As you have made your bed, so you must lie in it"; which again is simply a lie. If I have made my bed uncomfortable, please God I will make it again.' But nobody, reading these remarks, can fail to make sufficient retorts to them; and the soundness of Chesterton's opinion is lost in a fritter of nonsense.

Now what can we do about such men? All who enjoy debate and the flexible use of thought and language must delight in their adroitness. All who can stand outside the stream of current opinion must observe how many ideas of virtue and value continually appear in their work. But when all is turmoil, as it is today, it is too much to expect that the occupants of a backwater will receive the proper rewards of literary genius. That is what happened to Belloc and Chesterton. They were regarded as old gentlemen doing whatever is the nautical equivalent of fiddling while Rome burns. For this reason their gifts—finally separated for ever from their views and resistances—will not be fully realized and acknowledged until at least a century has passed. How great those gifts were, in my opinion, could be stated only in terms which would seem at this time extravagant.

HILAIRE BELLOC
On the Place of Gilbert Chesterton in English Letters

THE main points of what I have to say in my fragment upon the conditions of survival in Gilbert Chesterton's writings, may be tabulated as follows:

1. The leading characteristic of Chesterton as a writer and as a man (the two were much more closely identified in him than in most writers) was that he was *national*.

I will point out in a moment what the effects of this were upon his treatment of various subjects.

2. The next characteristic was an extreme precision of thought, such as used to be characteristic of Englishmen, though in modern times it has broken down and people have forgotten how native it was to the English mind in the past.

3. The third characteristic I note about his writing and thought is a unique capacity for *parallelism*. He continually illumined and explained realities by comparisons. This was really the weapon peculiar to Chesterton's genius. It was the one thing which he in particular had, and which no one else in his time came near to, and few in the past have approached. It is the strongest element in his writing and thinking, after the far less exceptional element of sincerity.

4. The structure upon which his work, like that of all modern men, had been founded, was historical: but it was only in general historical; it was far more deeply and widely *literary*. (I believe I notice this the more because with me it has always been the other way about; I have a very great deal of reading and experience upon history, far less upon literature.)

5. Charity. He approached controversy, his delight, hardly ever as a conflict, nearly always as an appreciation, including that of his opponent.

6. Lastly, there is that chief matter of his life and therefore of his literary activity, his acceptation of the Faith.

Let me now consider each of these six points separately.

1. Chesterton I say was national in every way it is possible for a writer and thinker to be national, or for a character to be national in private life apart from its public activity with pen or voice. It is one of the things, perhaps the main thing, in which Chesterton can be

bracketed with the great character to whom he was so often compared: Samuel Johnson.

Often in his conversation with me I have noticed some colouring adjective or short phrase which showed how all his mind referred experience to national standards. Here, as in other matters, he drew from sources older and far stronger than the perversion of nationalism which had afflicted Englishmen more and more since his own youth.

The characteristic writer of this degraded mood is Kipling. With the mention of that name we are reminded of the opposing fates that confront the future of England. The two public writers that stand for these two tendencies are Kipling and Chesterton. Each is national—but in very different ways. Chesterton is national in himself. He is English of the English. To follow his mind and its expression is an introduction to the English soul. He is a mirror of England, and especially is he English in his *method* of thought, as he is in his understanding of things and men. He writes with an English accent.

Kipling has little about him of England or indeed of these islands. He has no sense of the English past or of the things natively and essentially English; he is rather of Asia and of the transplanted. But he is a representative of modern England through his enormous audiences. All that which we call today 'The suburbs' is full of Kipling, and to Kipling the English urban middle class immediately respond. His verse especially appeals to it, far more than do his few powerful short stories.

All of us who have travelled can witness to the effect of Kipling upon the reputation of England abroad. Kipling's ignorance of Europe, his vulgarity and its accompanying fear of superiors (which modern people call an inferiority-complex) have profoundly affected and affected adversely the reputation of England and the Englishman throughout the world. This for two reasons:

(*a*) The fact that Kipling was un-English and Imperial threw him at once into contact with the New World, by which term I mean the United States of the Protestant tradition and Australasia.

(*b*) Fate would have it that Kipling should be translated into French by a genius, who makes him out in that language a far greater writer than he is. Now French being, after Latin, the universal language of civilized Western man, the Continent of Europe has approached Kipling and has been affected by him and has given him a reputation upon these lines.

Chesterton was the opposite of Kipling as a nationalist and

conservative.[1] Because his family and individual traditions were of an older and far more cultured time than our own, he could not appeal either to the New World or to the Continental world in a way either could easily understand. Chesterton's work has never been *properly* translated into French, and three-quarters of the ideas he had to put forward, were so unfamiliar to foreigners that they could hardly be understood by them. His most concise and epigrammatic judgements are often taken as mere verbal exploits, and the half-educated and uncultured, who are of the stuff by which modern opinion is ruled, use of him the term 'paradoxical', in that special meaning of their own which they give to this word, meaning 'nonsense through contradiction'; not the original and cultured meaning of 'paradox': 'illumination through an unexpected juxtaposition'.

This national character of Chesterton came out strongly in his appreciation of the English landscape. He had little experience of seafaring life; he saw the sea not from a deck but as it is seen from the land: from an English cliff. He was strongly impressed by its clean horizon and by its strength; even by the great voice of it, which is only heard when it comes in contact with its limits on the shore. But in this omission, as it were, of seafaring, he was more national than ever; for though the English have a strong political and literary passion for the sea, they have as a people little personal familiarity with it, save on its shores. Thus you do not find among them the knowledge of the blind impersonal force of the sea which some have called the cruelty of the sea. Nor do you find among them a regard of the sea as the road to things *other* than themselves. The sea in the modern English mind is a road to commerce and to their colonial fellow subjects or administrators.

The nationalism of Chesterton was providential, not only for his own fame but for its effect upon his readers. It formed a bridge or link between the English mind as it has been formed by the Reformation (and particularly the later part of the Reformation, during the seventeenth century), and the general culture of Europe which was created by, and can only be preserved through, the Catholic Church.

Chesterton by his intellectual inheritance from the high Unitarian English culture was highly sympathetic with the general classical culture of Europe. He could illustrate it and pass it on (often unconsciously), as could not a writer or a man who knew not the soul

---

[1] I do not use this word 'conservative' in the silly and now obsolete sense of a Parliamentary Party but in its true sense: 'Conservative of Tradition', especially of tradition social and national.

of that culture. He could not have conceived a world which should be of our civilization in a fashion and yet not based on Latin and Greek.

I remember, some years before he was received into the Church and before he ever visited America, his asking me, as one with a wide experience of the United States, whether it were true that the Latin and Greek classics were there of no effect. I told him this was increasingly so, save in a very few academic coteries and, of course, in the ubiquitous and very numerous Catholic clergy, and those influenced by them.

I had a private conversation with him walking in the lanes of Beaconsfield in which he said to me, some two or three years before being received into the Catholic Church, that an obstacle which always presented itself to him was the alien character of the Faith in the eyes of a modern Englishman. He said that I myself was cosmopolitan in experience, for I had often talked with him about the way in which I had torn myself up by the roots in my twentieth year and had gone all over America by myself and had later undertaken the adventure of service in a foreign army. Also I had a foreign name and certain ties of blood abroad.

I pointed out that among his intimate acquaintance apart from myself was the example of Maurice Baring. He answered with justice that Maurice Baring also was cosmopolitan in his experience and outlook. 'What I want,' he said (I recall his exact words, for they made a profound impression on me), 'is some-one entirely English who should none the less have come in.' The objection is one that has occurred to all Englishmen in a spiritual crisis of this kind. It is inevitable.

I may conclude this point by repeating that the *national* character of Chesterton is strongly marked in his style. The construction of his paragraphs and the sequence of his reasoning is so thoroughly national that efforts at translating him have, as I have said above, failed. I also here repeat that if he had been less national foreign nations of the Catholic culture would be more familiar with him today than they are.

One example of this national character and style is the way in which the word suggests the word in his writing: a thing not unconnected with the effect of the Jacobean Bible on the English mind since the seventeenth century, particularly the Epistles of St Paul. Renan has remarked on this characteristic in the Epistles. It is in this connection that the frequent use of puns, or, when they are not puns, plays upon words in Chesterton's writings, should be noted.

Lastly, there is the national character of high individualization, which some have also called 'localization'; the preference of concrete connotation to abstraction. Chesterton is in the full tradition of those creative English writers from Chaucer to Dickens, who dwell not upon ideas but upon men and women, and especially is he national in his vast survey of English letters, whereof Kipling is wholly ignorant.[2]

2. I have said that the second leading characteristic of Chesterton's mind and habit of writing is precision in reasoning. Many might superficially say that this talent clashed with his nationalism and was even contradictory thereto. I cannot agree.

The English, even the modern English, formed by the Reformation, have excelled in precision of thought. You see it in their vast volume of deduction in law, comparable to that of no other nation. You see it also in their special department of economics—a science which one may say was created in England.

This highly English tendency to precision has intellectual drawbacks as well as intellectual advantages, though the advantages can hardly be exaggerated. One drawback lies in this: that a man having exactly defined his terms and noted the weakness of an opponent's argument from the use of the same word in various senses, tends to verbalism. He is always in some danger of missing an opponent's just conclusions, when these have been arrived at from erroneous premises.

All thought is deductive. The opposition of 'induction' to 'deduction' is but verbal jingle. Induction is not thought at all. We have it in common with the animals. It is mere experiment and observation. Strictly speaking, when you think out a conclusion, that process must be deductive. For that very reason, you must be certain before you establish your conclusion firmly that your premises are sufficient. Now no premises other than the mathematical are wholly universal. So true is this that modern sophists have even sunk to attempting the reconciliation of contradictories. Great havoc have they made with this folly.

Take it by and large, Chesterton's passion for precision of thought was an overwhelming advantage for him over all his modern opponents in controversy, especially for his modern opponents of English speech, or rather of Protestant English culture.

In theology, for instance, he excluded Modernism with the nearest approach to contempt that a mind of such wide sympathy could achieve. He was impatient of all ambiguous nonsense, but it was in his

[2] He who would judge the *Nationalism*, the 'Englishry' of Chesterton's mind in contrast with the 'Imperialism,' the un-English tone of Kipling's should contrast their drawings as well as their writings. Chesterton's innumerable drawings of human expression are quintessentially English, distilled and double distilled English. Kipling's efforts in this line are as decadent as the French Latin quarter of a life-time ago, or as Aubrey Beardsley's exact and poisonous line.

nature to leave it aside rather than to transfix it with ridicule or invective. Had he indulged more in the latter, it is my judgement that his effect would have been stronger and perhaps more permanent: for it has been well said that the fame and effect of a man are buttressed by his enemies. As I shall come to say when I speak of his spiritual virtue, he made no enemies.

This habit of precision in thought and diction made of Chesterton a sort of what the French call a *revenant* on the highest phase of European and Christian thought. A *revenant* is one who comes back, one who reappears. Unfortunately, it also means a ghost. If you exemplify in your mind and also in your style an intellectual perfection which your contemporaries have lost, you will be a *revenant* and in danger of having less effect on those contemporaries. Although Chesterton's precision of thought and supreme talent for exact logic had much to do with his failure to conduct the mind of his contemporaries, he did influence that mind through the emotions. For indeed, his contemporaries of the Protestant culture live upon emotion and know of hardly any other process for arriving at conviction.

Fools are fond of repeating in chorus that the English are not a logical nation. They might as well say that the Italians are not an artistic nation, nor the Spaniards a soldierly nation. The English have been the special masters of logic in the past and still use it with a razor-like edge in circumscribed contemporary discussion, as you may see any day by listening to the pleadings in an English law-court, or by reading any one of the principal arguments advanced for theories in physical science by the English discoverers of the nineteenth century such as the great Huxley.

What men mean when they say that the English are not a logical nation is that nothing in their modern education makes them familiar with logic in the largest matters, whether these be political or the supreme matters of religion. It is a weakness, for with politics especially, but also with religion on its moral and practical side, an error in first premises is usually disastrous.

Thus you may predicate as a first principle the equality of man, which is an absolutely certain truth: men are only men by the qualities they have in common with their fellows. But unless you (*a*) get the term in its exact meaning and (*b*) supplement it by other equally certain general principles, you might be led into such absurdities as thinking that the two sexes have similar aptitudes for public life, or regarding the inexperienced as equally wise with the experienced, or denying the effect of wealth on the opportunities for acquiring experience and manners.

One effect of Chesterton's unique and exceptional precision of thought is the peculiar satisfaction his writing gives to men of philosophical training or instinct. But I use the word 'philosophical' here to mean the search for truth in the reasonable hope of attaining it; not a contemptible shilly-shally of opinion, or the still more contemptible practice of advocacy in defence of all theories or of none.

Men who are accustomed to the terse and packed rational process of the past, and particularly to the master mind of St Thomas, will always eagerly seek a page of exposition from the pen of Gilbert Chesterton. But men who cannot taste a truth unless it be highly seasoned with epigram and shock, will misunderstand his manner, because it will satisfy them for the wrong reason. Chesterton is perpetually pulling up the reader with a shock of surprise, and his pages are crammed with epigram. Neither the one nor the other is the heart of his style. The heart of his style is lucidity, produced by a complete rejection of ambiguity: complete exactitude of definition.

While there is in this, as I have said, a peril to his contemporary effect and to its permanence in one way, because he wrote in the English tongue and for a public melted into the last dilution of English Protestantism—a public therefore which was almost physically incapable of appreciating precision in the major matters of life—there is, on the other hand, a strong chance of permanence in another way. For your precise thinker stands unchanged: unaffected by the fluctuations of fashion in expression.

Here, as in many other connections, the permanent effect of Gilbert Chesterton's writing must largely depend upon our return or non-return to the high culture which we have lost. This means in practice the return or non-return of England to the Catholic Church. The English-speaking public, apart from the Irish race, is now Protestant. It has been strongly and increasingly Anti-Catholic for now 250 years. Through the effect of time it is today more soaked in Protestantism than ever it was before.

Here, as in every other matter, the permanence of Chesterton's fame will depend upon that very doubtful contingency—the conversion of England.

RONALD KNOX
G. K. Chesterton: The Man and his Work

My most cherished memory of the late Gilbert Chesterton is that of a luncheon-party with friends in Hertfordshire, after which he was asked if he would walk down to the end of the garden, so that a bed-ridden old lady upstairs might see, from her window, the great Mr Chesterton. He acceded readily enough, though it was with more difficulty that we persuaded him to remove the waste-paper basket with which he had modestly obscured his features. To relieve his self-consciousness, I suggested walking with him. 'Oh, do come,' he said, 'then you will look like *the ordinary person*.' He was (it is to be remembered) not only a fat man and proud of it, but very tall and broad. And I honestly think it was this physical greatness which he had the intention of parading, set off by contrast with the ordinary person, myself. But as I walked down such a garden path as he would have loved to describe, flaming with poppies and delphiniums by the side of an old mill-stream, I was vividly conscious that his intellectual greatness might have been set off, not by such an ordinary person as myself, but by almost any figure in contemporary life. Almost anybody was an ordinary person compared with him.

I call that man intellectually great who is an artist in thought. There have been artists in words who were content to borrow the thoughts of other men; there have been great thinkers who were content to express themselves anyhow. There are only a few whose thought seems to spring out of them clothed in words that adequately express it: Plato, for example, or Pascal. Chesterton was an artist in thought. He was an artist, in the sense of one who drew pictures, before he started writing; and most of us know how, in his pictures, a single figure, full of movement, stands out luminous from a vague background. So his mind saw things; it seized instinctively on the essences of them. When he writes about 'a primitive monster, with a strangely small head set on a neck not only longer but larger than itself: with one disproportionate crest of hair running along that neck like a beard in the wrong place; with feet each like a solid horn, alone amid the feet of so many cattle', it takes us aback at first, until we realize that it is a perfectly accurate description of the horse. So he saw, with a vision not given to many of us, that still stranger creature we call Man.

I call that man intellectually great, who can work equally well in any medium. I believe it is true that Chesterton walked into the office of his literary agent one day, and asked if there was any book the publishers wanted. 'Nothing in your line, I'm afraid: the last thing we heard of was the *Saturday Evening Post* wanting detective stories. 'Oh, well, I don't know', he said, and sitting down there and then wrote the first of the Father Brown stories. Detective stories, extravaganzas, poetry, drama, history, biography, essays, controversy—all came alike to him as his medium. He was not a careful craftsman in any of them; perhaps *The Ballad of the White Horse* is his most accurate piece of work: but always the luminous idea stood out—the idea we had never seen, looking at the facts a thousand times, because it was so simple.

I call that man intellectually great, who sees the whole of life as a coherent system; who can touch on any theme, and illuminate it, and always in a way that is related to the rest of his thought, so that you say 'Nobody but he would have written that.' Chesterton was such a man; the body of ideas which he labelled, rather carelessly, 'distributism', is a body of ideas which still lasts, and I think will last; but it is not exactly a doctrine, or a philosophy; it is simply Chesterton's reaction to life.

His work burst upon the world with an astonishing maturity of observation and of thought. By the time he was thirty, when he had written *The Napoleon of Notting Hill*, and his *Life of Dickens*, you would say he had not merely seen through, but lived through, everybody else's illusions. He wrote *Heretics* in 1905 as a man already tired of that tired aesthetic world in which he had grown up, as a man already too sophisticated for that sophisticated Liberalism which was then invading our politics, as a man already too disillusioned to believe in the incredulities of the late-Victorian scientists. And at this point, if I may be pardoned for a Chestertonian way of expressing myself, he grew up from manhood into boyhood. There was a boyish strain in him, as of one who has never quite got over reading *Treasure Island*. He owed much to Stevenson—RLS, we affectionately call him, just as we still talk affectionately of GKC. He borrowed from Stevenson, in spite of a wide difference of temperament, that aggressive optimism with which he proceeded, from 1905 onwards, to attack the winning side. He defended small nations at a time when we were being told to think imperially, defended private property when we were all playing with socialism; defended the small business and the small shop when everything was falling into the hands of the chain-stores; defended the home when women were going feminist; defended marriage when society had made up its mind to accept divorce. And yet, while he stood for very old things, he always seemed much younger than the people he was arguing with. His whole pose in controversy was that of the *enfant terrible* who cannot be stopped telling the truth. His general philosophy in *The Man Who Was Thursday* and *The Ball and the Cross*, his theology in *Orthodoxy*, his political views in *What's Wrong with the World*—they are thrown at you with a boy's light-heartedness.

The most boyish of his tricks was the little laugh he could not resist when one of his own impromptus amused him. You could hardly call it a chuckle or a giggle; it was more like a little neigh of excitement, high-pitched, impossible to reproduce. I can remember when I first heard it; it was at the meeting of an undergraduate society at Oxford, and the occasion of it was some controversy, then in progress between two towns in Suffolk, as to which was the original of Dickens's Eatanswill. Delighted as he was that Dickens should be in the news, he could not resist pointing out, as always, the obvious truth which everybody misses. 'The ideal of the modern man', he suggested (and here his own laugh came), 'is to be able to say, I have built my bungalow on the exact site of Sodom and Gomorrah.' But his whole manner in controversy was one—I do not know how else to describe it—of schoolboy impudence; he had the impish delight of the pupil who has found his master out in a mistake. Deeply as he cared about all that he stood for, an argument was always something of a game to him. I remember once, when I was criticising the theology of a peer of the realm who, for political reasons, had decided not to use his title, I asked Chesterton whether he thought it would be unfair, in writing about him, to give him his full title nevertheless. And the laugh came again as he said, 'It's a foul weapon, but we'll use it.'

It was in the year 1922, when his age was still short of 50, that Chesterton, if I may be allowed to pursue my paradox, grew up from boyhood into childhood, by a change of religion. To be sure there was always a childlike element in his character. I like the story of a small guest at a children's party in Beaconsfield, who was asked when he got home whether Mr Chesterton had been very clever. 'I don't know about clever', was the reply, 'but you should see him catch buns in his mouf.' He did not, like many grown-ups who are reputedly 'fond of children', exploit the simplicity of childhood for his own amusement. He entered, with tremendous gravity, into the tremendous gravity of the child.

It must be confessed, too, that, like other great intellects, he went about somewhat in need of a nurse. It may not be a true story, but it is certainly not an incredible story, that he once telegraphed to his wife, 'Am in Liverpool; where ought I to be?' But I mean something different when I suggest that, in the remaining years of his life, Chesterton reached the age of childhood. His thought was as vigorous as ever; and I am firmly of the opinion that posterity will regard *The Everlasting Man* as the best of his books. But his ideas seemed to grow even larger and more luminous; behind the tortuosities of his style you detected a vast simplicity of treatment. He contributed once to a broadcast series under the title of 'Six Days' Hard'; each speaker was to describe the events of a week, or his own experiences during the week, and choose his own method of approach. The rest of us talked about this and that; Chesterton devoted twenty minutes to the six Days of Creation.

The reason for this change was a simple one; he had found his home. Just as the hero of his own book *Manalive* walked round the world to find, and to have the thrill of finding, the house which belonged to him, so Chesterton probed all the avenues of thought and tasted all the philosophies, to return at last to that institution which had been his spiritual home from the first, the Church of his friend, Father Brown. He would, I think, have done so before, if he had not been anxious to spare the feelings of his wife, the heroine of all his novels, who followed him into the Church only four years later. Readers of his autobiography will remember how, in one of the earliest chapters he describes the chief figure in a peepshow which was the delight of his youth: a man who stood on a bridge, carrying a golden key. They will remember how, later in the book, he uses that figure for a symbol of one whom he came to recognize as Key-bearer and Pontiff, so many years afterwards. He felt himself that he had come home. He was still a fighter, and in some of the causes he fought for he did not carry with him the sympathies of all his co-religionists. But those religious ideas which were the deepest thing in him no longer made him an outlaw in a world of madmen; he had found companions at last, in the children of God's Nursery.

So, only a few years back, he went to Lisieux to visit the shrine of that Saint whose message to us all was to be converted and become as little children. He fell ill on his return, and a few days later we buried him in the new cemetery at Beaconsfield, the extension of that graveyard which covers the bones of Edmund Burke. A book which came out some time ago, under the title of *Premature Epitaphs*, summed up his characteristics in a tribute still applicable, though, alas, no longer premature:

> Chesterton companion his companions mourn;
> Chesterton Crusader leaves a cause forlorn;
> Chesterton the critic pays no further heed;
> Chesterton the poet lives while men shall read;
> Chesterton the dreamer is by sleep beguiled,
> And there enters heaven Chesterton the child.

THEODORE MAYNARD
Chesterton

It should be said at once that even the lesser Chestertons were still great. If GKC was not a good lecturer, that he lectured at all showed him to be a great man. He did so at personal sacrifice and in order to make money to subsidize his paper. And if he was not a good editor—as he very well knew—he accepted editorship as an inescapable duty, because he felt that only in this way could he continue his brother Cecil's work. Even the Chesterton who was a somewhat too facile humorist, with an insufficiently disciplined fancy, showed in the waste products of his exuberant imagination how rich that imagination was. Perhaps he should have curbed himself more than he did; but who except Chesterton had such an over-abundance that it needed to be curbed?

But the really great Chesterton—or the essential Chesterton—was the man of thought. In one of his love poems he described himself as the easy mirth of many faces, swaggering pride of song and fight ...

But under all that there was the man who might be described as a pure contemplative, the man wholly given to thought, and the loftiest kind of thought. His lecturing was only thinking aloud; his conversation was merely thinking in company. As Maisie Ward[1] truly remarks: 'For most people intensity of thought is much more difficult than action. With him it was the opposite. He used his mind unceasingly, his body as little as possible. I remember one day going to see them when he had a sprained ankle and learning from Frances how happy it made him because nobody could bother him to take exercise. The whole of practical life he left to her.'

In this connection there are three other quotations drawn from this book. To the somewhat testy H. G. Wells, with whom he managed to maintain his friendship to the end, he wrote in 1933: 'The fact that the other side often forget (is) that we began as free-thinkers as much as they did: and there was no earthly power but thinking to drive us the way we went.' And when Father Vincent McNabb described him as 'crucified to his thought', another friend protested, 'It was his lifelong beatitude to observe and ponder and conclude.'

But another thing should also be said. Their mother did not put it quite accurately when she said that Cecil had more heart than Gilbert,

---

[1] *Gilbert Keith Chesterton*, London and New York (Sheed & Ward), 1944.

though from one point of view this was true enough; Gilbert had little in him of the Virgilian sense of the 'tears of things'. He felt with his head rather than with his heart. That he was not without pathos his poem 'The Wife of Flanders' is sufficient to show. Yet even there it is indignation rather than pathos that is expressed. Utterly devoid of any class consciousness, and confiding once to the granddaughter of a Duke, 'You and I who belong to the jolly old upper middle classes', he had little actual knowledge of how the poor live, though he was always the champion of the poor. Yet the thing seen in the abstract and even felt in the abstract inspired him to write such passages as that which comes at the end of his *Short History of England*.

Not in any story of mankind has the irony of God chosen the foolish things so catastrophically to confound the wise. For the common crowd of poor and ignorant Englishmen, because they only knew that they were Englishmen, burst through the filthy cobwebs of four hundred years and stood where their fathers stood when they knew that they were Christian men. The English poor, broken in every revolt, bullied by every fashion, long despoiled of property, and now being despoiled of liberty, entered history with a noise of trumpets, and turned themselves in two years into one of the iron armies of the world. And when the critic of politics and literature, feeling that this war is after all heroic, looks around him to find the hero, he can point to nothing but the mob.

The lesser Chesterton, who set out to be funny, and who so frankly enjoyed his own jokes, was indeed often very entertaining. But he was at his best when his humour took the form of a deadly and controlled irony, and when his irony was incandescent with passion. Thus there is that overpowering passage on the Crucifixion in *The Everlasting Man*:

The grinding power of the plain words of the Gospel story is like the power of mill-stones; and those who can read them simply enough will feel as if rocks had been rolled upon them. Criticism is only words about words; and of what use are words about such words as these? What is the use of word-painting about the dark garden filled suddenly with torchlight and furious faces? 'Are you come out with swords and staves as against a robber? All day I sat in your temple teaching, and you took me not.' Can anything be added to the massive and gathered restraint of that irony; like a great wave lifted to the sky and refusing to fall? 'Daughters of Jerusalem, weep not for me but weep for yourselves and for your children.' As the High Priest asked what further need he had of witnesses, we might well ask what further need we have of words. Peter in a panic repudiated him: 'and immediately the cock crew; and Jesus looked upon Peter, and Peter went out and wept bitterly.' Has anyone any further remarks to offer? Just before the murder he prayed for all the murderous race of men, saying, 'They know not what they do'; is there anything to say to that, except that we know as little what we say? Is there any need to repeat and spin out the story of how the tragedy trailed up the Via

Dolorosa and how they threw him in haphazard with two thieves in one of the ordinary batches of execution; and how in all that horror and howling wilderness of desertion one voice spoke in homage, a startling voice from the very last place where it was looked for, the gibbet of the criminal; and he said to that nameless ruffian, 'This night shalt thou be with me in Paradise'? Is there anything to put after that but a full-stop? Or is anyone prepared to answer adequately that farewell gesture to all flesh which created for his Mother a new Son?

The famous libel action that Godfrey Isaacs brought and won against Cecil Chesterton has found its best treatment here, one very full and very just. It was written by Mr Sheed, whose wife turned this over to him because he is a lawyer. I have said before that I always thought that Cecil bungled that case very badly, and at that time I was seeing Cecil two or three times every week. But about the *New Witness* time in general Maisie Ward writes one of the least satisfactory parts of an otherwise eminently satisfactory book. Nor does she have a wholehearted sympathy with the rollicking Chesterton of Fleet Street or the convivial Chesterton of the songs of *The Flying Inn*. It is hardly to be expected that she should; if there were no other reason for this, it is to be found in the amusing but unreliable and malicious book that Cecil's widow wrote. There that side of GKC was somewhat exaggerated. And the definite accusation is brought against Frances that, by taking her husband away from Fleet Street to Beaconsfield, she removed him from his friends and so made it impossible for him to burn up in conversation the quantities of drink he consumed.

The thesis is of course absurd, but the facts do not substantiate it. No doubt in his Fleet Street days Gilbert did drink a good deal of beer and wine, but there was absolutely no trace of alcoholism. And for some years after his long illness in 1916 he remained a teetotaller. I can recall how at dinner in his house he would give me a bottle of claret and explain his own abstention with, 'My wife won't let me be a Christian.' But though he often wrote his articles in public-houses in his early days, it was Cecil who used to edit the *New Witness* in the 'George' or the 'Essex Arms', producing in the noisy bar his beautifully lucid prose.

The main purpose of 'The Chestertons' seems to have been to exalt Cecil at the expense of Gilbert, and still more to settle an obscure grudge against Frances. I could say a good deal about all this, as I knew everybody concerned. But perhaps it is enough to comment that if one wants a perfect illustration of the difference between cleverness and genius one finds it in the two brothers. As for Mrs Cecil's charge that Gilbert was a frustrated man because Frances refused to consummate her marriage, nobody who saw them together would ever put a moment's credence in such a story. As a matter of fact it was a great grief to Frances that she never had a child. In the hope of obtaining one, she submitted to an operation, regarding which Maisie Ward prints on page 244 of her book a letter from Mrs Chesterton's doctor. It might, however, be said that as documentation this would have had more weight had the doctor's name and the date of his letter been given. I do not have the slightest doubt that it does refer to this operation—about which Frances spoke to several of her intimate friends—but there is nothing in the letter that positively proves that it does. This little defect could easily be remedied, and I think should be, to put the accusation finally out of court. Mrs Cecil's conduct is all the more deplorable from the fact that she wrote after all the people involved were dead and that she had been left money by both Gilbert and Frances. A still greater obligation was created by Gilbert's having so loyally devoted himself to carry on what Cecil had begun.

L. A. G. STRONG
Chesterton's Drinking Songs

THE British public likes its writers to be consistent. When it thinks of them at all, it wants to know what to think. It requires a label, a pigeonhole. It needs to feel sure that Miss A's latest book will contain the mixture as before, and that Mr B's will once again expound the views that made his name.

This foible, while often soothing to Miss A and to Mr B. brings disadvantages. As long as the mixture and the views remain popular, all is well. But, should tastes change, Mr B and Miss A may find their labels turned to millstones. Even though they too have moved with the times, the utmost sincerity or adroitness may fail to free them from their accepted labels and persuade the British public to attach new ones.

Things are even worse if the original labels touch controversial matters; as in the case of G. K. Chesterton. Chesterton suffered from the start under our national love of labels. He had a very large following, as his publishers have good reason to know: but, in view of his tremendous gifts and his essentially English character, there were too many whom he might have called abstainers. Lately, his reputation has been in decline. Great numbers of potential readers are put off by a label which, from being a millstone, has come dangerously near to being a gravestone. They picture an obese and hearty figure banging a pewter pot upon a table and bellowing a paean in praise of beer and the Pope: and they abstain.

The picture is unjust, and wrongs a writer of vigorous, noble, and tender imagination. But for his Church, Chesterton would have been a pagan. He avows it, in deriding Mr Mandragon, the Millionaire, who did not know how to live, and was cremated:

> And he lies there fluffy and soft and grey
>   and certainly quite refined,
> When he might have rotted to flowers and fruit
>   with Adam and all mankind,
> Or been eaten by wolves athirst for blood,
> Or burnt on a good tall pyre of wood
> In a towering flame, as a heathen should,
> Or even sat with us here at food . . .

Even more explicit is the 'Song of the Strange Ascetic':

> If I had been a Heathen,
> I'd have praised the purple vine,
> My slaves should dig the vineyards
> And I would drink the wine . . .
> If I had been a Heathen,
> I'd have crowned Neœra's curls,
> And filled my life with love affairs,
> My house with dancing girls . . .

And he goes on to jeer at the 'poor old sinner' who 'sins without delight'; and, in conclusion, declares himself unable to read the riddle

> Of them that do not have the faith,
> And will not have the fun.

It would be fun, clearly; but Chesterton had the faith. He turned therefore to the celebration of those bodily pleasures on which the Church looked with tolerance: not the dancing girls, but eating and drinking. Claiming as his authority the miracle at the marriage feast at Cana in Galilee, he sang the praise of wine. A believer expresses his faith in the right use of the body, and the body thrives upon the right food and the right drink. Hurrah, then, for wine, for twopenny ale, for good red meat, for bacon and for pork: and be damned to vegetarian dishes, teetotallers, and every form of substitute and womanish victual and tipple.

> For the wicked old women who feel well-bred
> Have turned to a tea-shop 'The Saracen's Head.'

Yes, and worse.

> Tea, although an Oriental,
> Is a gentleman at least;
> Cocoa is a cad and coward,
> Cocoa is a vulgar beast,
> Cocoa is a dull, disloyal,
> Lying, crawling cad and clown,
> And may very well be grateful,
> To the fool that takes him down.
>
> As for all the windy waters,
> They were rained like tempests down
> When good drink had been dishonoured
> By the tipplers of the town;
> When red wine had brought red ruin,
> And the death-dance of our times,
> Heaven sent us Soda Water
> As a torment for our crimes.

I would like to have been by when someone offered him sherbet.

One of the difficulties about Chesterton was that he joked when he was in dead earnest. These quips and quiddities sprang from his deep ferocious faith. The average Englishman, who is not sure enough of his beliefs to jest about them, is apt to be embarrassed and put off Chesterton by what seems to him a lack of proportion. More than most writers, Chesterton was all of a piece. His imagination worked on many levels, from profound to trivial, but it was dominated always by the same enormous Facts. He would have said, it needed their discipline, for it was a strong, violent, gambolling imagination, disporting itself often in the the meads of fancy, but leaping sometimes to summits—or depths—where few have the stomach to follow it.

One should not build too much on these songs, which are incidental to a prose fantasia, *The Flying Inn*, and have the colour of their context. Chesterton's deepest simplicity is not to be found in them, nor the clarity of vision in which, a terrible child, he glances back to Blake and to Swift. He is in lighter mood here, but he tells us what he believes, and he believes what he tells us.

> My friends, we will not go again or ape an ancient rage,
> Or stretch the folly of our youth to be the shame of age,
> But walk with clearer eyes and ears this path that wandereth,
> And see undrugged in evening light the decent inn of death;
> For there is good news yet to hear and fine things to be seen,
> Before we go to Paradise by way of Kensal Green.

There is more than rhetoric here; the fourth line is memorable. Do you notice, by the way, the double parallel to Dr Johnson, in 'undrugged' and all the praise of inns with which the book is filled? Dr Johnson, who refused opiates at the last because they would cloud his faculties, and who said 'There is no private house in which people can enjoy themselves so well, as at a capital tavern'? The two men had more in common than bulk and gusto. They had awe, fear of sin, loud laughter, and great kindness. The Doctor's voice was the deeper, but Chesterton would not have so condemned the unfortunate lady whom Boswell sought to excuse—'The woman's a whore, and there's an end on't.'

> Great Collingwood walked down the glade
> And flung the acorns free,
> That oaks might still be in the grove
> As oaken as the beams above,
> When the great Lover sailors love
> Was kissed by Death at sea . . .

Each had a deep tenderness, but Chesterton's charity could open wider arms.

This does not mean that there was anything blurred or sentimental in his views on human conduct. He had what many moralists lack, a vision of evil, too appalling to admit of compromise. Few men's work can shudder as can Chesterton's. This awareness of evil as a force in the universe was quite distinct from the small strain of morbidity which was one of his weaknesses, and betrayed him now and then into a female note of hysteria (I use the adjective not in sex prejudice, but because no other will express the incongruity in this male and roistering literary character). It was the note of fear, shrill in Walpole, sullen in Belloc, camouflaged in Maugham. Kipling blustered at his fear, Joyce shivered, Yeats raged; Virginia Woolf faced it; it made Chesterton querulous. All of them knew it: it was endemic in their time. Chesterton was fortified against it by his faith.

On the technical side he was, like most writers of fertile imagination, an admirer of Dickens—an influence as strong, and as dangerous, as that of Dr Johnson. A Dickensian exuberance, the consolations of faith, and a clear-cut moral judgement on first principles were none of them aids to popular esteem in the late interval between two wars.

An added drawback to appreciation of his powers as poet was that Chesterton used the rhythms which the public had learned to associate with Kipling. If Chesterton was the truer poet, Kipling used them with unmatched brilliance and assurance: and the general reader, no stickler for niceties but no sort of a fool, decided that Chesterton was following the more famous writer, and at some distance. I am not maintaining for an instant that this is relevant to any proper consideration of Chesterton's merits, but it has gone to make a label, which is another name for a prejudice, and it is against prejudice that Chesterton's reputation has to fight today. He has, as I said before, a legion of admirers: but he would have two, three, four legions, if the old label could be removed, and new readers could be brought to look at him as he really is.

Those who love good writing care nothing about the swing of fashion and the ups and downs of reputations, but some may confess to a touch of partisanship: and, in any case, it would be fitting that a Christian writer of genius, who kicked hard on the backside all who would make Christianity prim and joyless, should have from a new generation of readers the welcome which his gifts deserve.

IVOR BROWN
A Multiple Man

THE chief greatness of Chesterton lay in his ability to fascinate and to keep in the friendliest mental play those most opposed to his beliefs. To H. G. Wells, prickliest of anti-Papists, he remained, in letters, 'Dear Old G.K.C.' (though Chesterton was the younger man). As a boy I gazed enchanted upon Chesterton's mastery of journalistic technique. (It was the briefer the better with him, his column articles being always even more brilliant that his books.) Detesting his first principles, his anti-Socialism, anti-Feminism, and scorn of reason, and infuriated by his torrent of puns and quibbles on the side of what I deemed to be out-moded angels, I none the less adored the man and craved to be a mimic of the method. In those bountiful days for a penny one had the Big Four, Shaw, Wells, G. K. C., and Belloc, all gloriously embattled week after week in *The New Age*. Wells and Shaw gave the young Socialists doctrine, but Belloc and Chesterton gave them style, company, and songs to sing. What an agile and friendly enemy was the 'Chester-Belloc'!

The Chester-Bellocian criticism of the Party System: that analysis of the Servile State and of the increasing failure of liberal democracy was a striking piece of prevision. This part of Chesterton's work was, to many of us, far more important than his word-jugglery was in the service of theology. Recently I tried to re-read *Orthodoxy* and found its antics on the flying trapeze of celestial paradox more tiresome than I had expected. But the political essays are of permanent value. Incidentally, he anticipated the Fascist revolt and the criticisms of capitalist democracy now pungently made by E. H. Carr and others. G. K. C. will certainly be remembered as a political poet. He ought to be remembered as a political prophet.

The wonder of the man was his intuition—if one dare use the word now. Apparently he had not read widely or deeply. But he divined perfectly what a book was about by just looking at the surface. His seeming omniscience was a kind of justified and triumphant bluff. He was pure Theorist, praising the life of a peasant from a flat in Battersea—did he ever dig an acre in his life?—generalizing about the Common man from Fleet Street Pubs, where only the uncommon abound, rhyming the glory of battle without ever being near a battlefield, and laying down the law about History with the sketchiest knowledge of the facts. From the scholar's point of view, many of his books were written in a glorious haze of ignorance. But he often saw more truth glistening through the mist than the learned could descry amid all their blaze of factual illumination. And what he saw thus shining he set down in words of no less lustre and animation.

His life was an orgy of contradiction. He cried for Freedom and angrily rejected most things which men regard as free. Free Thought, Free Churches, Free Love, were not at all to his taste. In theory he praised all things small and humble, whether farm, shop, or nation, but he chose to join the largest, the proudest, the most rigid, and most authoritarian of the sects, whose discipline is specially designed to limit man's range of thought and reading. He praised the Christian virtues, but was morbidly fascinated by weapons and blood, loving to play with a sword-stick, writing romantic rhodomontade about the loathsome homicide called war, and filling his poetry with a whole arsenal of blood and iron metaphors. He knew nothing of music, but he could set words dancing and choiring in his prose and verse as few have ever done.

'Do I contradict myself? Yes, I contradict myself. I am large. I contain multitudes.' So Whitman. It was even truer of G. K. C., who said that Dickens was not a man but a mob. Chesterton also was a huge and happy crowd in his own person. He was the Multiple Man, raging against the Multiple Store. Others will see him as a poet and politician and now, like Belloc, absurdly underrated in both roles. His poem about the English, 'The Secret People', reveals alike his power of divination and his mastery of phrase, 'Chuck it, Smith' stands at the pinnacle of angry laughter, and 'The Donkey' always, however often read, sends down the spine the shiver which for Housman was the impact of the greatest poetry. Above all, he was a great enjoyer, ready to find his purple hours on the platform of Clapham Junction, the genial giant who seemed to be Falstaff and Aquinas, Blake and Dr Johnson, Cobbett and Chaucer and Conan Doyle all rolled into one.

GRAHAM GREENE
G. K. *Chesterton*

CHESTERTON'S bibliography consists of one hundred volumes—Out of this enormous output time will choose. Time often chooses oddly, or so it seems to us, though it is more reasonable to suppose that it is we ourselves who are erratic in our judgements. We are already proving our eccentricity in the case of Chesterton: a generation that appreciates Joyce finds for some reason Chesterton's equally fanatical play on words exhausting. Perhaps it is that he is still suspected of levity, and the generation now reaching middle-age has been a peculiarly serious one. He cared passionately for individual liberty and for local patriotism, but the party which he largely inspired has an art-and-crafty air about it today. He was too good a man for politics: he never, one feels, penetrated far enough into the murky intricacies of political thought. To be a politician a man needs to be a psychologist, and Chesterton was no psychologist, as his novels prove. He saw things in absolute terms of good and evil, and his immense charity prevented him admitting the amount of ordinary shabby deception in human life. At their worst our politicians were fallen angels.

For the same reason that he failed as a political writer he succeeded as a religious one, for religion is simple, dogma is simple. Much of the difficulty of theology arises from the efforts of men who are not primarily writers to distinguish a quite simple idea with the utmost accuracy. He restated the original thought with the freshness, simplicity, and excitement of discovery. In fact, it was discovery: he unearthed the defined from beneath the definitions, and the reader wondered why the definitions had ever been thought necessary. *Orthodoxy*, *The Thing*, and *The Everlasting Man* are among the great books of the age. Much else, of course, it will be disappointing if time does not preserve out of that weight of work: *The Ballad of the White Horse*, the satirical poems, such prose fantasies as *The Man Who Was Thursday*, and *The Napoleon of Notting Hill*, the early critical books on

Browning and Dickens; but in these three religious books, inspired by a cosmic optimism, the passionately held belief that 'it is good to be here', he contributed what another great religious writer closely akin to him in political ideas, and even in style, saw was most lacking in our age. Péguy put these lines on man into the mouth of his Creator:

> On peut lui demander beaucoup de coeur, beaucoup de charité,
>   beaucoup de sacrifice.
> Il a beaucoup de foi et beaucoup de charité.
> Mais ce qu'on ne peut pas lui demander, sacredié, c'est
>   un peu d'espérance.

PATRICK BRAYBROOKE
Chesterton and Charles Dickens

I

FORTY years have elapsed since Chesterton wrote his critical book on Dickens. It was its author's best book, and possibly it is the best book ever written or ever likely to be written about the Victorian novelist who burst upon the world like a shattering thunderstorm.

Dickens was a great man and a great author. Chesterton asks us to be careful in the use of the word 'great'. In the case of Dickens, he was great because he was not only a careful artist, he was an exuberant artist, he painted with a zeal both excitable yet restrained: 'he could not make a monotonous man'. Chesterton, after a series of delicious dialectics tells us with a sudden fall into calm realism why Dickens was great:

> We are filled with the first of all democratic doctrines, that all men are interesting; Dickens tried to make some of his people appear dull people, but he could not keep them dull. The bores in his books are brighter than the wits in other books.

Because Chesterton admired Dickens was no reason to prevent him from being very fully aware of any of his weaknesses. He found him just a little inclined to be quarrelsome in literary matters; he made enemies when he might have made friends, he was a Victorian and really rather touchy about foibles. Chesterton sees him in a state of worry, in a 'stew', stamping about and writing passionate denials of matters, that even if they had been enumerated, could at worst be but annoying trifles. So Chesterton sums up the problem of this Dickens irascibility, this pique:

No one can say of him that he was often wrong: we can only say of him as of many pugnacious people that he was often right.

To Chesterton the very essential genius of Dickens lay in the fact that he was in reality something more than a novelist. He was not like so many modern authors, something less than a novelist. He was, in the judgement of Chesterton, a kind of manufacturer of literary symbols parading about, and parading about successfully, as ordinary men and ordinary women. But his art made them extraordinary in the almost forgotten sense of the word, not peculiar but out of the ordinary. There is a vast and fundamental difference. Thus Chesterton writes his definition of Dickens: something courageous and something not written before:

> Dickens was a mythologist rather than a novelist; he was the last of the mythologists, and perhaps the greatest. He did not always manage to make his characters men, but he always managed at the least to make them gods. They are creatures like Punch or Father Christmas. They live statically, in a perpetual summer of being themselves.

And the characters, so many of them are so jolly, how they do know how to be merry, an art we in modern England have almost completely neglected, but not I hope finally lost. *The Pickwick Papers* is almost a fairy story:

> Mr Samuel Pickwick is not the fairy: he is the fairy prince: that is to say, he is the abstract wanderer and wonderer, the Ulysses of comedy. He has set out walking to the end of the world, but he knows he will find an inn there.

II

Perhaps the most illuminating part of the book on Charles Dickens that Chesterton wrote, and by it advanced on his journey to fame, is that which deals with Dickens's amazing and unique popularity. He called it 'the great popularity'. At once we are told that this popularity was not just the fact that Dickens was a best seller or even was tremendously talked about, it was indeed something very much more, something that had not happened before, and almost certainly will never happen again. It was as though Dickens came to be some new and wonderful literary star, blazed a trail through the dolorous night of Victorian pessimism, and with his final passing from the earthly scene, left behind him a mass of permanent and indestructible idealistic and realistic fiction.

Men never grew tired of Dickens and with a queer kind of whimsical truth Chesterton writes:

Men read a Dickens story six times because they knew it so well. In short, the Dickens novel was popular because it was a real world; not a world in which the soul could live ... Men give out the airs of Dickens without even opening his books: just as Catholics can live in a tradition of Christianity without having looked at the New Testament.

III

It is manifestly impossible in an article to consider the Chestertonian position with regard to many of the famous characters of Dickens. I have said just a little about Mr Pickwick and let me say a little about Little Nell. I am not concerned here as to whether Little Nell was based on a real character Dickens had known or whether she was a kind of lovely symbol, too good to live, but also good enough to die. Chesterton gives us his critical judgement of her: and would her grandfather disagree? I think he would be the first to agree with the careful verdict given by Chesterton:

> Little Nell has never any of the sacred bewilderment of a baby.

She is indeed a grown-up child and I think she would have curtsied to the Victorian stage coach as it rolled by, with Mr Pickwick shouting 'all the way to Dingley Dell'.

So again to select at random another Dickens character. Of Mrs Gamp, Chesterton writes with penetrating insight:

> We can destroy Mrs Gamp in our wrath, but we could not have made her in our joy.

Many Dickensian critics have attacked Dickens on the grounds that his villains are so bad that they are not only inhuman but unhuman. They allege that Fagin is a little bit too bad, Sykes too shockingly rough with Nancy, Squeers an exaggeration. The critics have a right to their opinions; I have a right to disagree with them. I believe that the Dickens villains are by no means exaggerated; they are to be found in every modern walk of life. What does Chesterton say of them? He admits of no kind of compromise whatsoever. He is discussing the 'lurid villainy' of the bad people of Charles Dickens.

> He inherited, undoubtedly, this unqualified villain as he inherited so many other things, from the whole history of European literature. Nobody ever made less attempt to whitewash evil than Dickens. He crowds his stories with a kind of villain rare in modern fiction—the villain really without any 'redeeming point'.

To sum up on this problem of the mass of villains, Chesterton concludes:

His emphasis on evil was melodramatic.

In other words Dickens sailed into the regions of the stage villain of the Victorian melodrama, but he did not, I believe, sail out from the stormy seas of probability. This is, I think, important, and confirms the essence of the Chesterton critical approach to Dickens: that there was no kind of unreality either about him as a man, as a novelist or even as an ardent and indeed successful reformer.

He dealt with the world of reality, not as a Plato, but as a writer of remarkable genius. That is what can be found in a careful study of Chesterton's book.

IV

It is possibly with a feeling of melancholy that we approach to a consideration of what would be, for Chesterton, the future of Charles Dickens. I say melancholy for the pen of Chesterton cannot again create for us the exquisite pen portrait of the magnificent Victorian genius from whose pen came reforms no less important than those achieved by the greatest statesmen or scientists.

With a conclusion that seems almost commonplace Chesterton sees the certain and obvious permanence of Dickens.

That Dickens will have a high place in permanent literature, there is, I imagine, no prig surviving to deny.

But there is very much more to it than that. Permanence is not sufficient for Dickens. Many Victorians have achieved that kind of well-merited earthly immortality. Dickens has quite a different basis of fame. So Chesterton places Dickens an absolute and undisputed first in the enormous gallery of Victorian celebrities:

Some would now say that the highest platform is left to Thackeray and Dickens; some to Dickens, Thackeray and George Eliot; some to Dickens, Thackeray and Charlotte Brontë. I venture to offer the proposition that when more years have passed and more weeding has been effected, Dickens will dominate the whole England of the nineteenth century. He will be left on that platform alone.

With a kind of clarion cry of triumph, with a rational but exuberant peroration, with a masterly summing-up, with a declaration that even now seems to be as true as though it had been written in this year of 1945, Chesterton ends his critical study of Charles Dickens:

The hour of absinthe is over. We shall not be much further troubled with the little artists who found Dickens too sane for their sorrows and too clean for their delights. But we have a long way to travel before we get back to what

Dickens meant: and the passage is along a rambling English road, a twisting road such as Mr Pickwick travelled. But this at least is part of what he meant: that comradeship and serious joy are not interludes in our travel, but that rather our travels are interludes in comradeship and joy, which through God shall endure for ever. The inn does not point to the road: the road points to the inn. And all roads point at last to an ultimate inn, where we shall meet Dickens and all his characters, and when we shall drink again it shall be from the great flagons in the tavern at the end of the world.

If we would wish to take ourselves out of the rush and panic of the modern world in which men and women find it hard to know which road to take or down which winding way they have come, they cannot do better than take down from their bookshelves the volumes created by Charles Dickens.

And if they would wish further to consider in some detail the nature of the man and the formation of the mind that created these books, they may well, so many years after it was published, open the pages of the book by Chesterton, *Charles Dickens: A Critical Study*.

And in it they will discover most of what was Dickens, his hopes, his fears, his quarrels, his successes, his failures, his heroes, his heroines, his villains, and his reforms.

They will find the book though old in years is still young in outlook and they will be able to say at the end of it:

Let us raise our glasses to the immortal memory of Charles Dickens and the toast shall, even from the quiet of his grave, be proposed by this twentieth-century critic, who in 1906, published his critical study.

JAMES STEPHENS
The 'Period Talent' of G. K. Chesterton

A VERY strange thing happens when a writer dies. The strange thing is that his books tend to die with him. A month after his death his name is not mentioned again, for he is no longer in competition with the living, and twenty or more years will pass before criticism decides to take another look at him and his work and try to discover if there was a real person there, or only a living gramophone record engraved and wound up by his period. Shakespeare's poetry was dead for nearly two hundred years, and then, and only as by chance, it came to life again. This fact alone is a truly remarkable criticism on the generations that succeeded the Elizabethan Age.

THE 'PERIOD TALENT' OF G. K. CHESTERTON

It is, so, impossible to talk about Chesterton by himself. One reason is that he and his period are dead. The great majority of modern writers, of writers in general, do not, of course exist by themselves. They live with their own crowd: they are enwrapped in each other and in their period; they were born and they got married and they died. What does all this matter? They have scarcely any life of their own, and they are moved, from day to day and from book to book, by the daily happening and the daily acquaintance. They write in their crowd. They live in their crowd, and they die there.

G. K. Chesterton died only ten years ago: but subsequent events have so arranged it that those ten years ago are already fifty years away. He left a vast body of work behind him, both in prose and verse. There are a great many writers who consider that only by writing hundreds of books can they write against time and leave behind them the relative immortality which the French call '*La Gloire*' and we call 'Greatness'. Before Chesterton died there were a certain number of writers, but three in especial, about whom a curious but true thing is to be said. I am not now referring to the great writers, to, that is, Meredith and Hardy, Kipling and Moore and Yeats. The three I have in mind are Wells and Shaw and Chesterton. The thing to be said of these three is that every person of some cultivation who used the English language as a native language had come under the influence of these three writers, and recognized these three as the popular masters of their epoch, and as the guiding and energizing influences of their day. It is impossible to speak of one of these three without taking the others into account, for between them they summarize their epoch, and it is impossible not to recognize the enormous liberating and fertilizing world-influence of their work.

A question that must arise is, were these the greatest writers of their time? They were not. Popular acclaim is rarely given to the great writers of a period, but these three were the greatest public influences of their day, and their influence has not ceased to this day. The way in which the great writer differs from all his companions is that he is *not* a journalist, and the three I speak of were journalistic both in talent and in production. The talent of Wells and Shaw and Chesterton was, too, largely a period talent—it was more than that, but it was also that—they wrote eagerly, even angrily, the idea of the day: they wrote of the good, the bad, and the indifferent which the day and its newspaper brought them. But of human beings, of human intentions, they almost never wrote. Of the thing we call 'living' they had almost no experience: of the thing we call 'thinking' they thought in bulks of national and international exchanges. Of the person-in-himself, who is everywhere the poet and the religious, they had no experience and almost a deep dislike. Of the poetry which is the sole and lasting glory of England, Wells and Shaw took no account, and of the thing which we call 'passion' they took little account either: indeed they replaced it with anger and humour and violence and perpetual change for the sake of change. Change! We call it Science! There is scarcely room to blame them for this for their whole life was passed in a world that changed from year to year, and their work is a wonderful notation of this change. They saw money change and sex change, and speed change, and a change in any one of these is revolution indeed. They saw money become valueless, sex become promiscuous, and speed become monstrous.

In all this Chesterton was different from the others, although not greater. He did seek the solitude of religion—he became a Catholic, but he could not abide that solitude, or any solitude: he did seek the solitude of poetry, but he could not abide being alone with Urania. He did not know that poetry is a very private matter and is only communion with others by their merit, and by their identity with the solitary song. All poetry that is widely and immediately acceptable is a peasant poetry, and is almost a parody of everything that poetry intends and promises. For poetry intends your freedom, and promises you beauty.

Chesterton's poetry is shown at his best, or at his greatest reach, in 'Lepanto'. This poem evokes most interesting and even strange reflections. It is cast in the ballad-form, but, where every native reader can immediately understand an English ballad, only a scholar, and a Catholic scholar at that, can get the very remarkable poem which Chesterton put such an immensity of work into. This is the only English poem I know which is not English. Every English poem exists by reason of the under-song which is inseparable from English verse. This poem is, in fact, the only French poem in the English language, and in it the under-song that I speak of has been replaced by that amazing French quality which is called eloquence, and which, while it can be superb, can also be a pestilence and a blight to poetry.

In English or in French how remarkable this quality is! It can often be wonderful: but you don't go to bed with it. One goes to bed with the moon, and with poetry, but you don't go to bed with eloquence—in these two stanzas Chesterton is indeed magnificent:

> White founts falling in the courts of the sun,
> And the Soldan of Byzantium is smiling as they run;
> There is laughter like the fountains in that face of all men feared,
> It stirs the forest darkness, the darkness of his beard,

It curls the blood-red crescent, the crescent of his lips,
For the inmost sea of all the earth is shaken with his ships.
They have dared the white republics up the capes of Italy,
They have dashed the Adriatic round the Lion of the Sea,
And the Pope has cast his arms abroad for agony and loss,
And called the kings of Christendom for swords about the Cross,
The cold queen of England is looking in the glass;
The shadow of the Valois is yawning at the Mass;
From evening isles fantastical rings faint the Spanish gun,
And the Lord upon the Golden Horn is laughing in the sun.

Dim drums throbbing, in the hills half heard,
Where only on a nameless throne a crownless prince has stirred,
Where, risen from a doubtful seat and half attainted stall,
The last knight of Europe takes weapons from the wall,
The last and lingering troubadour to whom the bird has sung,
That once went singing southward when all the world was young,
In that enormous silence, tiny and unafraid,
Comes up along a winding road the noise of the Crusade.
Strong gongs groaning as the guns boom far,
Don John of Austria is going to the war,
Stiff flags straining in the night-blasts cold
In the gloom black-purple, in the glint old-gold,
Torchlight crimson on the copper kettle-drums,
Then the tuckets, then the trumpets, then the cannon, and he comes.
Don John laughing in the brave beard curled,
Spurning of his stirrups like the thrones of all the world,
Holding his head up for a flag of all the free.
Love-light of Spain—hurrah!
Death-light of Africa!
Don John of Austria
Is riding to the sea.

Chesterton was a divided man—he wished to be a successful and voluminous writer, and he was that: he also wished to be the solitary, the poet, the religious, and he made as desperate an attempt to be a poet and a saint as any man has ever made. But how can the impatient writer ever become a writer? How can the poet-in-a-hurry ever come on the intense quietude which makes of itself the wing, the singer, and the song? Or how can the practice of the presence of God be written, as it were, daily for a newspaper? He was in his crowd.

Chesterton used to publish books at the rate of six or seven a year. He was a critic of anything and everything; he was a daily journalist: he would write you, at a moment's notice, the biography of anyone you asked for: he would throw off a couple of novels in his spare time: he would write so many short stories that you couldn't count them—they were excellent short stories: he would engage in written controversy with anyone on almost anything: he was a political and social theorist, and a public lecturer, and he would throw off at any moment a couple of dozen poems if he thought that would please the neighbour. He was not dishonest in this, nor even wrong. He did please everyone, I think, and he was truly a most pleasant person. He was of an unshakeable kindness. He was modest in despite of popularity. He was willing to work himself to death if that would help anyone. He would give, I think, up to the very limit of his possessions to whomever or whatever was in need. He was a good and a gallant man and writer. He knew that something was all wrong in his age, that the values had departed and that everything was at odds and at stake, so he went political, and like Mr Gandhi, he proposed that we should retreat some hundreds of years into the Middle Ages, or even further back into the Dark Ages, and that we should all become glorious peasants again and live happily ever afterwards. Perhaps that will happen, and perhaps the revolution in money and sex and speed will see that it does happen.

Meanwhile, Chesterton and his companions of the pen put all the blame on the immediate past. They had inherited the oddest legacy of hate that has been bequeathed from one generation to another. They were all devoted to the total destruction of something or other which they referred to as the Victorian Age. They declared war against Victorian comfort, Victorian complacency, Victorian morals, Victorian customs, against anything and everything which would be thought of as Victorian, and they have helped lead us to where there aren't any customs left to us, very little of morals that one can recognize, and nothing of comfort in being or in prospect. The growth of violence in the years I speak of is very strange and is truly the remarkable fact of the past age. Under the term 'detective-story', literature was a continuous and vast tale of murder, sex, theft and treachery. In every country these matters were the favourite writing and the favourite reading, these were the universal day-dreams, and when the 1914 Great War arrived every land was already soaked in the idea of sensuality and robbery and slaughter. For our 'popular' writers always see to it that their people are psychologists ready for any abomination that can arrive, and they are not neglecting that duty now.

C. S. LEWIS
On Stephens on Chesterton

OPENING *The Listener* a few days ago I came upon an article on Chesterton by Mr James Stephens—an article which seemed to me ungenerous and even unjust. There were two main charges made against Chesterton; the one, that he was too public (for on Mr Stephens's view poetry is a very private affair) and the other, that he was 'dated'. The first need not, perhaps, be discussed here at very great length. Mr Stephens and I find ourselves on opposite sides of a very well-known fence, and Mr Stephens's side is, I must confess, the popular one at present. It still seems to me that the burden of proof rests on those who describe as 'private' compositions which their authors take pains to have multiplied by print and which are advertized and exposed for sale in shops. It is an odd method of securing privacy. But this question can wait. It certainly would not have worried Chesterton. Nor would the maxim that any poetry which is immediately and widely acceptable (like that of Euripides, Virgil, Horace, Dante, Chaucer, Shakespeare, Dryden, Pope, and Tennyson) must be merely 'peasant' poetry have offended a man who desired nothing so much as the restoration of the peasantry. But the question of 'dating' remains.

It is very difficult here to resist turning the tables, to ask what writer smells more unmistakably of a particular period than Mr Stephens himself. That peculiar mixture of mythology and theosophy—Pan and Aengus, leprechauns and angels, re-incarnation and the sorrows of Deirdre—if this does not carry a man back into the world of Lady Gregory, AE, the middle Yeats, and even Mr Algernon Blackwood, then the word 'period' really has no meaning. Hardly any book written in our century would be so nearly dated by internal evidence as *The Crock of Gold*. Even Mr Stephens's curious suggestion that detective stories (of which Chesterton was notoriously guilty) somehow helped to produce the first German war could be retorted. It would be just as plausible to trace the Nazi ideology to the orgiastic elements in Mr Stephens's own work; to the cult of Pan, the revolt against reason (symbolized by the Philosopher's journey, imprisonment, and rescue) or the figure of the Ugliest Man. And it might easily be maintained that the theological background of Chesterton's imaginative books has dated a good deal less than the blend of Celtic Twilight and serious occultism (Yeats claimed to be a practising magician) which we cannot help surmizing for Mr Stephens.

But though this would be easy it would not be worth doing. To prove that Mr Stephens had dated would not be to prove that Chesterton was perennial. And there is, for me, another reason for not answering Mr Stephens with this *argumentum ad hominem*. I still like Mr Stephens's books. He holds in my private pantheon a place inferior to Chesterton's but quite as secure. It is inferior because the proportion of dead wood in *The Crock of Gold*, *The Demigods* and *Here are Ladies* (there is no dead wood in *Deirdre*) seems to me higher than in *The White Horse*, *The Man Who Was Thursday*, or *The Flying Inn*. I think the long paragraphs of what used, at Boston, to be called 'Transcendentalism' which we find in Mr Stephens are bad, sometimes even nonsensical. But then they always were bad: dates have nothing to do with it. On the other hand the gigantic (and, *in the proper sense*, Rabelaisian) comic effects—the arrest of the Philosopher or the postmortal adventures of O'Brien and the threepenny bit—are inexhaustible. So is the character of that admirable *picaro* Patsy MacCann. So is the Ass. So is the painting of nature; the trees that stood holding their leaves tightly in the wind, or the crow that said 'I'm the devil of a crow'. I cannot give up Mr Stephens. If anyone writes a silly, spiteful article to say that Mr Stephens was only a 'period' talent, I will fight on that issue as long as there is a drop of ink in my pen.

The truth is that the whole criticism which turns on dates and periods, as if age-groups were the proper classification of readers, is confused and even vulgar. (I do not mean that Mr Stephens is vulgar. A man who is not a vulgar man may do a vulgar thing: you will find this explained in Aristotle's *Ethics*.) It is vulgar because it appeals to the desire to be up to date: a desire only fit for dressmakers. It is confused because it lumps together the different ways in which a man can be 'of his period'.

A man may be of his period in the negative sense. That is to say he may deal with things which are of no permanent interest but only seemed to be of interest because of some temporary fashion. Thus Herbert's poems in the shape of altars and crosses are 'dated'; thus, perhaps, the occultist elements in the Celtic school are 'dated'. A man is likely to become 'dated' in this way precisely because he is anxious not to be dated, to be 'contemporary': for to move with the times is, of course, to go where all times go. On the other hand a man may be 'dated' in the sense that the forms, the set-up, the paraphernalia, whereby he expresses matter of permanent interest, are those of a particular age. In that sense the greatest writers are often the most dated. No one is more unmistakably ancient Achaean than Homer, more scholastic than Dante, more feudal than Froissart, more

'Elizabethan' than Shakespeare. *The Rape of the Lock* is a perfect (and never obsolete) period piece. *The Prelude* smells of its age. *The Waste Land* has 'Twenties' stamped on every line. Even Isaiah will reveal to a careful student that it was not composed at the court of Louis XIV nor in modern Chicago.

The real question is in which sense Chesterton was of his period. Much of his work, admittedly, was ephemeral journalism: it is dated in the first sense. The little books of essays are now mainly of historical interest. Their parallel in Mr Stephens's work is not his romances but his articles in *The Listener*. But Chesterton's imaginative works seem to me to be in quite a different position. They are, of course, richly redolent of the age in which they were composed. The anti-Germanicism in the *Ballad of the White Horse* belongs to a silly and transitory historical heresy of Mr Belloc's—always, on the intellectual side, a disastrous influence on Chesterton. And in the romances, the sword-sticks, the hansom cabs, the anarchists, all go back both to a real London and to an imagined London (that of *The New Arabian Nights*) which have receded from us. But how is it possible not to see that what comes through all this is permanent and dateless? Does not the central theme of the *Ballad*—the highly paradoxical message which Alfred receives from the Virgin—embody the feeling, and the only possible feeling, with which in any age almost defeated men take up such arms as are left them and win? Hence in the very nadir of the late war a very different, and exquisite, poet (Miss Ruth Pitter) unconsciously and inevitably struck exactly the same note with the line:

> All but divine and desperate hopes go down and are no more.

Hence in those quaking days after the fall of France a young friend of mine (just about to enter the RAF) and I found ourselves quoting to one another stanza after stanza of the *Ballad*. There was nothing else to say.

So in the stories. Read again *The Flying Inn*. Is Lord Ivywood obsolete? The doctrinaire politician, aristocratic yet revolutionary, inhuman, courageous, eloquent, turning the vilest treacheries and most abominable oppressions into periods that echo with lofty magnanimity—is this out of date? Are the withers of any modern journalist quite unwrung when he reads of Hibbs However? Or read again *The Man Who Was Thursday*. Compare it with another good writer, Kafka. Is the difference simply that the one is 'dated' and the other contemporary? Or is it rather that while both give a powerful picture of the loneliness and bewilderment which each one of us encounters in his (apparently) single-handed struggle with the universe, Chesterton, attributing to the universe a more complicated disguise, and admitting the exhilaration as well as the terror of the struggle, has got in rather more; is more balanced: in that sense, more classical, more permanent?

I will tell Mr Stephens what that man is like who can see nothing in these stories but an Edwardian 'period' piece. He is like a man who should look into Mr Stephens's *Deirdre* (the one unmistakably great and almost perfect book among its author's many good books) and having seen the names (Connohar, Deirdre, Fergus, Naoise) should mutter 'All the old Abbey Theatre stuff' and read no more. If Mr Stephens is too modest to reply that such a man would be a fool, I will do it for him. Such a man would be a very notable fool: a fool first for disliking the early Yeats; second, for assuming that any book on the same theme must be like the early Yeats; and a fool thirdly for missing some of the finest heroic narrative, some of the most disciplined pathos, and some of the cleanest prose which our century has seen.

EVELYN WAUGH
The Man Who Was Thursday

THE need for a secret society is one that many story-tellers have felt. Most good stories are, in some fashion, the conflict between Good and Evil, and whereas it is easy enough to pile up virtues until some fairly plausible hero has been created, villains, however black the crimes attributed to them, tend to remain limp in their makers' hands. Iago is wicked, but the reasons he gives for his actions are so fatuous that few readers, offhand, can remember them. Bosola is the unwilling agent of men of diabolic character; it is impossible to take seriously their statement that they are impelled by avarice. And the Elizabethans were far nearer to the springs of Evil than Miss Agatha Christie.

The writers of detective stories, indeed, are in a peculiar difficulty. Their concern is with the mechanics of crime and the logic of its discovery, rather than with Good and Evil. But convention demands that the crime be murder. This is the accepted token-coin of extreme wickedness, yet it is one of the few sins which the civilized man can regard quite dispassionately. However misanthropic, he has never been tempted to its commission as he has been tempted, say, to adultery or suicide, and he recognizes, as perhaps he would do with other sins, could he regard them with equal coolness, that the dangers and exertions are appalling and the rewards trifling. Moreover, conscience, silenced before more alluring transgressions, is here plainly audible, interposing its voice and debunking the tempter: murder for profit—'What shall it profit a man if he gain the whole word and lose his own soul?'; murder for revenge—'Vengeance is mine, saith the Lord'; murder for love—'Love suffereth long and is kind'...

Only writers like Mr Graham Greene and M. Bernanos—how glibly the words come! What writers, in fact, are there like these two—who plumb the human spirit at depths where few venture, can create villains. More humble writers, who know they are qualified to deal only with rather ordinary characters, and who know that such people do not kill, are constrained, often, to exonerate their murderers either by inventing circumstances—blackmail, the threat of corruption of a loved one, etc.—in which the victim is an intolerable aggressor and his removal justifiable, in which case the forces of law are deprived of the reader's sympathy and only his intellect is engaged in the process of solution, or by borrowing from the psychologists. Schizophrenes, those rare and often harmless creatures, have become an accepted device of stage and fiction, such as identical twins were to an earlier age. The modern 'psychological-thriller' has, properly, no villain, for the crimes are not acts of free will.

But there is a third resource, the group villain—the gang, the spy-ring, the subversive organization, the secret society—which doubly commends itself to the critical reader, first by titivating the conspiracy-mania which is latent in most of us, and secondly by emphasizing the deep moral truth that men in association are capable of wickedness from which each individually would shrink.

The simplest case is the gang, the army of outlaws. Though there is a commercial 'racket' at the origin of it and wads of 'grands' disbursed, the essence of the gang, as it appears in fiction, is not the making of money, but an organized war against society; they take their orders, observe their own loyalties, and 'shoot it out' with the police without hope of victory. In advance of this is the organization which aims not merely at an independent existence in a law-abiding universe, but at the actual subversion of the social and moral order.

The conviction that such a conspiracy is feasible may be traced in history—for example in the Albigensian suppression and in the pogroms inflamed by the spurious 'Protocols of the Elders of Zion'. It is even more frequent in fiction. Conan Doyle, Agatha Christie, John Buchan, Francis Beeding, Edgar Wallace—almost every writer of crime stories has at one time or another made use of this expedient.

The classic example, of course, is in Chesterton's *The Man Who Was Thursday*, only in that case, significantly, the whole thing turns out to be moonshine.

Significantly of Chesterton's character and age, Chesterton's cheerfulness was redeemed, but abundantly redeemed, from vulgarity only by his innocence. He was so sweet and virtuous a man that crime appeared to him as ingenious, schoolboy mischief and sin as something remote and palpably perverse, diabolic, scarcely human at all. Dr Fu Manchu was his type of villain. And he lived in an age when it was possible to retain this purity untarnished. *The Man Who Was Thursday* ends in a great chase when the Common Men rise and take arms; it is thought in cosmic anarchy; it is discovered in defence of order and truth; and the dionysiac Sunday is revealed as the beaming, tutelary, Cheeryble Brother, god of the hearth.

Could Chesterton have written like that today, if he had lived to see the Common Man in arms, drab, grey and brown, the Storm Troopers and the Partisans, standard-bearers of the great popular movements of the century; had he lived to read in the evidence of the War Trials the sickening accumulation of brutality inflicted and condoned by common men, and seen, impassive on the bench, the agents of other criminals, vile, but free and triumphant?

Chesterton was the poetic and romantic child of a smug tradition. It is typical of his age and class that the dawn of sweetness and light in which are dissipated all the night-fears of *The Man Who Was Thursday* should be the discovery that the secret society, so long hated and dreaded, proves to comprise nothing but policemen. For Chesterton the police were the angelic hosts in action; the corpulent blue figure under the Kensington lamp-post represented Justice and Order and held his commission from the innate and inalienable sanctity of popular good sense. For half the world today 'the police' are the Gestapo and the NKVD, and a very macabre parable might now be written of a poetic anarchist whose associates one by one unmask and reveal themselves as policemen, for *they* are the new secret society, so often foreseen in shadows, that conspires against the social and moral order.

H. MARSHALL McLUHAN
Where Chesterton Comes In

TODAY the Chesterton public remains very much the public which read his books as they appeared. And for these readers he inevitably represents a variety of literary attitudes and manners which have begun to 'date' in a way which bars many younger readers from approaching him. So that, for example, even in Catholic colleges books by Chesterton are not commonly on the reading-lists, nor do many of the present crop of students read anything more by him than an occasional 'Father Brown'.

The specific contemporary relevance of Chesterton is this, that his metaphysical intuition of being was always in the service of the search for moral and political order in the current chaos. He was a Thomist by connaturality with being, not by study of St Thomas. And unlike the neo-Thomists his unfailing sense of the relevance of the analogy of being directed his intellectual gaze not to the schoolmen but to the heart of the chaos of our time.

The Catholic teaching of philosophy and the arts tends to be catechetical. It seeks precisely that Cartesian pseudo-certitude which it officially deplores, and divorces itself from the complex life of philosophy and the arts. This is only to say that the Catholic colleges are just like non-Catholic colleges: reflections of a mechanized world. The genuine critical discoveries, on the other hand, made by T. S. Eliot and F. R. Leavis, about how to train, simultaneously, aesthetic and moral perceptions in acts of unified awareness and judgement: these major discoveries are ignored by Catholic educators. Rather in the rationalistic and dialectical patterns of Buchanan and Adler they imagine that there is some Thomistic residue which is to be trusted.

That is where Chesterton comes in. His unfailing sense of relevance and of the location of the heart of the contemporary chaos carried him at all times to attack the problem of morals and psychology. He was always in the practical order. It is important, therefore, that a Chesterton anthology should be made along the lines indicated by Mr Kenner. Not an anthology which preserves the Victorian flavour of his journalism by extensive quotation, but one of short excerpts which would permit the reader to feel Chesterton's powerful intrusion into every kind of confused moral and psychological issue of our time. For he seems never to have reached any position by dialectic or doctrine, but to have enjoyed a kind of connaturality with every kind of reasonableness.

So very impressive is this metaphysical side of Chesterton that it is always embarrassing to encounter the Chesterton fan who is keen about *The Ballad of the White Horse* or the hyperbolic descriptive parts of Chesterton's prose. In fact, it might be the kindest possible service to the essential Chesterton to decry all that part of him which derives so obviously from his time. Thus it is absurd to value Chesterton for that large and unassimilated heritage he got from William Morris—the big, epic dramaturgic gestures, riotous colour, medieval trappings, ballad themes and banal rhythms. Morris manages these things better than Chesterton ever did: and nobody wants to preserve William Morris.

There is also a lot of irrelevant pre-Raphaelite rhetoric in Chesterton. From Rossetti came those pale auburn-haired beauties who invariably haunt his stories. The tiresome alliteration is from Swinburne. From Edward Lear came the vein of anarchic nursery wisdom which served the Victorians as a strategy for keeping sane. By acting insane in a childish way, a kind of temporary equilibrium was maintained: but it was also an evasion of that world of adult horror into which Baudelaire gazed with intense suffering and humility. For the Victorians the nursery was the only tap-root connecting them with psychological reality. But for Chesterton the rhetoric and dimensions of childhood had also their true Christian vigor and scope. He was never tempted into the cul-de-sac in which the *faux-naif* of the Christopher Robin variety invariably winds up.

Nevertheless, there is in Chesterton a considerable aroma of the desperate jauntiness and pseudo-energy of the world of Stevensonian romance: enough to make it desirable to give back to Stevenson the things that are Stevenson's rather than to try to make this dubious adolescent rhetoric appear to be of equal value with Chesterton's metaphysical intuition of being. From Stevenson's master Henley, Chesterton adopted the note of professional heartiness—a journalist's strategy for debunking the aesthetish despair of the 1880s and 1890s. It has led to Kipling and Bulldog Drummond. Henley fathered the optimistic reaction to the intellectual langour of the later Victorians, and 'Wine, Water and Song' is typical of Chesterton's sympathy with that sort of lugubriously self-conscious jollity. But just how unessential it was to him is plain from the fact that Chesterton really was happy. Henley and Kipling never were.

One Victorian feature of Chesterton's which is more closely allied to his real strength he got from Oscar Wilde: rhetorical paradox and epigram. Pater's Marius the Epicurean awoke to 'that poetic and, as it were, moral significance which surely belongs to all the means of daily life, could we but break through the veil of our familiarity with things by no means vulgar in themselves'. Wilde made much of this basic paradox in his life and art, as when in 'The Decay of Lying' he proved that social and artificial things are more exciting than the 'nature' of the romantic poets, and that 'Life imitates Art far more than Art imitates Life.'

The way in which Whitman and Browning and others appear in Chesterton is even more obvious. But the conclusion which it seems necessary to draw from these Victorian aspects of Chesterton is simply that he was not sufficiently interested in them to make a genuinely personal fusion of them. Had they been necessary to his primary awareness of things, he would have been obsessively limited by them in that drastic way in which a Stevenson or a Pater is limited and 'dated'.

In a word, Chesterton was not a poet. The superstition that he was is based on the vaguely uplifting connotations of 'the poetic' prevalent until recently. He was a metaphysical moralist. Thus he had no difficulty in imagining what sort of psychological pressures would occur in the mind of a fourth-century Egyptian, or a Highland clansman, or a modern Californian, popping himself inside of them and seeing with their eyes in the way that makes Father Brown unique among detectives. But he was not engaged in rendering his own age in terms of such varied experience, as the artist typically is. The artist offers us not a system but a world. An inner world is explored and developed and then projected as an object. But that was never Chesterton's way. 'All my mental doors open outwards into a world I have not made', he said in a basic formulation. And this distinction must always remain between the artist who is engaged in making a world and the metaphysician who is occupied in contemplating a world. It should also relieve the minds of those who from a sense of loyalty to Chesterton's philosophical power have felt obliged to defend his rhetoric and his verses as well.

It is time to abandon the literary and journalistic Chesterton to such critical fate as may await him from future appraisers. And it is also time to see him freed from the accidental accretions of ephemeral literary mannerisms. That means to see him as a master of analogical perception and argument who never failed to focus a high degree of moral wisdom on the most confused issues of our age.

HUGH KENNER
The Word and the World

THE essential Chesterton is the man with the extraordinarily comprehensive intuition of being .... There is a sense in which his enormous literary production is a by-product; what must be praised in Chesterton is not the writing but the seeing. The reader who has ... seen that Chesterton's wildest parallels and metaphors are not excogitated illustrations of the vision but ingredients of it has gotten nearly all a commentary can give him. Our task is to examine the nature and quality of Chesterton's writing, defining as we do so a third kind of paradox which may be labelled the aesthetic. It is essential to consider him as an artist, however inartistic he may be, because his vision is after all manifested in language, and his every excursion into language brings him up against certain problems of the artist.

We must first of all enforce a distinction ... between art as making in general and art as a significant expansion of sensibility. This may be most directly done by determining exactly what kind of merit can be claimed for Chesterton's poetry. Like most poets he is praised today by his admirers for the wrong reasons. Charity would suggest leaving the uncritical Chestertonian to his illusions, but prudence insists that the bubble must be pricked, because of the curiously exclusive nature of misguided praise. If a man is praised for the wrong reasons he will almost certainly not be praised for the right reasons. This is attested in the present instance by the abject refusal of Chestertonians to see that interest in their idol as a significant figure must centre not on his cleverness or heartiness but on his perceptivity.

Consider, for example, the opening of one of the best of Chesterton's poems, 'Gloria in Profundis':

> There has fallen on earth for a token
> A god too great for the sky.
> He has burst out of all things and broken
> The bounds of eternity:
> Into time and the terminal land
> He has strayed like a thief or a lover,
> For the wine of the world brims over,
> Its splendour is spilt on the sand.

The paradoxes here are perhaps ... directly and explicitly rooted in the Incarnation . . ., and a little thought will justify any of them. Our concern here is with the mode of their poetic realization, and the judgement must be that the realization is not poetic at all, but intellectual. The alliteration (via Swinburne) and the hearty rhythmic thump (via Kipling) exert a hypnotic influence in their own right and direct attention *away from* the intellectual content. There is no development of imagery: one must pause, shutting one's ears to the sound, to think out the aptness of the thief, the lover, and the wine-cup as analogues for Christ; and each image exists in isolation, without connections before and after. The latter is also true of each stanza; the four stanzas of the poem may be arranged, without serious confusion, in any one of twenty-four possible orders. In sum, the reader is confronted with a cluster of epigrams while a brass band drums at his ears.

The reader who will compare this poem, or any Chesterton poem he likes, with, say, the fourth part of T. S. Eliot's 'East Coker' (in *Four Quartets*) will have no difficulty perceiving the radical difference. Eliot, for example, writes, in a passage equally replete with paradox:

> The whole earth is our hospital
> Endowed by the ruined millionaire,
> Wherein, if we do well, we shall
> Die of the absolute paternal care
> That will not leave us, but prevents us everywhere.

Here the operative word is 'die', finely enforced by its initial position in the line. Expectancy of the rhyme with 'hospital' slows down the reading of 'we shall'—a process accentuated by the parenthetical interruption 'if we do well'—and further isolates the key-word 'die', enforcing the paradoxical contrast with 'doing well'; which contrast in turn suddenly expands the convalescent associations of 'if we do well' to a moral well-doing implicated in the moral irony of the nature of man. These hints on the way one word in the stanza is made to function could be carried on indefinitely to involve all the rest of the poem, although written analysis is at best a clumsy demonstrative instrument. Enough has been said, however, to show that in the Eliot stanza all the poetic devices are enlisted behind, and not at cross-purposes with, the 'meaning'; which meaning is not a detachable intellectual thing but consists in one's total response to the entire stanza.

This radical difference in the mode of working of the two poets cannot be brushed aside by calling the demonstrable differences merely finicky or ascribing Eliot's superiority to more laborious craftsmanship. Indeed, the latter argument tells exactly the other way. Chesterton is simply uninterested in the job a serious poet undertakes. The merit that can be claimed for his verse, once the careful reader has shut his ears to the sound-effects and deciphered the relevance of the array of images, is simply the merit owing to any triumphal celebration. Read in this way, as celebrations of cosmic fact, his poems take on their full meaning; but it is a philosophical, not a poetical meaning, and a noisy rather than a perceptive celebration.

That he should write in this way is the inevitable consequence of the way he perceived. The conflicts reflected in the language are not in his mind but out in front of him, in the things; he admires them, he does not feel involved in them. His analogical vision was both total and in an odd way painless. It unfits him for poetry; it equips him admirably and beyond question for philosophy and exposition.

The next thing to be said about Chesterton as an artist is that his poetic failure carries with it no moral imputation. He fails because he is so constituted that certain gears will never mesh: not because he misconceives the moral basis of making. In a moral sense, in fact, and within the limits of a less intense definition that includes under 'art' all making, even the making of expositions, he is unsurpassed in his time. He never fumbles to reach a position, because he never needs to reach a position. He occupies a central position all the time. And he never fumbles in stating some truth drawn from contemplation of the nature of things, because his statements are so intimately bound up with his perceptions that the central clarity of the latter induces an authoritative finality in the former. It cannot be too often repeated that his gifts and habits were such as to fit him pre-eminently for philosophical discourse. In a better age, with greater incentive for scholarship and less pressure for immediate, continuous, and dissipating journalistic action, he might have been a principal ornament of the medieval Sorbonne. It is doing him the fullest possible homage to call him a splendid anachronism: the operative word is splendid.

Yet he was by no means an anachronism in any moral or political sense. Indeed it is curious to compare his continual centrality with the centre to which the most recent of the intelligent socially conscious are turning. The comparison emphasizes even further Chesterton's firm roots in a timeless philosophy. For the post-war world is cluttered with the hulks of disillusioned Marxists, and some few of them are suddenly and earnestly discovering the kind of cultural conditions and the kind of conception of man which Chesterton had been celebrating since the turn of the century.

THE WORD AND THE WORLD

From a September 1946 speech by André Malraux, for example, I extract the following numbed recognition that the nineteenth-century progressive dream that nourished Marxism is at last pragmatically dead:

At the end of the Nineteenth Century the voice of Nietzsche took up the classical refrain, 'God is dead', and gave it a new and tragic sense. Everyone knew that the death of the deity meant the liberation and deification of man.

The question which faces us all today on this old European earth is whether not God but man is dead . . .

Europe, ravaged and bloody, is not more ravaged and bloody than the picture of mankind which in the pre-war days it hoped to create.

It is with pardonable satisfaction that one now turns to *Heretics*:

The modern man says, 'Let us leave all these arbitrary standards and embrace liberty.' This is, logically rendered, 'Let us not decide what is good, but let it be considered good not to decide it.' He says, 'Away with your old moral formulae; I am for progress.' This, logically stated, means, 'Let us not settle what is good, but let us settle whether we are getting more of it.' . . . Never perhaps since the beginning of the world has there been an age that had less right to use the word 'progress' than we.

Malraux is lamenting the passing of an ideal in which he had for a long time invested everything. Chesterton did not need to have any such ideal beaten out of his head by two bloody wars. *Heretics* was published in 1905.

The April 1946 issue of *Politics*, to take another example, contains some 10,000 anxious words of socialist self-searching under the general title, 'The Root is Man'. The author is a sincere man, and a responsible one. Yet to follow him in his valiant, hesitant, fumbling approach to the Chestertonian position ('All I attempt here is to explain, as coherently as possible, why the Marxian approach to socialism no longer satisfies me, and to indicate the general direction in which I think a more fruitful approach may be made') is to realize most forcibly, while applauding a new political hopefulness, the fact that Mr Dwight Macdonald is merely groping after the most elementary principles of *What's Wrong With the World*, which Chesterton dashed off in 1912. This is not to wish that Chesterton had been listened to long ago: too facile a mass conversion would have been unearned and subject to relapse. It is rather to wish that there had been, contemporary with him, twenty men of his unique gifts.

Such comparison of his spontaneity with the hampered and stammering effort of other men emphasizes the easy inevitability of phrase and analogue which made his vast output possible, and which is usually wrongly accounted for. What, to take a concrete instance, is the exact explanation of that Chestertonian habit which has so distressed so many sensitive reviewers, the habit of allowing a categorical conclusion to issue from a play upon words? Mr Maurice Evans, for example, is concerned with

> The illegitimate use to which Chesterton frequently puts his admirable command of word and image, so that proof appears, where, in fact, none exists. Analogy is obviously a very dangerous weapon in this respect, for what begins as an illustration may, after sufficient development, be accepted as proof... For example, he observes that the American mentality is child-like and loves 'to watch the wheels go round' (*Generally Speaking*). Then taking the metaphor literally, he argues from it: 'watching the wheels go round' implies that they will return back to the same place, or if they move on, they will move in a rut. Therefore, Americans are conservative. This may be the case, but there is no logical connection in the argument.

This may be the case, one must reply, but there is in fact no argument present. Mr Evans does not realize that he has been told a parable.

The parable, as Belloc had the sagacity to observe, is Chesterton's chosen form:

> His unique, his capital genius for illustration by parallel, by example, is his peculiar mark.... No one whatsoever that I can recall in the whole course of English letters had his amazing—I would almost say superhuman—capacity for parallelism.
>
> Now parallelism is a gift or method of vast effect in the conveyance of truth.
>
> Parallelism consists in the illustration of some unperceived truth by its exact consonance with the reflection of a truth already known and perceived....
>
> Thus if some ass propounds that a difference of application destroys the validity of a doctrine, or that particulars are the enemies of universals, Chesterton will answer: 'It is as though you were to say that I cannot be an Englishman because I am a Londoner.' ...
>
> Always, in whatever manner he launched the parallelism, he produced the shock of illumination. He *taught*.
>
> He made men see what they had not seen before. He made them *know*. He was an architect of certitude, whenever he practised this art in which he excelled.
>
> The example of the parable in Holy Writ will at once occur to the reader. It is of the same origin and of similar value. The 'parable' of the Gospels differs only from pure parallelism in the artifice of introducing a story in order to capture the reader's mind. But in essence a parable is the same thing as a parallelism.
>
> Let us remark in conclusion that parallelism is of particular value in a society such as ours which has lost the habit of thinking. It illustrates and thereby fixes a truth or experience as a picture fixes a face or landscape in the mind.

When Christ says, 'Salt is good; but if the salt have lost his saltness, wherewith will ye season it? Have salt in yourselves, and have peace one with another', no one accuses Him of a *non sequitur*. The parable is obviously a parallelism, the illumination of the unknown by its exact consonance with a truth previously perceived. There is no attempt to argue from an observation about salt to an injunction about peace, which is the kind of argument Mr Evans and other critics accuse Chesterton of attempting. Yet they never so accuse Christ: not merely, it may be suspected, because it is patently not the job of Divinity to argue. One reason Christ is sacrosanct is probably to be found in a popular tradition that He is solemn: as if He had never produced the wine for a feast. If Christ had playfully turned the lost sheep of the parable of the ninety and nine into a stray from a herd of kangaroos, it would have been all up with Him as a theologian. Apropos of solemnity Chesterton observes:

If you say that two sheep added to two sheep make four sheep, your audience will accept it patiently—like sheep. But if you say it of two monkeys, or two kangaroos, or two sea-green griffins, people will refuse to believe that two and two make four. They seem to imagine that you have made up the arithmetic, just as you have made up the illustration of the arithmetic. And though they would actually know that what you say is sense, if they thought about it sensibly, they cannot believe that anything decorated by an incidental joke can be sensible.

The joke and the parable are not so far apart as they seem; for properly speaking even the statement, 'two sheep and two sheep make four sheep' is a parable; a common corporeal phenomenon corresponding exactly to the unfamiliar, almost mystical idea that two and two make four. Sheep as a parallel to an abstract idea are fascinating and fantastic enough to make one wonder why the solemn critic should boggle at kangaroos. It is still more important, however, to recognize the almost irresponsible fantasy by which the word 'two', either as a set of wriggly marks on paper or as a man-made noise at once abrupt and cooing, is made to correspond to that same abstract idea, the idea of twoness. The word is certainly not the reality; it is only something analogous to the reality. Apropos of allegorical painting, Chesterton asks:

But what does the word 'hope' represent? It represents only a broken instantaneous glimpse of something that is immeasurably older and wilder than language, that is immeasurably older and wilder than man: a mystery to saints and a reality to wolves. To suppose that such a thing is dealt with by the word 'hope', any more than America is represented by a distant view of Cape

Horn, would indeed be ridiculous. It is not merely true that the word itself is, like any other word, arbitrary; that it might as well be 'pig' or 'parasol'; but it is true that the philosophical meaning of the word, in the conscious mind of man, is merely a part of something immensely larger in the unconscious mind, that the gusty light of language only falls for a moment on a fragment, and that obviously a semi-detached, unfinished fragment of a certain definite pattern on the dark tapestries of reality. It is vain and worse than vain to declaim against the allegoric for the very word 'hope' is an allegory and the very word 'allegory' is an allegory.

Language is not thought, and thought is not reality, any more than figures in a ledger are money, or money is human wealth. Yet one must always use language and thought, as the bookkeeper must always use figures and coins. One can never, in short, escape parallelism; and we never speak but in parables.

'I doubt whether any truth can be told except in parable,' Chesterton makes one of his characters say; and the proposition is accompanied by its Chestertonian corollary: 'I doubt whether any of our actions is really anything but an allegory.' That these observations were self-evident to him in the light of his metaphysical intuition of being may be guessed from the central lines of the poem 'Ubi Ecclesia':

> Where things are not what they seem,
> But what they mean.

One who saw the world as a vast inter-reflecting organism saw language implicated in that reality along with every other ingredient. When Chesterton writes, in short, the very words he uses are part of the vision he exploits; his facility in word and image derives from a real analogical relation, of which he was keenly aware, between language and the other parts of reality. To tax him with verbalism is to deny the existence of analogy, to deny that anything is like any other thing, to deny therefore that connecting things by thinking has any metaphysical meaning: all that. It is rank nominalism.

He was a sturdy realist; and his vision of things showed him very clearly what he could most readily do. His favourite logical device was the *reductio ad absurdum*, because that line of argument springs most readily, with the least possible degree of abstractness, from a direct metaphysical perception. When you see that something is absurd, you are in touch with reality.

It is true for the same reason, though insufficient, to say that he was more concerned with stating cases that proving them. It is still more accurate to say that he strove above all else to *show* men what he saw, on the principle that a thing once seen is its own proof.

False religion ... is always trying to express concrete facts as abstract; it calls sex affinity; it calls wine alcohol; it calls brute starvation the economic problem. The test of true religion is that its energy drives exactly the other way; it is always trying to make men feel truths as facts; always trying to make abstract things as plain and solid as concrete things; always trying to make men, not merely admit the truth, but see, smell, handle, hear, and devour the truth.

Hence Chesterton's purple patches, his parallelisms, his vivid wordplay. Hence his perpetual reiteration of concrete imagery, concrete argument: his avoidance of language which unhappily no longer keenly indicates reality, having been abstracted to death. In *Orthodoxy* he says

If you say, 'The social utility of the indeterminate sentence is recognized by all criminologists as a part of our sociological evolution towards a more humane and scientific view of punishment', you can go on talking like that for hours with hardly a movement of the grey matter inside your skull. But if you say, 'I wish Jones to go to gaol and Brown to say when Jones shall come out', you will discover, with a thrill of horror, that you are obliged to think. The long words are not the hard words, it is the short words that are hard. There is much more metaphysical subtlety in the word 'damn' than in the word 'degeneration'.

On these principles, in the kind of argument we have caught Mr Evans deploring, Chesterton translates the jargon of a Swiss professor about the conscience into short words which force men to think, and then reproduces the pattern of absurdity presented by those naked words with a corresponding pattern secured by putting for the word 'conscience' the word 'nose'. He does not argue, he need not argue, that the statement about conscience is as absurd as the statement about noses. It was always absurd. One does not even need to know the meaning of the word 'conscience' to see the absurdity. The law of logic has been transgressed, and it is a logical, not a factual, flaw that is being exposed. Unfortunately, logic is a strangely unfamiliar tool: equally unfortunately, no one has ever seen a conscience, though all men have seen noses. Hence it is that nonsense talked about the conscience has a fair chance of passing muster, though corresponding talk about noses fairly shrieks its own falsity.

Having grasped all this, the reader of the following passage will not accuse Chesterton of irresponsible play:

The first argument is that man has no conscience because some men are quite mad, and therefore not particularly conscientious. The second argument is that man has no conscience because some men are more conscientious than others. And the third is that man has no conscience because conscientious men in different countries and quite different circumstances often do very different things. Professor Forel applies these arguments eloquently to the question of human consciences: and I really cannot see why I should not apply them to the question of human noses. Man has no nose because now and then a man has no nose—I believe Sir William Davenant, the poet, had none. Man has no nose because some noses are longer than others, or can smell better than others. Man has no nose because not only are noses of different shapes, but (oh, piercing sword of scepticism!) some men use their noses and find the smell of incense nice, while some use their noses and find it nasty. Science therefore declares that man is normally noseless; and will take this for granted in the next four or five hundred pages, and will treat all the alleged noses of history as the quaint legends of a credulous age.

The nose-pattern repeating the conscience-pattern is the type of all Chesterton's writing. The truths with which he deals are not those of a rarefied kind which the normal mind can only discover at the further end of a wearisome logical process; they are mostly elementary truisms which have only to be *seen*. 'It is the paradox of human language', he says of the fundamental convictions, 'that though these truths are in a manner past all parallel hard and clear, yet any attempt to talk about them always has the appearance of being hazy and elusive. The best that language can do is indicate them, and the best language for the purpose is that which indicates most sharply. It follows that Chesterton's concern throughout his writing will be to frame paragraphs which first, correspond with the reality whereon he has fixed his eye, and second, shout for attention. 'We try', he says, 'to make our sermons and speeches more or less amusing, . . . for the very simple and even modest reason that we do not see why the audience should listen unless it is more or less amused.'

Chesterton's humility here underrates his achievement. . . . Word, image, and epigram cooperate to do superbly something that could not otherwise be done at all, when, as too rarely, he disciplines them rigidly in the service of metaphysical statement. The example of the nose and the conscience cited above is really excessively simple; as an instance of the precision and flexibility Chesterton was capable of when he chose, it is worth while examining the working of the paragraph which develops the comparison between the mystical mind and the dandy's dressing-room. As usual, he opens with a specific example, the wooden post:

> When (our contemporary mystics) said that a wooden post was wonderful, they meant that they could make something wonderful out of it by thinking about it. 'Dream; there is no truth', said Mr Yeats, 'but in your heart.'

The quotation from Yeats recalls to the reader any number of similar statements, and so places Chesterton's simplification in relation to the entire tradition he is attacking. With the next sentence the controlling image is introduced:

> The modern mystic looked for the post, not outside in the garden, but inside, in the mirror of his own mind. But the mind of the mystic, like a dandy's dressing-room, was entirely made of mirrors. That glass repeated glass like doors opening inwards for ever; till one could hardly see that inmost chamber of unreality where the post made its last appearance.

The word 'dandy' reflects on 'mystic' as much as on 'dressing-room'; and at the same time the stock image of the mirror of the mind is subtly transmuted into a vivid, pejorative image of a room lined with mirrors. 'Dressing-room' brings to mind the triptych mirrors at tailors', where everyone has had experience of infinite multiple reflections; and the additional comparison of doors opening inwards for ever gives additional concrete force to the idea. In the final clause, 'inmost chamber of unreality' gives new precision to the comparison of the mystic's mind to a room: and the sequence ends where it began, at a deeper level of penetration. In the next sentence, mirrors return, with a difference:

> And as the mirrors of the modern mystic's mind are most of them curved and many of them cracked, the post in its ultimate reflection looked like all sorts of things . . . etc.

It is perhaps unnecessary to point out that 'cracked', applied to the mirror and to the mind, has double force. Another functional pun turns up immediately afterwards:

> But I was never interested in mirrors; that is, I was never interested in my own reflection—or reflections.

'Reflection' is, of course, supplied by the mirror-imagery, with an overtone of vanity versus humility; Chesterton is never far from the moral implications of metaphysics. And the pun introduced by 'reflections' brilliantly refocuses the entire enquiry on the operations of the mind, preparing for a statement of the positive conclusions:

> I am interested in wooden posts, which do startle me like miracles. I am interested in the post that stands waiting outside my door to hit me over the head, like a giant's club in a fairy tale.

The giant's club recalls the episode of bumping into a post which was the initial stimulus of the essay; but it here functions locally as a physical image of the metaphysical surprise evoked in 'posts which do startle me like miracles'. In the next sentence the door outside which the post stands introduces a transition to the doors of the senses which open on that mental room with which the preceding passage has been concerned: and the peroration after so much preparation carries enormous force:

> All my mental doors open outwards into a world I have not made. My last door of liberty opens upon a world of sun and solid things, of objective adventures. The post in the garden; the thing I could neither create nor expect; strong plain daylight on stiff upstanding wood; it is the Lord's doing and it is marvellous in our eyes.

Careful study along these lines of the way in which the transition from image to image is made in similar passages will reinforce the constant theme of this chapter: that Chesterton's writing at its best is concerned with fixing exactly a statement of a metaphysical vision, by indicating relationships of word and example within that vision. He is not inventing illustrations, he is perceiving them. The conventional patristic divine, Jeremy Taylor, summarizes the kind of analogical perception that this writing is exploring:

> Thus when (God) made the beauteous frame of heaven and earth, he rejoyced in it, and glorified himself, because it was a glasse in which he beheld his wisdom, and Almighty power: . . . For if God is glorified in the Sunne and Moon, in the rare fabric of the honeycombs, in the discipline of Bees, in the œconomy of Pismires, in the little houses of birds, in the curiosity of an eye, God being pleased to delight in those little images and reflexes of himself from those pretty mirrours, which like a crevice in a wall thorow a narrow perspective transmit the species of a vast excellency: much rather shall God be pleased to behold himself in the glasses of our obedience. . . .

Gerard Manley Hopkins puts it more succinctly: 'This world then is word, expression, news of God.' Chesterton would interpret that news. Perception of this fact reduces to a simple manifestation of humility his claim to be a journalist rather than an artist. If he was not a creative artist, he was, when he took the pains, an extremely competent workman, framing intricate analogies to interpret the supreme analogy which he saw all around him. He was in this sense an artist because he was the highest kind of journalist, having as his object truth.

He suggests in *William Blake* the way in which the analogist's art must be called in to present truth:

> In the modern intellectual world we can see flags of many colours, deeds of manifold interest; the one thing we cannot see is the map. We cannot see the simplified statement which tells us what is the origin of all the trouble. How shall we manage to state in an obvious and alphabetical manner the ultimate

query, the primordial point on which the whole modern argument turns? It cannot be done in long rationalistic words; they convey by their very sound the suggestion of something subtle. One must try to think of something in the way of a plain street metaphor or an obvious analogy. For the thing is not too hard for human speech; it is actually too obvious for human speech.

The plain street metaphor or the obvious analogy are for Chesterton the simple key to the problem of conveying reality, short-circuiting as they do the fore-doomed attempt to trace the contradictory labyrinth of being with any continuous rational thread. 'Long rationalistic words . . . convey by their very sound the suggestion of something subtle'; and being is the reverse of subtle. It is simple, though the principle of analogy shows it to be paradoxically complicated as well. A locomotive is both simple in essence and complicated in detail; one would scarcely call a locomotive subtle. And one should beware of trying to describe a locomotive to the uninitiated by rationalistically describing its workings, beginning with the vaporization of heated water, lest its puffing power come to seem very subtle indeed. One does better to call it an iron horse.

No one who has finally grasped these points will press the question, why Chesterton's prose is so intricate. Too wise to try to explain the obvious, he drew pictures of it; and his pictures, like those of God with whom the artist is often audaciously compared, took on life; a life of their own; a life of alliteration and epigram, of sudden unexpected correspondences, of accidental patterns writhing and weaving with all the crawling energy of the Gothic architecture which was his craftsman's ideal.

We have shown in analysing the mirror-passage how Chesterton in his best work manipulates his images functionally, to control the reader's response towards a total meaning which cannot itself be briefly and exactly stated. We have stated further Chesterton's explicit view that brief and exact statement of an analogical reality is in fact a priori impossible. These principles may be tidied up in a new statement of the ubiquitous necessity of paradox: for in paradox is the practitioner of art, even expository art, perpetually landed. The reason is that the thing, the work of art, that he is constructing must both hang together itself and be consistent with the reality on which his eye is fixed. Insofar as it hangs together itself, insofar as it obeys its own artistic laws, it will have being, which is analogical. Its statements, to put it another way, will in only a relative sense be logically interdependent. Insofar as it is consistent with that other being whose shadow it is, it will not only tend to be twisted out of coherent shape, but it will partake of the paradoxicality of its prototype. The law that all being is intrinsically analogical operates here with a double vengeance. Things are paradoxical, and art performed in homage to those things is doubly so. If I say, for example, that there is a pinkish man in the room, my exact transcript of reality lands me in a paradox of language: it is customary to speak of a white man. If I say that there is a white man in the room, I obey the laws of language, but only by convention the laws of fact: a genuine white man would be monstrous. The relationship between statement and fact is analogical: you have paradox, whatever you do.

The artist, especially the artist like Chesterton with his eye on a ready reception, must be constantly in this way adjusting the strain between inner and outer consistency; constantly striving after words that will say something and at the same time say the right thing. And the more adequately his words proclaim both the unity of speech and that of being, both the contradictions of art and those of being, the more he will forge a chain of paradoxes. There is scarcely a great mystical poem in the language that is not, at the merely literal level, stark nonsense. Hence—and it is their most triumphant justification—hence the paradoxes of Gilbert Chesterton.

It is illuminating to notice how thoroughly paradox enters into the workmanship of that most conscious aesthetician among modern writers, James Joyce. The method of Joyce's masterwork, *Finnegans Wake* concealed as it is by the use of some dozen different languages, is simply to fold paradoxes back upon themselves in such a way as to utter both contrasting halves simultaneously. When Joyce writes 'phoenish', he is telescoping alpha and omega, the end and the beginning, *finish* and *phoenix*. When he writes, 'For nought that is has bane', he said simultaneously, 'Nothing that exists is evil', and 'The appearances of evil have no permanence; they were not and they shall not be'; simultaneously posing and resolving the problem of pain.

The oddly esoteric vocabulary of *Finnegans Wake* represents a final straining attempt to overcome the basic paradox of art and make the Thing identical, beyond any possibility of separation, with its verbal vehicle. It utterly defeats paraphrase. It is not a little startling to see how this audacious, almost blasphemous attempt to re-utter the world-generating Word, to achieve a totality corresponding to the totality which is of God, achieves its object—insofar as it does achieve it—by virtue of multi-layered paradoxes whereby a river is all rivers, riverdom, woman, all of life, and but half of life; and a stone is an innkeeper, a dreamer, the fount of life, quite dead, and both food and feeder at his own funeral feast.

Joyce, as we shall see, has other affinities with Chesterton as a myth-maker. It is surely a demonstration of the contemporary critical muddle to find the most advanced experimenter of his time building upon the same first principles, and exploiting the same kind of analogical perception, as the man whom avant-garde critics decry as the very type of hearty Toryism.

Indeed it is the analogical perception which makes it possible for *Finnegans Wake* to be taken seriously; though Joyce, who, it is true, offers far more temptation to the inept than does Chesterton, has suffered from essentially the same charge of verbalism. Chesterton's insistence that the artist keep his eye on the object finds in the career of Joyce a particularly ironic vindication.

We have shown that Chesterton's eye never wandered from the object, from an especially intricate simultaneous perception. It is helpful to remember that he elevated this principle into a positive prescription. It is, to begin with, scarcely necessary to point out that keeping one's eye on the object does not mean copying the externals of the object. Rather it means knowing what the object is, knowing from the inside. This interior knowledge when it can be obtained is a guarantee against the errors introduced by falsely framed concepts. The surest way to find out that the 'economic man' doesn't exist is to try to draw a picture of him. It was the attempt to draw pictures of things that existed only as concepts that vitiated the later poetry of Blake; it took his art into that unreal otherworld of doubts and riddles that has for the past thirty years been the playground of a certain kind of critic. On the failure of Blake Chesterton commented,

> No pure mystic ever loved pure mystery. The mystic does not bring doubts and riddles: the doubts and riddles exist already. . . . The mystic is not the man who makes mysteries, but the man who destroys them. The mystic is one who offers an explanation which may be true or false but is *always* comprehensible. The man whose meaning remains mysterious fails, I think, as a mystic.

The early Blake, he says, like every great mystic, was also a great rationalist. In this sense, another great rationalist is Chesterton's own Father Brown. It is startling to count the Father Brown stories which turn on the war of reason with mystification. Father Brown, the professional supernaturalist, is constantly at war with the sham supernatural. In *The Arrow of Heaven* there is talk of a curse, misdirecting attention from a simple stabbing. In *The Perishing of the Pendragons* a family doom and a supernaturally flaming tower are reduced to mere arson and shipwreck. In *The Doom of the Darnaways* an ancient interdiction boils down to a very modern murder-plot. These stories are repeated parables of the true function of the artist and seer; their wildly paradoxical solutions are true; the straightforward, frequently supernatural explanation is falsification. The cloak of evil, Chesterton seems to be saying, is the false paradox; the trap of truth is the incomplete paradox.

Equally the trap of truth is the word written in the void, the writer's eye not firmly fixed on the object. A parable of this principle is presented in the contrasting peasants of the following passage:

> Knowing nine hundred words is not always more important than knowing what some of them mean. It is strictly and soberly true that any peasant, in a mud cabin in County Clare, when he names his child Michael, may really have a sense of the presence that smote down Satan, the arms and plumage of the paladin of paradise. I doubt whether it is so overwhelmingly probable that any clerk in any villa on Clapham Common, when he names his son John, has a vision of the holy eagle of the Apocalypse, or even of the mystical cup of the disciple whom Jesus loved. In the face of that simple fact, I have no doubt about which is the more educated man; and even a knowledge of the *Daily Mail* does not redress the balance. It is often said, and possibly truly, that the peasant named Michael cannot write his own name. But it is quite equally true that the clerk named John cannot read his own name. He cannot read it because it is in a foreign language, and he has never been made to realize what it stands for. He does not know that John means John, as the other man does know that Michael means Michael.

Chesterton's acute awareness of this danger was one of the things that led him to prefer journalism to a more cloistered if less distracting life among the 'pure artists'. For it is patently true that for one such genius as Eliot, 'the most conscious point in his age', there are a hundred poseurs of Bloomsbury with their eyes turned inward upon their egos. In exoneration of his trade he wrote.

> A poet writing his name upon a score of little pages in the silence of his study may or may not have an intellectual right to despise the journalist: but I greatly doubt whether he would not morally be the better if he saw the great lights burning on through darkness into dawn, and heard the roar of the printing wheels weaving the destinies of another day. Here at least is a school of labour and of some rough humility, the largest work ever published anonymously since the great Christian cathedrals.

He preferred journalism because it kept him constantly in touch with real work and real problems. What troubled him about the efforts of the emptier modern artists was that they had their eyes on no object: they meant nothing. The heresy of Realism, which celebrates things for what they seem and not for what they mean, he presents under the parable of a gigantic Gothic cathedral revisited by a priest who has lost his memory:

He saw piled in front of him frogs and elephants, monkeys and giraffes, toadstools and sharks, all the ugly things of the universe which he had collected to do honour to God. But he forgot why he had collected them. He could not remember the design or the object. He piled them all wildly into one heap fifty feet high; and when he had done it all the rich and influential went into a passion of applause and cried, 'This is real art! This is Realism! This is things as they really are!' . . .

The finest lengths of the Elgin marbles consist of splendid horses going to the temple of a virgin. Christianity, with its gargoyles and grotesques, really amounted to saying this: that a donkey could go before all the horses in the world when it was really going to the temple. Realism means a lost donkey going nowhere.

He closes this essay 'On Gargoyles' with an illuminating reference to his own work:

These monsters are meant for the gargoyles of a definite cathedral. I have to carve the gargoyles, because I can carve nothing else; I leave to others the angels and the arches and the spires. But I am very sure of the style of the architecture and of the consecration of the church.

Journalist or no, gargoyle-carver or no, he nevertheless knew what his words meant, and the contradictions into which language leads; he knew what being meant, and the contradictions implicit in it; and he did not shrink from the baffling task of making the latter visible through the former.

Keeping in mind all the criteria we have considered: Chesterton's insistence that art be responsible to truth and rooted in the perceptions of the artist; the scope and explicitness of Chesterton's metaphysical perception, within which he moved so freely; yet disabling both, his patent incapacity to realize particular conflicts seriously enough to produce significant poetry: keeping all this in mind, what are we to make of his output of novels and stories? What kind of relevance have they to his lifelong moral and metaphysical concern?

The novel that is not simply documentation owes its vitality to the epigram at its heart: it works by expansion.

Chesterton's novels expand his elsewhere concisely developed perceptions, function in the same way, and have the same kind of value: but with (as Belloc said of the scriptural parable) a story to capture the interest of the reader. The reader who has followed the analysis above of the richly allusive passage on the mirrors of the mystic's mind, with its shifts of imagery and expanding and contracting symbolism, can see that the movement of ideas is exactly like that of a Chestertonian story; and the experienced reader can readily imagine the story Chesterton might have made of it. The reader familiar with the Father Brown collection will know the story he did make of it: *The Man in the Passage*.

Like a paragraph of vintage Chestertonian exposition, the Chestertonian novel or story constructs a web of analogies. Its value is ultimately moral: the value of any parable. His novels, like his poems, are the products of a born philosopher, not of a born dramatist. The Father Brown stories, for example, with all their machinery of murder and repentance, and all the genuine moral interest in the fact of human sin that makes them unique among detective stories, are patently devoid of the intense dramatic life of *Crime and Punishment*. This is not to say that they exist, like the ordinary mechanical detective story, only as neat constructions: rather they exist as ingenious analogues of psychological facts. Chesterton knew perfectly well, and repeatedly asserted, that as human documents they are trifling; he took them seriously enough to write them because they reflect, like everything else he wrote, the unique metaphysical intuition it has been the purpose of this book to explore.

To say that the characters exist as abstractions, that the life of the stories is conferred entirely by the continual local brilliance of the writing and that they function ultimately as expansions of the philosophic conflicts in his paragraphs of moral and metaphysical paradoxes, is to say that Chesterton's fiction is not drama but parable; on a large scale, as in *The Man Who Was Thursday*, it is allegory: myth. It is unnecessary to recall the tradition of Christian allegory in which they are rooted: *Pilgrim's Progress* may be cited as a late example, springing from the tradition that had flowered in the morality plays. It is more fruitful and suggestive to point to two men whose perceptions tended to be, like Chesterton's, of a detached and philosophical kind, and much of whose output is explicitly on the level of myth: William Blake and James Joyce.

Much of the recent re-emphasis on Blake is based on the appetite of a collapsing civilization for sustaining myths, and to a current belief that the artist fulfilling his supreme function assumes a sort of priestly character and becomes myth-maker. That the myth tends to become dehumanized is counted no demerit by the modern taste for the abstract. At best, the myth-maker erects a pantheon and brings it to life; and so bringing the universe to life, he presents that life under the figure of something living: a man: hence the approach of the most ambitious philosophical speculation to the ancient conception of the macrocosm: the gigantic man who is all things. This conception is everywhere present in Christian thought: to say that in Adam all die and in Christ all are made alive is literally to think in terms of the fall of one all-subsuming human form and the redemption effected by an all-sustaining human God.

Men capable of thinking with any comfort in terms of such magnitude have been few; one is William Blake, another is James Joyce; a third is Gilbert Chesterton. Blake in the nineteenth century and Joyce in the twentieth represented the pattern of the cosmos by the figure of a gigantic man, or by the eternal recurrence of a gigantic circle or wheel; or contemplating the persistence of the unfallen state as an eternal reality lying behind the fallen, by both together.

Such figures were Chesterton's, and they were the logical fruit of his talent for metaphysical perception dramatized on a large scale. One of his themes, developed in *The Man Who Was Thursday*, is the figure of fallen and scattered men conceived as parts broken off the whole and perfect man, free according to their limited being to recapture some analogical image of that former wholeness by pushing to the limit such virtues as now lie within their powers: an image of the isolation of soul from soul. It is in this sense that he sees the supernatural goodness of saint after saint arising to union with God and yet intrinsically imperfect because its emphasis is on goodness of one kind. It is in obedience to this principle that 'St Francis, in praising all good, could be a more shouting optimist than Walt Whitman; St Jerome, in denouncing all evil, could paint the world blacker than Schopenhauer.' Saints may contradict one another's virtues and be right, because saints live in a fallen world. The best man develops only a corner of his potential virtue; he is but a fragment of the unfallen Adam.

Following from and completing this idea is the corollary conception of good men everywhere seemingly at odds, breaking each other's heads in the name of good, yet ultimately fighting all on the same side, the warring members of the cosmic man.

At the end of *The Napoleon of Notting Hill* there stands a passage pushing this idea to the utmost of which Chesterton was capable, a passage pointing back to the celestial wars of Blake and forward to the cosmic paradoxes, as yet unuttered, of Joyce. There comes out of the silence and darkness that followed the settling of the dust upon the last battlefield of Notting Hill a chill voice saying:

> So ends the Empire of Notting Hill. As it began in blood, so it ended in blood, and all things are always the same.

And another voice replies out of the ruins,

> If all things are always the same, it is because all things are always heroic. If all things are always the same, it is because they are always new. To each man one soul only is given; to each soul only is given a little power—the power at

some moments to outgrow and swallow up the stars. If age after age that power comes upon men, whatever gives it to them is great. . . . We who do the old things are fed by nature with a perpetual infancy. No man who is in love thinks that anyone has been in love before. No woman who has a child thinks that there have been such things as children. . . . Yes, oh, dark voice, the world is always the same, for it is always unexpected.

. . . Wherein the experienced reader will hear the soft Irish voice of James Joyce: 'Teems of times and happy returns. The seim anew.' Then the first voice retorts again that all is dust and nothingness, and again the second voice carries forward its theme:

> Men live, as I say, rejoicing from age to age in something fresher than progress—in the fact that with every baby a new sun and a new moon are made. If our ancient humanity were a single man, it might perhaps be that he would break down under the memory of so many loyalties, under the burden of so many diverse heroisms, under the load and terror of all the goodness of men. But it has pleased God so to isolate the individual human soul that it can only learn of all other souls by hearsay, and to each one goodness and happiness come with the youth and violence of lightning, as momentary and as pure. And the doom of failure that lies on all human systems does not in fact affect them any more than the worms of an inevitable grave affect a children's game in the meadow. Notting Hill has fallen; Notting Hill has died. But that is not the tremendous issue. Notting Hill has lived.

But the first voice laughs on, scoffing at Notting Hill as vanity. Then they know one another: Auberon Quin, who gave Notting Hill its charter for a joke, and Adam Wayne, who fought for that charter as a creed. And Wayne finishes:

> The equal and eternal human being will alter (our) antagonism, for the human being sees no antagonism between laughter and respect, the human being, the common man, whom mere geniuses like you and me can only worship like a god. When dark and dreary days come, you and I are necessary, the pure fanatic, the pure satirist. We have between us remedied a great wrong. We have lifted the modern cities into that poetry which everyone who knows mankind knows to be immeasurably more common than the commonplace. But in healthy people there is no war between us. We are but the two lobes in the brain of a ploughman. Laughter and love are everywhere. The cathedrals, built in the ages that loved God, are full of blasphemous grotesques. The mother laughs continually at the child, the lover laughs continually at the lover, the wife at the husband, the friend at the friend . . . Let us go out together. . . . Let us start our wanderings over the world. For we are its two essentials. Come, it is already day.

That, as the conclusion and summation of his earliest novel, shows clearly the abstract and mythological conception on which it is based.

The 'equal and eternal human being' was to become Sunday, the fantastic anarchist whose face in the last wild chase of *The Man Who Was Thursday* is concealed from sight, and who turns out finally to be the chief of police: Sunday, 'huge, boisterous, full of vanity, dancing with a hundred legs, bright with the glare of the sun, and at first, somewhat regardless of us and our desires', Sunday—'Nature as distinct from God.'

The reconciliation of that antagonism between him who scoffs and him who worships is accomplished in *The Man Who Was Thursday*; for the antagonist of them both turns out to be the leader of them both. He is like the cosmic man of so much quasi-mystical speculation: the stupendous figure through whose limbs circle the stars. In him is transcended the isolation of soul from soul, which begets both loneliness and its blood-brother courage. The cosmos has the pattern of a man, which is one of its two traditional ultimate patterns; the other being the wheel, the unending cycle, the serpent with its tail in its mouth, which Chesterton also perceived and abominated, summing it up through countless scattered passages in the restless, formless patterns of Turkish carpets, the restless, pointless cycle of Nirvana, and the annihilistic self-contemplation of the East. The cosmos has become a man, a man of will and energy and fantastic beauty, a man and therefore a cross. And when, in the final sentences of *The Man Who Was Thursday*, the last mask is torn off the face of Nature, there is displayed the older face of God: 'Can ye drink of the cup that I drink of?'

The restless brother-battle consequent on the Fall and resolved in a transcendental resurrection was a myth that Chesterton arrived at early in life: saw, embodied in a hasty novel subtitled *A Nightmare*, and passed over. He said, with acute self-penetration, that he was a journalist because he could not help being a controversialist, and hence never a novelist. Had he been a novelist he might well have lingered with that single vision, and elaborated it as it deserved to be elaborated, for his largest talents lay towards myth and allegory, and that vision, or rather the perception underlying that vision, underlay everything that he was later to write, in however scattered or fragmentary a form. Only twice in English letters has that vision been perceived and elaborated towards its perfection: in the Apocalyptic vision of *Jerusalem* and in that other nightmare of the dreamer of *Finnegans Wake*. Sunday is the gigantic Albion of Blake, the nameless panheroic HCE of Joyce. The Two Voices disputing amid the failing firelight of Notting Hill are the rebellious Orc and the sunlit Los of Chesterton's great predecessor, the scoffing Shem and the conserving

Shaun of his great contemporary. Had he given himself to his art as did these men, he might have been received into their trilogy. It was as well that he did not. Myths tend to be sterile; Blake's reputation, after the flurry of symbolic interpretation has died down, will probably rest on his early dramatic lyrics, and Joyce's on the inevitable discovery that his myth is vitalized by an intense personal conflict. That Chesterton's potentiality, had he chosen to be an artist, lay in the direction not of drama but of myth, is another way of saying that with his secure metaphysical perception he would have found his true fulfilment as a great philosopher. The times, however, and his sense of immediate duty, were against him; that he preferred instead to be a practical mystic whose vast moral vision was to be placed at the daily service of immediate political and educational ends, is matter both for gratitude and regret. That he preferred loosing a thousand lightning-strokes to achieving the calm sunlight of a single perfect work is in the simplest sense a true summary of his career; and he would have justified it by his faith in the final paradox of *The Everlasting Man*: 'The lightning made eternal as the light.'

HUGH KINGSMILL
G. K. Chesterton

ROUND about 1930 G. K. Chesterton was the chief catholic apologist writing in English. Nowadays, Herbert McLuhan says in his introduction to Hugh Kenner's revaluation of Chesterton[1] books by Chesterton are seldom included in the reading lists of catholic students. Chesterton, says McLuhan, had 'an unwavering and metaphysical intuition of being ... a kind of connaturality with every kind of reasonableness'; but this element in his work has been swamped by the extravagances of 'the Toby-jug Chesterton of a particular literary epoch'. From Stevenson, McLuhan continues, Chesterton borrowed a desperate jauntiness and pseudo-energy; from Henley, a note of professional heartiness which he employed to debunk the pessimism of the 1880s and 1890s; from William Morris, dramaturgic gestures, medieval trappings, ballad themes and banal rhythms; from Rossetti, the pale, auburn-haired beauties who haunt his stories; from Swinburne, a tiresome trick of alliteration; from Edward Lear, the vein of anarchic

---

1 *Paradox in Chesterton*, by Hugh Kenner.

nursery wisdom which helped the Victorians to keep sane. The influence of Whitman and Browning on Chesterton seems to McLuhan too obvious to need any more than a mention, and he does not even refer to the still greater influence of Dickens, about whom Mr Kenner also maintains an unbroken silence.

Having claimed that Chesterton had an unwavering insight into being, and then presented him as swayed by every passing fashion in thought and expression, however shoddy, McLuhan concludes abruptly, 'It is time to see him freed from the accidental accretions of ephemeral literary mannerisms', and hands on his smoky torch to Mr Kenner, who snatching it blithely from him speeds buoyantly on his way. Chesterton, Mr Kenner says, was not so much great because of his published achievement as great because he was right. He scarcely left a page that is not in some way blotched and disfigured, his perceptions are metaphysical rather than æsthetic; he never achieves a great poem, for his poems are compilations of statements not intensely felt but only intensely meant and have a philosophical not a poetic meaning. Celebrations of cosmic fact, they are a noisy rather than a perceptive celebration. The conflicts they deal with are not in Chesterton's mind but out in front of him, in the things; he does not feel involved in them; they are in an odd way painless, for he never experienced the self-distrust mirrored in some of Gerard Hopkins's verse, 'There is a sense in which it is to his praise that he could not be a poet.'

It would be useless to enquire how a man can be right whose work is blotched and disfigured on nearly every page; how he can mean intensely without feeling intensely; how though noisy rather than perceptive, he can yet be philosophically sound; or how he can resolve mental conflicts which take place outside his mind, and bring great truths painlessly to birth in a kind of twilight sleep. Mr Kenner is, what Chesterton became, a Thomist. Thomas Aquinas, the most gifted of medieval dialecticians, has been revived of late years in order to supply sceptics anxious for a faith with a complete set of answers to persons willing to engage with them in arguing about what can neither be established nor disproved by mental processes. The Thomist seeks refuge from the complexity of things in the more manageable complexity of words. With the inexhaustible verbal arsenal of Aquinas to draw upon, he is able at the same time to preserve the Christian virtue of humility by affirming nothing of himself, and to claim that he is always right because Aquinas is never wrong. To this type the free exercise of the imagination is a presumptuous attempt to by-pass the theological route to truth, and careless or tawdry writing such as Mr McLuhan and Mr Kenner note in much of Chesterton's work, a pleasing sign of indifference to the artist's ideal of perfection. From Mr Kenner's standpoint the trouble with the Toby-jug Chesterton is not that he is unreal, exaggerated, rhetorical, but merely that he obscures the later Chesterton, the master of analogy and paradox whose course even from the earliest years, was set towards the haven of Aquinas. In rescuing Chesterton from Dickens and Browning and Stevenson and aligning him with such verbal artificers or contortionists as Gerard Hopkins and T. S. Eliot, James Joyce and Gertrude Stein, Mr Kenner may well have made him more palatable to catholic students, grateful for an up-to-date flavouring in their theological dish. But his Chesterton is no more authentic and much less amusing than the mythical figure of twenty years ago, the roaring, beer-swilling swashbuckler at whose name capitalists and cocoa-drinkers turned pale. His book, however, contains many illuminating quotations which could give some idea of Chesterton to a careful and reflective reader.

In his autobiography Chesterton narrates that in his late teens he passed through a phase of mental disease bordering on madness. He had horrible fancies which he used to put into drawings. They were not, he says, of a homosexual nature, but he tells us nothing else about them, though from the prevalence in his stories and poems of blood and slicing swords, one may surmise that they expressed the craving for violence which permeated his otherwise kindly and pacific nature. In a poem to his schoolboy friend E. C. Bentley, he speaks of 'the sick cloud upon the soul when we were boys together', and defines it as an exhalation from the decadent 1880s and 1890s, when 'science announced non-entity and art admired decay', the age of Wilde and Whistler and Haeckel. He and his friend were upheld in their despair by Stevenson and Whitman; sent 'far out of fish-shaped Paumanok some cry of cleaner things', and 'Truth of Tusitala spoke'. Then 'God and the good Republic came riding back in arms', and they 'found common things at last, and marriage and a creed'.

The sick cloud was not an exhalation from the age, it was within Chesterton himself. Whitman was hardly a sure prophylactic against Wilde, nor the author of *Dr Jekyll and Mr Hyde*, with his cry 'Shall we never shed blood?' a breath of fresh air after the far less subterranean Whistler. But Stevenson and Whitman inevitably attracted Chesterton, who like them longed for health, and like them hoped to achieve it not only by facing his nature but by externalizing it in the outer world as a conflict between health and disease, with himself as the protagonist of health. It is in his imaginative work that one most easily perceives how little of daylight and freshness he had by nature. Much of his poetry and more of his fiction are overshadowed by the evil twilight of a child's nightmares, swelling to a horror which only some sudden act of violence can dissipate. This internal tension he resolved, both in his critical and imaginative work, by simplifying every situation into a conflict between an evil oppression and a liberating champion. His Dickens is the English embodiment of the French Revolution, a man who 'panted upwards on weary wings to reach the heaven of the poor'. His Browning is a plain man who made poems out of the simple things of life, shaming the æsthetes at their dubious diversions in velvet-curtained rooms. His King Alfred is Free Will and Christian Hope wiping the floor with the pagan Guthrum and his determinist Danes. And so on. He applied the same simplifying process to countries. In *An Alliance*, written, as will easily be credited, before he knew Hilaire Belloc, good embodied itself in 'the Saxon lands', England and America, evil in Spain, who had recently been defeated by the United States—'Four centuries doom of torture, choked in the throat of Spain' is Chesterton's way of putting it. Later, Spain was replaced by Germany ('O thousand folk and frozen folk'), by the Orient ('land of purple and passion and glamour'), and finally by all non-Catholic countries, including England, of whom, a few years before she was left as the sole defender of all he valued, Chesterton wrote: 'This noble and generous nation which lost its religion in the seventeenth century has lost its morals in the twentieth.'

Chesterton was at his best in the first decade of this century, when his conversion to Roman Catholicism lay many years ahead. Work and success had lifted the cloud of his adolescence; his wit was keen, his fancy wonderfully fertile, his sense of reality intermittently active, and his quibbling and rhetoric with which in his last years he tried to dispel his fretful uneasy gloom still far from being in complete control of his mind. The progressives, led by Wells, and the æsthetes, from Pater to George Moore, were his chief targets. Wells's happiness in the thought that just as the motor-car was quicker than the coach, so something would be quicker than the motor-car, Chesterton echoed and amplified in a Dr Quilp, who foretells a machine on which a man could circle the earth so quickly that he could keep up a talk in some old world village by saying a word of a sentence each time he came round. George Moore served Chesterton as a type of self-conscious artist who values everything only as a setting for himself—*The Grand Canal with a distant view of George Moore, Effect of George Moore seen through a Scotch mist, Ruins of George Moore by moonlight*. Religion in those years still meant more to Chesterton than creeds and institutions; and to measure the change between what he then was and what he later became, one need merely place side by side his criticism of

Pater's view that we must enjoy the passing moment for its own sake and his defence of the wealth and pomp of Rome. Moments of love or of any other great emotion, he says, in reply to Pater, are filled with eternity: 'These moments are joyful because they do not seem momentary. Man cannot love mortal things. He can only love immortal things for an instant.' Visiting Rome after his conversion he concedes that the Catholic faith 'might have scored in some ways if it had remained absolutely austere and unworldly; as poor as the birth in the stable; as naked as the Victim on the cross'. But, he continues, unless it could have been kept at the last extremity of severity, it was right to rush to the last extremes of splendour: 'The Pope is the Vicar of Christ, and when he goes splendid in white and silver and gold, with the ostrich plumes and the peacock fans borne before him, he is only making the approximate attempt that every picture makes, to symbolise a sort of vision. Rome had to decide whether it would express the simplicity of Christ in simplicity or the glory of God in glory.'

It would not have taken the early Chesterton long to reply that the glory of God is in the simplicity of Christ not in the magnificence of Caesar.

GEORGE ORWELL
Great is Diana of the Ephesians

TEN or twenty years ago, the form of nationalism most closely corresponding to Communism today was political Catholicism. Its most outstanding exponent—though he was perhaps an extreme case rather than a typical one—was G. K. Chesterton. Chesterton was a writer of considerable talent who chose to suppress his sensibilities and his intellectual honesty in the cause of Roman Catholic propaganda. During the last twenty years or so of his life his entire output was in reality an endless repetition of the same thing, under its laboured cleverness as simple and boring as *Great is Diana of the Ephesians*. Every book that he wrote, every paragraph, every sentence, every incident in every story, every scrap of dialogue, had to demonstrate beyond possibility of mistake the superiority of the Catholic over the Protestant or the pagan. But Chesterton was not content to think of his superiority as merely intellectual or spiritual: it had to be translated into terms of national prestige and military power, which entailed an ignorant idealization of the Latin countries, especially France.

Chesterton had not lived long in France, and his picture of it—as a land of Catholic peasants incessantly singing the 'Marseillaise' over glasses of red wine—had about as much relation to reality as 'Chu Chin Chow' has to everyday life in Baghdad. And with this went not only an enormous overestimation of French military power (both before and after 1914–18 he maintained that France, by itself, was stronger than Germany), but a silly and vulgar glorification of the actual process of war. Chesterton's battle poems, such as 'Lepanto' or 'The Ballad of Saint Barbara', make 'The Charge of the Light Brigade' read like a pacifist tract: they are perhaps the most tawdry bits of bombast to be found in our language. The interesting thing is that had the romantic rubbish which he habitually wrote about France and the French army been written by somebody else about Britain and the British army, he would have been the first to jeer. In home politics he was a Little Englander, a true hater of jingoism and imperialism, and according to his lights a true friend of democracy.

Further Comments on Chesterton

G. K. Chesterton courageously opposed the Boer War, and once remarked that 'My country, right or wrong' was on the same moral level as 'My mother, drunk or sober'.

Chesterton's vision of life was false in some ways, and he was hampered by enormous ignorance, but at least he had courage. He was ready to attack the rich and powerful, and he damaged his career by doing so.

Chesterton's Introductions to Dickens are about the best thing he ever wrote.

Chesterton, another cockney, always presents Dickens as the spokesman of 'the poor', without showing much awareness of who 'the poor' really are. To Chesterton 'the poor' means small shopkeepers and servants. Sam Weller, he says 'is the great symbol in English literature of the populace peculiar to England'; and Sam Weller is a valet!

Comic verse has lost all its vitality—there has been no English light verse of any value within this century, except Mr Belloc's, and a poem or two by Chesterton.

Henry Miller ... displays an attitude not very different from that of Chesterton.

ANONYMOUS (FROM THE *TIMES LITERARY SUPPLEMENT*)
Chesterton as Essayist: The Mirror of a Silver Age

ALL his life G. K. Chesterton wrote essays. His friend and schoolfellow, E. C. Bentley, describes the Chesterton home in Warwick Gardens.

In the sitting rooms of that house, in the hall, in the conservatory, on the table under the big tree in the garden, one was always likely to find at least one manuscript book with a few or many of its pages filled by his scrawling energetic writing, the embryo of that remarkable calligraphy which was to become so well known in the world of letters.

And when, following the general fashion, GKC joined the ranks of those who dictated, the essay remained his favourite and most natural literary form. Dictating came naturally to him because he first began to address his fellows as a debater; and it remains a clue to the whole of his subsequent career to understand that it was all an extension, a kind of permanent session, of the JDC, the Junior Debating Club of St Paul's School.

The debater has also to be an entertainer. How he speaks matters as much as what he says. He is one performer, bringing his individual contribution to the joint achievement of a good evening's debate. He is not a statesman or a reformer, speaking with any nice choice of words in the light of the particular meeting, and he only feels responsible for himself and his own opinions. In after-years critics both friendly and unfriendly noted this about Chesterton, that he made up his mind, considering that a man should begin with an open mind because that is the prelude to being able to make it up: that he fought for his causes vigorously and robustly enough, fulfilling his own dictum that what our age requires is the mixture of intellectual ferocity with personal amiability, which is the formula of a good debate: and yet that he seemed curiously detached from any expectation of any practical results. No one could ever call him defeatist or over-awed by the forces ranged against him. But in a time when other men like Shaw and Wells and Belloc made their literary work part and parcel of a wider public activity, were active in societies and leagues, Chesterton, even when, after his brother Cecil's death, he was editing the journal which was then renamed *G. K.'s Weekly*, remained all the time a literary man; so that when he was writing on current political controversy he was writing essays, not leading articles, and writing them with a continual consciousness of form, seeing them as things which had to be entertainments in their own right.

We are constantly reminded through all his range as an essayist of his beginnings as a youth much more interested in poetry and the arts than in the Second Home Rule Bill. He went not to Oxford but to the Slade. He began writing about Robert Browning and Watts, and when he fell into the lively Fleet Street world of the *Speaker* and the *Daily News* and found himself caught up in all the fierce feeling which the Boer War generated, he managed to retain intact his own conception of the essay. He was surprised and glad to find that the talent for vivid, wildly unexpected but intellectually relevant and often telling metaphors and parallel illustrations, with which he had delighted himself and his school cronies, was wanted by the new halfpenny Press for the entertainment of the readership which had grown up after 1870 for Fleet Street to gratify. He followed the excellent rule—not that he could have followed any other—of being himself, as he had been in the JDC. And although the years were quite quick to disclose how little he had in common with the special public of the *Daily News*, so Puritan and Nonconformist, his first experiences in Fleet Street confirmed him in the resolution to treat all newspapers alike, not to try to write in their tradition or for their readers, but to treat each of them as one more society, even as one more soap-box, from which to hold forth for the benefit of anybody who cared to listen.

His essays always had this immediacy. They had to be in the first person, a very definite first person, communicating personal impressions, experiences and judgements. They were the work of a man for whom an idea found in a book was an event, and an encounter, as much as anything that could happen in human intercourse, and it was one element in his wide popularity that he recognized no frontiers, never supposed literature or art to be subjects set apart in worlds of their own to be studied and enjoyed there, but treated the poets and artists and the philosophers as people all taking part in the same jolly interminable discussion, all members of the same huge Sixth Form, wondering and arguing about the character of the world and the meaning of life.

He had thus two great advantages, that he could establish an immediate personal, almost confidential, relationship with the reader, and that he could make the third party to the discussion, the subject-matter, also immediately intimate as well as relevant. He was very much the least didactic of the controversial writers of his time, and the young readers who knew that H. G. Wells was determined to change them, and getting impatient at their slowness, and that Mr Shaw and

Mr Belloc both had their very clear intentions for the country, found Chesterton much more content to treat the relationship of writer and reader as the reader wanted to treat it. The reader was to be diverted and interested, given something to think about, but not followed up afterwards to see how far he had thought about it, and what difference it had made. Chesterton could communicate enjoyment, and even when he was writing about evils and their reform, he managed not to write as a reformer in the great tradition. He never treated the essayist's space in the daily or weekly Press as a pulpit, and this was remarkable enough, for the pulpit for centuries had been the main setting for expression and left its mark on the approach and style even of the men in whose programme for their countrymen the abolition or desuetude of pulpits had a high place. When he was once asked to contribute to a symposium to be called 'If I had one Sermon to Preach', his own contribution was against pride as the first of the Seven Deadly Sins. It was this clear realization of the central place of pride in human ruin which determined his own literary style. The antics—what was called 'standing on his head to attract a crowd', and more harshly, and theologically, from the Deanery of St Paul's, 'crucifying Truth head downwards'—were seen by the performer as fulfilling a double purpose. They arrested attention, but they also offset any suggestion that the speaker was above his audience and laying down the law—in a word, pontificating.

He saw himself as a journalist, a man whose activities are not to be surrounded with any liturgy or ritual, because a journalist is a man talking to his fellows, and no more than that. Church and State had to have their great offices, and had to be staffed by human beings, but all human beings were inevitably in a false and unnatural position as soon as they had to assume the functions of authority. Therefore they had to have special uniforms, a special setting, and had to use a special vocabulary. But he had no sort of desire for his own part to go inside, or even very near to, all that great apparatus by which men are enabled to live together. When he wrote about the State, notably in the work that most brings his views to a focus, *What's wrong with the World*, it was to proclaim something which all humanity was to learn more and more painfully in every decade after 1910: that all this apparatus is always threatening to become an end in itself, to develop a set of interests of its own, to produce human types of its own, with a special interest which is really hostile to the general interest of the great unofficial majority; and that all public action ought always to be measured and tested in the light of abiding human standards. There is, he argued, such a thing as human nature, with its norms which every reformer and every specialist will be tempted to distort, unless the broad sanity and clarity of public opinion restrains them.

So he conceived himself as a journalist, engaging in something which was ultimately of more consequence than anything else, the condition of public knowledge and public feeling. The journalist was himself an ephemeral creature, and the more he sought to separate himself by acquiring and boasting of special knowledge, the more he sought to put distance between himself and the ordinary man he was addressing, the more he would lose. This determination to be listened to and understood immediately meant that the essays had to be journalism, that their writer had to be in no ivory tower but a Fleet Street man, entering into, and feeling strongly about, the topics of the hour. There was no effort involved in this for Chesterton; he asked nothing better, and the men he admired and whose friendship he sought were men so immersed.

> We have known smiters and the sons of thunder,
> And not unworthily have walked with them.

he was to write many years later, in one of the few tributes he ever paid himself.

His work was accordingly a curious blending of the expression of attitudes and judgements on late Victorian and Edwardian England with permanent subjects and judgements on the nature and destiny of man which really preoccupied him. This has had the curious effect on today's generation of young people who are given Chesterton to read, that he can be at one and the same time a very wholesome spiritual influence and a mixed sociological one. He was a liberal reformer with an immense distrust of the State machine, the bureaucrat and the visiting official. Like the young Dickens a lifetime before him, he combined a passion for reform and an indignation at the oppressions and injustices he saw around him, with a conception of radicalism as a moral force to diminish, and not to rely upon or extend, the action of the State. To the young Chesterton, as to the young Dickens, the State was part, and a great part, of the organized injustice of society, and to invoke it was to seek to cast out Beelzebub by Beelzebub. If men were unjust and cruel when they had the advantage of economic superiority they could only be expected to be more unjust and more cruel when they had the much greater superiority conferred by political power: and, in fact, it could be taken for granted that the forces of oppression would come to an understanding among themselves, that the State would be the rich and the rich the State.

This was the central thesis of a book which Chesterton, although he had not written it, wholeheartedly accepted, Mr Belloc's *The Servile State*. The Chesterbelloc grew up in a world of apparently unshakeable private giants; Gog and Magog, the titular guardian giants of the City of London, looked impregnably strong at the turn of the century. If there was war in South Africa, it was because the City wanted war. If there was no Anglo-German *entente*, it was because of trade competition, and if there was no Home Rule, it was because of the capital holdings in Ireland. The Edwardian plutocracy was in its day so great and strong a reality that few if any foresaw how weak it would prove, or would discern that it ruled not by its own authority, but because public opinion accepted its dogmas, but that if it were ever to lose its ascendency over opinion, it had no powers in itself to dominate the State. Chesterton, defending the Common Man, conceived him as denied his elementary rights and frustrated in his normal instincts by the power of anonymous finance, wasting the countrysides and defiling the towns. The Common Man had to suffer all this, and then on top of it had to bear the last oppressions of the rich in the form of philanthropy and social legislation. It was to Chesterton no accident but the most natural consequence in the world, although a tragically bad one, that the successful capitalist families of the Victorian age, manufacturers like the Potters or financiers like the Samuels, should then turn up in the next generation as advanced political reformers, and that the legislation issuing from them, the blueprints of the Webbs and the Samuel Act for keeping children out of the public houses, should reflect the same lack of fundamental respect for men and women, if they were poor, which the manufacturers and financiers showed in business towards the common and resourceless man.

He objected to the whole trend of legislation which came in with the Liberal administration of 1906 because it was not nearly radical enough, in the sense of really going to the roots. What was wrong with the world was that men did not stop to ask what was right, what ought to be, and accepted far too complacently an inherited complex of false relations and customs, and then superimposed on them what they called progressive and reforming legislation which really only made matters worse. Poverty prevented the ordinary working man from being a fully responsible husband and father, and the new legislation, instead of increasing his responsibility and giving him the means to live up to it, further diminished it by taking from him and entrusting to officials the decisions about his health or his savings or the upbringing of his children which should be, pre-eminently and essentially, his. Chesterton's own prescription, like his diagnosis, was the same as Belloc's, and largely derived from his friend; it was the spread of ownership and, with ownership, of independence and responsibility, the great middle-class virtues, which held the secret why British Government in Victorian England had been so far more honest and scrupulous, and limited, than any Government in history. The nineteenth century was a golden age for independent men. But there were fewer at the end than at the beginning, and more proletarians, although better off.

Chesterton was the contemporary of Northcliffe, and of the great expansion of commercial advertising in the popular Press. He was always being confronted with private industry in its largest units and most vulgar forms, so that he came more and more to write of the capitalist structure simply in terms of its biggest and most aggressive manifestations. The reader of his many essays on the economic structure of England would not have imagined that there were, through all his time, some 50,000 business firms, most of them quite small, and embodying a type of human relationship of men associated for a specific purpose which had great and proved merits. When he once debated with the late Gordon Selfridge at the Essex Hall, Selfridge said afterwards that but for himself everyone in the room had been a socialist, and the remark, repeated among Chesterton's circle, was hailed as one more proof of the abysmal obtuseness and mental confusion of the giants of modern shopkeeping. There was, all the same, a measure of truth in Selfridge's dictum, and a good deal of unconscious socialism in the Distributist movement, precisely because it had come into existence, not against the Planned State but against irresponsible plutocracy.

The Distributists wanted political power in order to carry legislation designed to equalize property and encourage the small man. But they underrated the extreme difficulty of using the State as an agent to watch the distribution of property and adjust it, without, in fact, spreading the idea that all property is to be held by permission of the State and during its good pleasure. The central thesis of Mr Walter Lippmann's book, *The Good Society*, was that Americans had lost sight of the basic truth that all social arrangements are permissive, that all can be revised, that, in short, there are no natural rights which do not derive from society, that the authority of the State is just as much behind case-law as behind statutes, behind what is very old and taken for granted as much as it is behind what is brand-new legislation.

The difficulty for Chesterton and his school was how to preserve that for which they chiefly valued property, its power to instil into men feelings of self-confidence and responsibility which the dependent retainer could not know, if property was taken out of the old setting in which it was a sacrosanct conception. Nowhere had property been more sacrosanct than in England after 1688, when men of great property controlled the State. The law guarded property with an extreme ferocity as the foundation of society, and men could argue whether it was the sturdy independence of the English character which had so enthroned property or whether the security of property had been the chief influence in fostering and strengthening that great national characteristic. Once the conception of the State rearranging the pattern of possession is accepted, property in such a society becomes at once a different and a lesser thing, not a free animal but a domesticated one, with an inevitable shadow behind it, an element of contingency and the disappearance of the old absolute confidence. Sir William Harcourt's death duties were not very onerous, but they were philosophically significant. It was with much aptitude that their author said: 'We are all Socialists now', because the general collectivist trend included the idea that the State should regulate the amount of property individuals might possess, and should legislate not merely to provide for public purposes, but to ensure greater equality between individuals.

To the charge of revolutionary innovation the Socialists had plenty to answer historically—that through most of history the ruler had disposed of the property as of the lives of his subjects, that in the Middle Ages all land had been held from the Crown with no absolute possession, and so on. They had less ready answers to make to the statement that historical experience had shown that it was the mark of civilization to accept individualism and to keep enactment in close accord with natural law, and that under natural law there is an absolute right to ownership. There is consequently no operation harder to perform than land reform or any other public process by which some property rights are set aside without the essential nourishing conception of property as 'mine' being debased and weakened, so that the new beneficiary has instead the mentality of a tenant-at-will, an employee or pensioner, who must be careful not to offend. Chesterton was over 50 when he first visited the United States, a country whose constitution and history offer more in the way of guidance and illustration than any other for the would-be legislator about property relations, and about the intertwined relationship of the common man and the sense of equality to capitalism; for the good and the bad are but the obverse and reverse of the same medal.

It was right and necessary at the time that the plutocracy in its Park Lane heyday should have been mocked and derided as Belloc and Chesterton derided it. But it is not useful that today young men should come from reading so much good invective in prose and rhyme and fail to appreciate how completely our society has been transformed, so that there is now great need for the large company and the intermediate unit between the Government and the mass of private citizens of very small resources. This is perhaps the chief field in which Chesterton's occasional writing has dated. On his other front, against the instrusions of the Welfare State, what he wrote has gained in force and relevance by the passage of time. He saw and continually illustrated the direct descent of Socialism from capitalism as materialist creeds which could not but degrade men while helping them materially; and the abiding value of his most occasional pieces is that he never wrote about men without instinctively placing them in a large setting.

The dignity of man as man has never had a more eloquent expositor. But the large setting was a religious setting; for the great moral Chesterton set himself to bring home to his fellows was the duty of taking things with gratitude and not taking them for granted. The last thing he wanted was to be treated as championing the common man simply because of his commonness, although the effect of discovering so much in everybody was to diminish the sense of difference, and to make quality seem less important. But what Chesterton really thought of quality comes out as soon as he writes of individual poets or novelists or artists, and he can never really be claimed for 'the Quantity' against 'the Quality', merely because he found a transcendental quality in man as such. He found so much in the natural world that to him every man was rich if he possessed his five senses, and he combined the feeling of Walt Whitman of a human equality that comes from the immense richness of everybody, with the Christian sense of man's profound equality in immortality. On both grounds it seemed to him absurd that men should exploit or oppress each other, and that they should not all have something much better to do.

KENNETH M. HAMILTON
G. K. Chesterton and George Orwell: A Contrast in Prophecy

G. K. CHESTERTON begins his *The Napoleon of Notting Hill* by describing the game of 'Cheat the Prophet'. This game is, in fact, not practised by humanity at large, which, far from ignoring the prophets and thus confounding them, is enormously attracted to false prophets and gives the true ones much angry attention, though seldom a hearing.

The matter of 'Testing the Prophet', however, is conscientiously performed by history itself for mankind's edification. The decline of the position of the prophets in Israel has been attributed to the failure of the majority to survive the judgement of history.[1] Nearly half a century has elapsed since *The Napoleon of Notting Hill* claimed to see a hundred years into the future. It is time enough to see how this particular prophecy has worn and how its message compares with present-day foretelling of the future.

One limitation of Chesterton's vision appears at the outset of his story, when he says the world to be would be little changed in appearance. He failed to appreciate the situation created by modern technology. In trying to laugh away H. G. Wells's optimistic faith in science, he did not distinguish between Wells's true conviction that the motor-car's pace would be superseded as the horse's had been and the shallow assumption that the wonderful increase in speed of travel must be wonderfully good. Chesterton's 'Dr Pellkins' who held that the largest pig in the litter must some day become larger than an elephant was correct, so far as breeding-habits of scientific method were concerned, and here Chesterton's jibe sounds flat in modern ears. It is highly dangerous in a prophet to ignore the material conditions within which the spirit must operate; the prophetic word must reach men as they are and where they are. Having nothing to say of the truth (which Marxists have made into an idol) of the determination of culture by the means of production employed, Chesterton's *Napoleon* forfeits from the start the claim to comment profoundly upon society and becomes a romantic fable. A fable which does not try to reproduce reality in its complexity may yet have a deep wisdom of its own. But here the fable is romantic in the worst sense. It is not true to life in its externals—a small matter. More important, the ideals it exalts in the shadow world of its creation are not controlled by the moral realities of any conceivable universe and appear attractive only in proportion to the unreality of the setting.

Like *News from Nowhere*, *The Napoleon of Notting Hill* presents a dream-world from which economic and kindred problems have been banished. In all the upheavals of the Notting Hill Empire there is plenty of tea in the shops and gas in the Gas-works. Morris's leisurely scene of pastoral placidity was drawn by one who hated industrial society and followed the Marxist analysis, naïvely hoping that by economic action economics would cease to be a problem; sharing, indeed, the Marxist illusion that Satan is the only one who can cast out

---

[1] See Aubrey R. Johnson, *The Cultic Prophet in Ancient Israel* (1944), chap. iv.

Satan. Chesterton was also in revulsion against the civilization which nineteenth-century capitalism had created. But he preferred to ignore its causes and concentrated on describing a society transformed, without working out any rational notion of how societies are created or maintain themselves. Morris postulated a revolution as a prelude to Utopia, a revolution which was like the real thing, nasty, but (in his view) necessary. Chesterton's revolution is as impossible (and as entertaining) as the schoolboy comic's account of the school where the boys take charge. In a London which exactly reproduces the London of 1904, but which is ruled 'by a popular despotism without illusions', because 'the people had lost faith in revolutions', Auberon Quin as head of the state introduces halberdiers and other medieval trappings for the Boroughs, as a joke. Adam Wayne of Notting Hill accepts the joke seriously. He takes up arms on a question of local rights and in a series of street battles defeats all the rest and establishes an Empire on the basis of medieval chivalry. It is all fine and frolicsome, yet under the legerdemain the writer is putting his case and displaying his values. We are easily persuaded that the philosophy underlying it is as pleasant as the plot and we may forget that prophecy must be tested, not only upon the pulses, but also upon complex recording-tape of history.

Many shared Chesterton's dissatisfaction with a world that 'had lost faith in revolutions'—real revolutions which men desired in their hearts and fought for with their hands, not the souless slavery to the machinery of commerce which was the consequence of the 'Industrial Revolution'. Back then to the Middle Ages when life was single and beliefs really mattered and art was not the drudge of money-power! For the romantically inclined, these Middle Ages were of course Golden Ages containing only values we had seemingly lost and lacking only what we could do without in the interests of a simpler life. We could well jettison the gun, the aeroplane, the lounge suit, and perhaps as a gesture (one must show one is in earnest) plumbing, electricity, and surgery as well. Others who knew we could not put the clock back materially, thought we should do so spiritually. Berdyaev proclaimed that we were standing at the threshold of a New Middle Age when the bourgeois values of liberalism must vanish and belief once again take central place. He recognized that this was also the coming of a 'spiritual night' and would bring misery before better ways were born again among men. Chesterton saw no blackness in the picture. And his medieval Utopia was built on belief indeed, but on a kind of belief suited to the uncomplicated mind of a Tarzan. Though Christian terms are repeatedly used, the values they propose are mainly pagan ones, where love as *agape* has no place, and mind and spirit little function except that of waiting upon the emotions. A strongly felt loyalty is absolute, although the worth of the object of loyalty is unexamined. Religion appears chiefly as a sanction for self-assertiveness. Everywhere feeling is paramount: 'Whatever makes men feel young is great: a great war or a great love-story.' Even the plot takes a pagan form. Adam Wayne's early victory over great odds and ultimate gallant defeat under great odds shows the basic pattern of Germanic mythology, to which the gods themselves conform. Adam Wayne is nearer to Beowulf than to any other ideal figure. Physical violence is almost *the* good in life. It is to the credit of the Empire of Notting Hill that 'as it began in blood, so it ended in blood'. To take the sword and perish by the sword is the glory and the purpose of existence. A Christian values in humble gratitude the shed blood of martyrs and of every witness to truth and righteousness, for their sacrifice is taken up into the sacrifice of the One who shed His blood for us all. For Chesterton, shed blood has value in itself, because it is emotionally satisfying: 'Blood has been running, and is running, in great red serpents, that curl out into the main thoroughfare and shine in the moon.' The words are Auberon Quin's, the great jester at last impressed by something seriously interesting. There is no hint of what this fine sight means in human terms of pain and mutilation, savagery, and bereavement. The fighting in Notting Hill is seen always as a spectacle; never too near, so that unpleasant details are lost; never too distantly so that the consequences are ignored. There are no women in the book. It is not to be thought that war brings with it famine, disease, and a legacy of fear. Since this is old-fashioned hand-to-hand fighting and not modern mechanized warfare it must be gentlemanly. (It might be instructive to illustrate *The Napoleon* with Goya's *Disasters of War*.)

Chesterton's pagan romanticism is shown at its height in the dedicatory poem to Hilaire Belloc, where he grows warm to 'your tall young men'—the adjectives obtrude for short, middle-aged soldiers would spoil the picture—who 'drank death like wine at Austerlitz' and looks forward to the omnipotence of emotion:

> The drums shall crash a waltz of war
> And Death shall dance with Liberty . . .
> And death and hate and hell declare
> That men have found a thing to love.

It is a flamboyant revulsion from the thought of 'what cold mechanic happenings must come'. The prophecy has come true, for we have seen the conversion of millions to a perfervid belief in revolutions. And in Fascism we have seen the enthronement of feeling as the arbiter of values. We have seen blood in the streets and death and hate and hell written across the nations because men found a belief to cherish with passionate intensity. We have seen it as a romantic outburst, complete with banners and dressing up and torchlight in true Notting Hill style, but within the setting of a civilization ordered by modern technology, of continually increasing power to manipulate the environment—and man himself—in units of increasing size and complexity. Hitler was as single-minded and as humourless as Adam Wayne and as much more dangerous as scientific knowledge had made him. Yet even Adam Wayne, flapping his archaic flag, was more dangerous than his creator allowed him to appear. Lip service to religious ideals does little to make the conduct of the successful strongman different from the openly cynical tyrant; much blood that has run under the contemporary moon has been shed in the name of Christian loyalty.[2] Beside Wayne stands Auberon Quin—'a man who cares for nothing except a joke. He is a dangerous man.' After their death the two became reconciled, as being two sides of the same medal. The totally irresponsible man and the blind fanatic are certainly one at bottom and the world will suffer from them, either separately or blended in one personality, as long as human nature surrenders itself to false values. Quin without Wayne, however, does little damage. It is the revival of blind fanaticism whether based on emotionalism, as in fascism, or on a quasi-rationalism, as in communism, which has brought the world to the edge of the abyss.

The Napoleon charms because of its exotic escapism—swords, cloaks, rhetoric, water-towers plated in silver, and grocers who learn to speak like some one out of *Hassan*. Its teaching seems to suggest that a return to a strong, simple, basic living will result from a destruction of decadent, money-controlled, megalopolitan civilization. It seems so easy, since all that is needed is a resumption of local patriotism and the ethics of the strong right arm. National Socialism made use of this idealistic appeal too, but it equally used the darker enticement of the permission of cruelty in the strong. There is nothing of this in Chesterton, whose basic convictions, in spite of everything, were still liberal, humanitarian, and nineteenth-century. Fundamentally, he also did not believe in revolutions. Whether under the gaudy feudalism of Wayne or under the full despotism of pre-Wayne, the essential landmarks of democratic London and its characteristic habits of mind remain. A jester on paper is not dangerous, except where his jokes

---

[2] See, for instance, a Catholic's reaction to the 'Spanish Crusade': George Bernanos, *Les Grandes Cimetières sous la Lune* (1938).

strike at simple goodness, mercy, and truth. Chesterton's background of Victorian tolerance remained even when his dislike of what the Victorians has achieved led him to approve of intolerance. This can be seen in his ambiguous attitude to Italian fascism. He could not condemn the Abyssinian war. Yet he would have been appalled at the suggestion that Vittorio Mussolini's pleasure in burning a village of four thousand souls had his full approval. There is no essential difference, all the same, between the dictator's son's reaction to slaughter and Auberon Quin's. Vittorio's words, 'It was all extremely interesting', sum up the two—and he even achieves in his descriptions some of Quin's pictorial effects. But Chesterton wanted to play with children, not with incendiary bombs; even his sword-stick was a gesture, not a weapon. He looked at the Middle Ages, both Old and New, through the spectacles of suburban English tastes.

George Orwell's *Nineteen Eighty-Four* brings us right into the New Middle Ages, forcing us to look into all that Chesterton would rather ignore. It is characteristic of the pace of our day that this new prophecy does not look a hundred years ahead, or even fifty. It predicts Berdyaev's 'spiritual night' from the knowledge that spiritual twilight is already upon us. It is not romantic, in that it envisages a possible situation, even if the possibility is a nightmare. It acknowledges the fact that it is technological advance which lays down the conditions of society—seeing also the sole limit that can be set to its course, the regimentation of specialist research. But while this prophecy is realistic in intention and tone, its central pre-occupation is religious. *The Napoleon of Notting Hill*, like Hitler, often spoke about God while retaining a pagan level of thought. Orwell, like the Communists, does not deal in religious phrases, but Judaic-Christian patterns of thought are only just below the surface.

Nineteen Eighty-Four imagines the world divided into a few great rival totalitarian Powers, continually at war. This external war is not a threat to the existence of any, being rather a condition of the continuance of all. The real war is an internal one, the retention and extension of the Party power over its members, a war against the emergence of freedom. The book deals with the attempted revolt of one citizen, Winston Smith, against the order, and its suppression. As O'Brien, the Inner Party member who superintends this small but all-important campaign, points out, physical obedience is not the essential thing. Technology has made that a relatively simple matter. The Thought Police are not concerned to stamp out disobedience, but heresy. Killing the body is a power which tyranny has always possessed. The new campaign is waged against the personality of man, lest he should think his soul his own. This makes clear to us what faith in revolution really means and what happens when belief is taken seriously. This is the New Middle Ages, which sees in the authoritarianism of its predecessor a pale version of its proper task. It is not simply that the new way has an evil intent, while the old purposed good. 'The Party' does not only prefer to rule in hell than to serve in heaven; it sees clearly that it is only in hell that one can rule. To enjoy power is to enjoy causing suffering, therefore the aim of the powerful is to rule by promoting strife, pain, and misery, by stimulating revolt in order to crush it and by turning every instinct into unreflecting loyalty to authority. That Winston Smith must learn to *love* the Party leader is the command of the authority which torments him. 'Whatever makes men feel young is great', said Chesterton. O'Brien enjoys an extended youth by identifying himself with the Party whose agent he is and whose stability in tyranny modern technology has ensured. No 'cold mechanic happenings' here, but the full enjoyment drained from the wine of death; Quin-Wayne watches the curling of the serpent of humanity's blood for his delectation; Vittorio Mussolini exclaims 'This is extremely interesting.'

Nineteen Eighty-Four speaks to our condition. It reflects the situation which a large part of the human race have to endure at the present moment, the unholy alliance of modern scientific power with the medieval demand for the submission of individual liberty to a spiritual authority punishing heresy with death and degradation. Even those scientists, whose thought is tied to the illusion that knowledge and progress are synonymous terms are today entertaining doubts. Every discovery of science is a weapon for the torturer as well as for the healer, as we see when psychology and the social sciences are brought in to support guns and drugs in 'conditioning' 'enemies of the state'. As a prophet Orwell is far more clear-sighted than Chesterton, because he has seen the real cancer of our age. He has had the advantage of seeing the evil fruit of tendencies which earlier seemed promising growths—the enthusiasm that would die for, and kill for, a faith and the substitution of passionate feeling for tolerant reason. Orwell finds that the pagan virtues of the strong right arm and loyalty to one's own clan, turn to cold cruelty and blind submission. Pride, the spiritual sin of rebellion against God and the desire to become as gods, is laid bare as the root of our misery. But just because Orwell has no residual nineteenth-century illusions about the decency of man when he had power and no responsibilites, so the humanity and optimism that abound (and sometimes irritate) in Chesterton are absent. The nineteenth-century tolerance and kindliness and all that has been labelled 'liberal' which many have been so anxious to sweep away in the name of the Faith, either Christian or anti-Christian, was not a transient foible of the age. It was the fine flowering of centuries of Christian witness. If the Victorians thought the flowers were enough by themselves, cut off from their roots, or imagined that no further cultivation was required, that did not affect the worth of what was achieved. Orwell has no message of assurance. It might be objected that the scope of the book precludes any such thing, the moral by implication being: take warning; cherish the humane values or expect the coming judgement. But form is not accidental. *Nineteen Eighty-Four* invites us to identify ourselves so fully with Winston Smith that we share his spiritual defeat. The form Orwell has chosen meets us at every turn in an age which has shed its illusions without achieving faith, an age which knows evil but not God. In literature its purest expression is found in Kafka's *The Trial* (of which *Nineteen Eighty-Four* reads like a revised version, in which the action is merely lifted out of the interior world and put into the objective plane); but it is the staple theme of the greater part of 'serious' fiction today, quite apart from such romantic outpourings as Huxley's *Ape and Essence*. In fashionable philosophy it is systematized in Existentialism. In history it confronts us in the tragedy of Benes and Jan Masaryk. It is a pattern which must be broken, if the world is to make its way, painfully as it must, to sanity instead of suicide.

Romantic escapism, a return to sub-Christian values or to an external authority which identifies itself with Christian truth is no solution. The prophetic word which will speak to us through the 'spiritual night' of our age must speak to our desperate sickness. It must also maintain: *'So God be with us, who can be against us!'*

MICHAEL ASQUITH
G. K. Chesterton: Prophet and Jester

I SUPPOSE the vivid impressions of our childhood are rather like certain Italian primitive paintings. They show the same kind of disproportion, the same crudity—sometimes they have the same kind of charm. My own memories of G. K. Chesterton are something like this. They are vivid but subjective—a sketch rather than a portrait—and probably a very crude sketch at that. During the time I knew him I was still a child, so it is perhaps not surprising that my first impression is not of him but of his trousers. (As a matter of fact, I think anyone would have found them impressive.) I should perhaps explain that the first time I met them they were without Chesterton. They were being carried reverently along the passage by the butler, in a country house where I spent a large part of my childhood. This house in the Cotswolds belonged to my grandfather: but during the summer months it was regularly taken by Sir James Barrie, the dramatist. Among his guests was Chesterton, who came to stay on two or three occasions. The butler, whom I knew well, disentangled himself fold by fold from an enormous black garment, and held it out for my inspection. 'Mr Chesterton's evening trousers', he explained briefly; 'it reminds you of going down the underground.'

I need hardly say that this introduction to his trousers led me to expect a great deal of their owner. I was not disappointed. I had been told Mr Chesterton was a very great man. When I saw him I realized he was also a very big man—the biggest I had ever seen. When I came in, he was sitting in a deep armchair, but on seeing me he politely set himself in action to rise from it. This laconious act of courtesy to a child of 13 would have been impressive in any man nearing 60. But in Chesterton's case, you can imagine, it was far more so. You could not help feeling that the energy needed to raise him would take an ordinary man several times up and down a steep flight of stairs.

While he was getting up, I had plenty of time to observe his massive head with its impressive mane of grey hair, the pince-nez perched insecurely on his nose, his drooping walrus moustache, his rumpled, shapeless suit with its bulging pockets. I remember that I liked him at once, and he gave me a most unusual feeling that we were really contemporaries. Looking back now I think that of all his remarkable qualities, that impressed me the most. I think he gave all children and young people this feeling. And probably the reason was that Chesterton, in a very real and very rare sense, was everyone's contemporary. Talking, acting, playing games, or reciting nonsense rhymes, you never felt with him that cordial and friendly efforts were being made to bridge a gap between you. You simply felt it was not there. Perhaps this was one reason why children were so fond of him. Certainly he loved them, and it was clear that he was completely at their mercy. I do not know if any child ever asked him to stand on his head. But I have no doubt that if asked he would have tried—that is, if his wife had not been there to restrain him.

I remember how Mrs Chesterton kept an anxious eye on him, and particularly on the chairs he was about to sit down in. She seemed to be judging their powers and resistance with an expert eye, and now and then she would voice her anxiety. 'I wonder if it's *wise* for Mr

Chesterton to sit on that chair?' she said once when the more robust chairs were occupied and her husband was bearing down purposefully on one of the smaller ones. I don't think Chesterton was ever in any way what is called a practical man. He had little or no sense of money, and he would often say that the only way to catch a train was to miss the one before. After his marriage, Mrs Chesterton took charge of this side of his life. While he was travelling all over the country on lecture tours she would keep his engagement book, and she once got a telegram from him saying: 'Am in Birmingham. Where ought I to be?'

Chesterton was the only man I have ever seen stuck in a door. It was entirely typical of him that he seemed greatly to enjoy this experience. I think this was perhaps another reason for his great popularity with children—his willingness to make himself ridiculous in public. 'Willingness' is really too weak a word. He positively delighted in it. He was in fact so far the reverse of pompous that you might almost say he was always standing on his *in*dignity. Others might proclaim their triumphs and recite their success stories; it was in his failures and fiascos that GKC preferred to glory. 'Only man', he writes, 'can be absurd, for only man can be dignified.' Always the staunchest champion of human dignity, he never ceased to delight in the absurdities of man—and in no man's more than his own.

It was one of his maxims that 'what was worth doing was worth doing badly'. This was his attitude to games, and I seem to remember him discussing the subject with Barrie one evening over the dinner table, which I was quite often allowed to attend. Neither man had the build of a natural athlete. But Barrie, who was ambidextrous and had a good eye, took pride in his prowess at various sports. He had a life-long passion for cricket, and excelled at quoits, skeeball, and golf-croquet. Chesterton, who was not addicted to violent exercises, took equal pride in his incompetence at all forms of sport. Boasting of his blunders, he stoutly maintained that to go on doing things badly showed a disinterested love of them for their own sake, unspoilt by any selfish satisfaction that you might derive from doing them well. Afternoon croquet was a regular feature of the day; and I seem to remember one glorious occasion when these two met on the lawn in single combat. I can see Barrie in his shirt-sleeves, crouching tense as a tiger over his ball with mallet cunningly poised for the execution of some deadly stroke. Chesterton, meanwhile, his great bulk balanced on his incongruously small feet, takes the long handle and lashes out with a fine quixotic abandon. He loses the game, and probably his ball, but wins a tremendous ovation from the gallery. And both are equally pleased with the result.

Either Barrie or Chesterton, alone, was the kind of man you would look at more than twice. Together, and particularly when they were talking over the dinner table, the combination was startling. In my mind's eye I have several vivid pictures of Chesterton. I see him rising ponderously from his armchair, or squeezing himself, inch by inch, into the back seat of a small saloon car. I fancy I can also see him at some game of hide-and-seek, his familiar form bulging from behind a curtain. But it is with Barrie at dinner that I shall always remember him best. Barrie, small and somewhat melancholy behind his large bulldog pipe, his chair drawn back from the table as he sits with one leg tucked under the other, wreathed in a dense haze of tobacco smoke. Chesterton, vast and genial, probably a napkin under his chin and a glass of red wine in his hand; the great white expanse of his shirt-front sweeping down, like some gentle ski slope, to the edge of the dining-room table. In a voice bubbling with enthusiasm, and rising at moments of excitement to a squeaky falsetto, he pours out his long, effortlessly flowing sentences, punctuated from time to time by the sharp crackle of verbal fireworks. I remember, too, how he would often pause in the midst of the most serious discussion to recite a comic verse, or recall some snatch of the music-hall songs he was so fond of. 'I believe firmly', he wrote, 'in the value of vulgar notions, and especially of vulgar jokes ... the men who made the jokes saw something deep which they couldn't express, except by something silly and emphatic.' He was not only a good talker but also an inspired listener. Heard by Chesterton the most trivial commonplace was transfigured and shone with new meaning, until you found to your surprise that you had said something clever. Like Falstaff, whom he also resembled in figure, he was 'not only witty in himself, but the cause that wit was in other men'. No one laughed louder than Chesterton at his own jokes, but as they were usually at his own expense, it did not spoil them.

Self-ridicule, as I have said, was his favourite gambit. Brilliant in debate, he could hit very hard when he wished, and on any subject near his heart he was easily provoked to the hottest indignation. But though never lukewarm, he was never bitter in argument; and even his most piercing sarcasms were somehow softened by a kind of geniality, which gave it the atmosphere of a mock battle, fought with wooden swords and padded trousers, by clowns in a circus. Like his great friend Belloc, he loved wine. But few men can have had less need of any external aid to intoxication. His zest for life was enormous, and nothing provoked him more than the attitude of those who took life for granted, without feeling, as he thought they should, a perpetual gratitude for the experience.

Chesterton had a very highly developed sense of fun. I remember once, during his last visit, how he devoted a whole day to the preparation of an elaborate game to be played that evening. During the morning he made a shopping expedition into Evesham, and returned laden with several parcels. All the afternoon he was closeted in the library, completely absorbed in his mysterious business. He emerged at tea-time with a confident smile, and announced that he had invented a new kind of murder game. The problem was not to find the murderer, but to find the victim, who proved to be a well-known lord of the press. Chesterton, a very talented artist, had made a magnificent effigy of him in cardboard, painted it, dismembered it, and hidden the fragments all over the house. He had also written in rhyming couplet a number of clues which not only enabled us to find the corpse piece by piece, but also provided some very interesting notes on its past life. Last of all we discovered the head, complete with coronet. The dismembered peer was finally assembled, and I have never seen a better caricature. It might perhaps be invidious to mention names, but I imagine that he was not one of Chesterton's heroes. I cannot remember exactly how we scored at this game; but I rather think that I won it.

I hope these scrappy memories of Chesterton will give no one the impression that he was primarily an entertainer, still less a sort of amiable buffoon with a talent for writing. A jester of genius he certainly was. But, as his friend and biographer Emile Cammaerts has consistently pointed out, he was first and foremost a prophet. The cap and bells were only a means to capture his audience. This done, they were thrown aside, and the jester got down to his real business, which was not entertainment but salvation.

As a prolific journalist, poet, pamphleteer, and theologian; as an untiring champion of Christian orthodox principles, crusading for the unpopular causes which he so brilliantly defended, Chesterton has often been compared in spirit to a medieval knight-errant, tilting, sometimes a little recklessly, sometimes a little impetuously, at the dragons of modern industrialism. After his death, his friend Walter de la Mare wrote of him:

> Knight of the Holy Ghost, he goes his way,
> Wisdom his motley, truth his loving jest.
> The Mills of Satan keep his lance at play,
> Pity and Innocence his heart at rest.

But I think my own childhood recollections of Chesterton are best concluded with a very different kind of verse. It was one of his favourite music-hall rhymes, and he once wrote it out for me in his flourishing, decorative hand. He used to recite it with Cockney gusto:

> When we goes up to London Town,
> We likes to drown our sorrers:
> We likes to go to the waxwork show,
> And sit in the Chamber of 'Orrers.
> There's a lovely image of Mother there,
> And we do enjoy it rather:
> We likes to see 'er 'ow she was,
> The night she strangled Father.

DOROTHY L. SAYERS
Chesterton's The Surprise

To the young people of my generation, GKC was a kind of Christian liberator. Like a beneficent bomb, he blew out of the Church a quantity of stained glass of a very poor period, and let in gusts of fresh air, in which the dead leaves of doctrine danced with all the energy and indecorum of Our Lady's Tumbler. No doubt, in her substance, the Church was, then as ever, glorious and immutable, but she was, at the beginning of the century, somewhat unfortunate in her accidental presentation, seeming, to the man in the street, to offer little choice except between a somewhat complacent Protestant morality and a rather simpering Catholic piety. And since a great many clever people were busily engaged in undermining the foundations both of the piety and the morality, Christians were popularly felt to be somewhat in the position of death-watch beetles, chewing their blind way through rotten timbers which were due to come down with the building.

It was therefore stimulating to be told that Christianity was not a dull thing but a gay thing, not a stick-in-the-mud thing but an adventurous thing, not an unintelligent thing but a wise thing, and indeed a shrewd thing, for while it was still frequently admitted to be harmless as the dove, it had almost ceased to be credited with the wisdom of the serpent. Above all it was refreshing to see Christian polemic conducted with offensive rather than defensive weapons.

To be sure, the weapons were sometimes felt to be offensive in more senses than one. The style was accused of being more florid and more flippant than was compatible with either dignity or profundity; and there may have been some justification for a contemporary Oxonian satirist who complained of conditions in the University:

> Where Chestertonian youths five things revere:
> Beef, noise, the Church, vulgarity, and beer.

But a man is not responsible for all the indiscretions of his followers: and I think it was not always realized, either by the critics or the followers themselves, how sound a theological basis underlay all the flourishes of paradoxical wit. Nor indeed was the fundamental accuracy confined to the theology. The ordinary reader, brought up on little books about logical Positivism and the inevitability of scientific 'laws', probably snorted with contempt when he came upon the assertion:

> It is no argument for unalterable law ... that we count upon the course of things. We do not count on it; we bet on it.

He took it for a perverse piece of mystification, if not plain humbug. Today, having read other books (or more likely 'digests') he accepts it as commonplace, for reputable physicists have said so. Similarly, a number of prophecies about the ultimate issue of certain highly approved trains of thought have, in recent times, been somewhat painfully fulfilled in our ears.

To those within the tradition, it is perhaps not so surprising after all. If a man is dealing with 'a truth-telling thing', one would rather expect him to hit on the truth more often than not. But GKC did surprise his contemporaries and that must have pleased him, for he thought that truth should be perennially surprising and that surprise did people good. 'Blessed is he that expecteth nothing, for he shall be gloriously surprised.' He complained that men had lost 'the innocence of anger and surprise', and he fought indefatigably for a recovery of a frame of mind he was afraid to call 'romantic'—a term in those days not yet degraded to the status of a mere pejorative.

It may be argued that in the present time our need has changed; that the depression of spirits from which we now suffer has been caused, not by security but by shock—we have had so many surprises, mostly of an unpleasant kind, that we have been battered into insensibility. It is all very well for the Poet in the play (*The Surprise*) to tell the Princess (the ardent exponent of the Welfare State) that the people

> do not want sufficiency or security. They want surprise.... You say they can always find a pig at the pig-trough and ale in the ale-cask. If ever, one fine morning, they found the pig in the ale-cask and could drink ale out of the pig-trough—they would think they were in a fairy tale.

We have of late become too accustomed to finding trough and cask both empty and the pig missing, and would rather have an extra rasher of bacon in the right place than all the fairy tales in the world. But with that Chesterton would cordially agree—provided that the rasher was welcomed as a glorious windfall instead of being morosely demanded as a right. And he saw, correctly enough, that it is the sense of insecurity which gives to the simplest possessions (as to Robinson Crusoe's two guns and one axe) the triumphant value of something saved from shipwreck. There is, indeed, nothing in the doctrine of 'surprise' that is necessarily critical of social security as such; what is criticized is the attitude of mind which takes security, happiness, or life itself to be no more than what our own merit entitles us to.

But the title of *The Surprise* refers primarily to a salvage more astonishing and a shipwreck more universal than anything to be found in the pages of *Robinson Crosoe*, or in the annals of global warfare. The theme is one of peculiar fascination to any writer who deals with the creation of characters and situations, and was adumbrated by GKC himself as long ago as in *Orthodoxy*.

According to most philosophers, God in making the world enslaved it. According to Christianity, in making it He set it free. God had written, not so much a poem, but rather a play; a play He had planned as perfect, but which had necessarily been left to human actors and stage-managers, who had since made a great mess of it.

The 'Author's' cry on behalf of his puppets:

They only existed because I wanted to get them out of my mind. I wanted them separated from me and my life and living lives quite different and entirely their own.

is an echo of the earlier theological assertion:

Love desires personality; therefore love desires division. It is the instinct of Christianity to be glad that God has broken the universe into little pieces, because they are living pieces. . . . The world-soul of the Theosophists asks man to love it only that man may throw himself into it. But the divine centre of Christianity actually threw man out of it in order that he might love it. . . . No other philosophy makes God actually rejoice in the separation of the universe into living souls. But according to orthodox Christianity this separation between God and man is sacred, because this is eternal.

Dante had said the same, in an epigram more concentrated and more august.[1] And given this theme, and given the free-will of the human actors and stage-managers concerned, there remains the problem of the Author's action, which forms the 'surprise' of the final curtain-line.

---

[1] *Paradiso*, xxix, 13–18.

ALFRED NOYES
The Centrality of Chesterton

THE sixteen years that have elapsed since the death of Chesterton have left his memory very much alive, and he is now seen in a clear perspective by his biographer Maisie Ward. He has often been compared with Dr Johnson for the way in which his personality actually counted for more than his books. There was a grain of truth in this, but it has been greatly exaggerated, except in the fact that his personality was one of the most extraordinary of his time. It is true that he was physically massive, but he was not in the least like Samuel Johnson. He was in fact more like Father Christmas. Where the author of *The Vanity of Human Wishes* was majestically solemn, Chesterton was either bubbling over with jests, or writing satirical ballads about Mr Mandragon the Millionaire. Where Johnson continued in thought and form and feeling the eighteenth-century classicism of Pope, Chesterton in poems like 'Lepanto' and 'The Ballad of the White Horse' carried romanticism to a point where he might almost be identified with his own 'Wild Knight', a twentieth-century Quixote tilting at what Blake called its 'dark satanic mills.'

His own play on Dr Johnson, however, is perhaps his masterpiece in prose, incomparably finer than his more frequently mentioned 'Magic'. In the splendid rhetoric he puts into the mouth of Dr Johnson, when he tells the American critic of the old world what the dangers of the new may be, there is a depth of thought and a dramatic thunder of prophecy, coruscating with lightning flashes of humour, before which the epigrams of Mr Shaw and even the importance of his earnestness dwindle into mere persiflage.

It has not been noticed how much Chesterton owes to the methods of Lewis Carroll. *You might as well say*—the phrase so familiar to readers of *Alice in Wonderland*—clinches Chesterton's arguments again and again with some aptly absurd illustration of a logical fallacy. It has been said that *Alice in Wonderland* and *Alice Through the Looking-Glass*, edited with commentaries and notes, might serve as an admirable textbook on logic, by the brilliant way in which they exemplify and destroy many familiar varieties of fallacious reasoning. The regret of the Mad Hatter that his watch had stopped although he had greased its wheels with the best butter is a classic example. The arguments of the mad tea-party ('You might as well say that I see what I eat is the same thing as I eat what I see'; 'I like what I get is the same thing as I get what I like,' and 'I breathe when I sleep is the same thing as I sleep when I breathe') probably educated the youthful reader as efficiently as Euclid could, and might be described as the prolegomena to the philosophy of Chesterton. He seized on the method and applied it with a new depth and brilliance to politics, literature, ethics, and religion. There is, for instance, a real critical value in his amusing remark about free verse: 'You might as well call sleeping in a ditch free architecture.' It covers everything. It does not deny that sleeping in a ditch may have its own charm. The sleeper may have the free wind blowing over him, a friendly frog on his chest, and is he not lodging at the Sign of the Beautiful Star? The one thing he is not entitled to do is to call sleeping in a ditch architecture.

Chesterton's remark about certain misuses of the 'relativity' theories was equally cogent and equally Carroll. 'Mathematically', he said, 'it may make no difference whether a man is running towards a tree or the tree running towards the man; but I notice it is the man who is out of breath.' Alice might have said something like that to the Red Queen when they were both running at top speed to keep in the same place. The subtle metaphysical suggestions in Alice's argument with Tweedledum and Tweedledee as to whether she was only something in the Red King's dream, have their counterpart, with a deeper significance, in many pages of Chesterton. The Wild Knight of his poem might be described as a transfiguration of the White Knight in *Alice Through the Looking-Glass*. Alice, in that golden book of wonder, remarks, 'It's a great huge game of chess that's being played all over the world'; and Chesterton himself, in his essay entitled 'The End of the Moderns', almost continues the remark by saying,

> To almost all the modern moral and metaphysical systems, as stated by the moderns themselves, I should be content to add the comment, 'Mate in three moves.'

The influence of Lewis Carroll is occasionally present in what was perhaps Chesterton's greatest gift—that unspoiled sense of wonder to which every object in the world around us appears as the miracle it really is. He wakes in the morning and sees 'incredible rafters overhead'; and through his almost overwhelming apprehension of the mystery of being, the fundamental paradox of an impossibility that is somehow a most certain fact, he finds himself in the presence of the supernatural. In one of his early poems he says, with the depth and simplicity of Blake,

> Speller of the stones and weeds,
> Skilled in nature's crafts and creeds,

> Tell me what is in the heart
> Of the smallest of the seeds?
>
> God Almighty, and with Him
> Cherubim and seraphim,
> Filling all eternity.

There is a legend growing up about Chesterton which the biography by Maisie Ward should help to dispel by its insistence on the solid philosophical ground of Chesterton's best work. In this book, as in the recently published *Return of Chesterton*, the daughter of the distinguished biographer of Newman shows that she has inherited her father's great gift. In her earlier book, *Young Mr Newman*, the characters live and move as they would in a novel by Jane Austen; and the two portraits of Chesterton give the same sense of life. The biographer has brought together much new material, and her own sense of humour has enabled her to select from this some of the most humorous verses Chesterton ever wrote. His parody of an early poem by Swinburne is more than a parody. In the form of a jest it gives the complete philosophical answer to a very ancient fallacy:

> I am sorry, old dear, if I hurt you,
> No doubt it is all very nice,
> With the lilies and languors of virtue
> And the raptures and roses of vice.
> But the notion impels me to anger,
> That vice is all rapture for me,
> And if you think virtue is languor,
> Just try it and see.

The biographer's task was not an easy one, for in Chesterton's immense journalistic output (in which he was not always at his best, and often repeated himself), together with his achievements in poetry, fiction and political and religious polemics, the trees made it exceptionally difficult to see the woods. As he himself wrote,

> The million forests of the earth
> Come trooping in to tea,

and

> The sea had nothing but a mood
> Of vague ironic gloom,
> With which to explain its presence in
> My upstairs drawing-room.

Whatever else may be said of Chesterton, he certainly had one of the most original minds in the Europe of his generation, and this originality was manifested, not in a rebellion against the laws of God and nature, but in a rediscovery of their grandeur. This in fact is the only true originality, for it draws its strength from the origin and end of all things. It is the only road upon which

> All joy is young and new all art.

The others necessarily end in defeatism, bitterness, and despair, as much of our contemporary literature abundantly demonstrates.

There is no paradox in the very old assertion that only through those laws can the perfect freedom be found, just as they are the only means of establishing a right relationship (which we call religion) with the ultimate Reality. It is the old sun that makes the new morning; this is the central fact that in Chesterton reconciles originality and tradition.

It was a mere chance that decided whether Chesterton should write articles for the radical *Daily News* or the conservative *Blackwood's Magazine*. Many years ago I was told by William Blackwood that Chesterton wanted to write the monthly *Musings Without Method* for Maga [the *Magazine*], but an arrangement had already been made with Charles Whibley. Chesterton would have been simply himself, brilliantly central, wherever he wrote. It is worth noting that, although he bitterly satirized the flag-waving of the jingoes, he out-Kiplinged Kipling in one powerful early poem, asserting that whatever differences or even hatreds might divide the English-speaking nations, the ultimate test would find them united. The racial terms of his poem are unsound and cannot be taken literally, but their spirit is symbolical of something that has happened twice since the poem was published in 1902, and may yet save our civilization from red ruin:

> This is the weird of a world-old folk,
>   That not till the last link breaks,
> Not till the night is blackest,
>   The blood of Hengist wakes.
> When the sun is black in heaven,
>   The moon as blood above,
> And the earth is full of hatred,
>   This people tells its love.
>
> In change, eclipse, and peril,
>   Under the whole world's scorn,
> By blood and death and darkness
>   The Saxon peace is sworn;
> That all our fruit be gathered,
>   And all our race take hands,
> And the sea be a Saxon river
>   That runs through Saxon lands.

The two biographies by Maisie Ward do not, of course, achieve the impossible. No biography can ever give the completeness of any man; and some of those who knew him have said that her portrait was not quite the Chesterton they knew. But the most important things are all here, and though she idealizes him, it is the kind of idealization which probably comes nearer to the man as God knew him. There is no suggestion of the false perfection against which Chesterton wrote that tremendous little dialogue between a dead man and his sanctified image, showing us the lean dead hands tearing the white figure down from its shrine with the cry,

> Where is my good, the little real hoard,
> The secret tears, the sudden chivalries?

Indeed that secret treasure-house is unlocked in many pages of the *Return of Chesterton*. The first biography, a full-length study, is admirably supplemented by the second, a record of the impression made by Chesterton upon all sorts and conditions of men, women, and children outside the literary world.

The centrality of the man is his most important characteristic, and, in the legend that has been growing up about Chesterton, there is some danger of losing sight of this in paying too much attention to stories which would make him appear a kind of gargantuan eccentric. His physical stoutness has been greatly exaggerated. His jokes about it gave it a kind of false jollity, for it was due to a real ailment with which his mind and spirit had to wage a laborious but always cheerful war. In his paradoxes one sometimes felt that 'the whole world turned over and came upright'. When the paradoxes became too mechanical the effect was lost, but this does not alter the fact that when they were not, they brought about a real 'renascence of wonder' in the mind of his generation.

It is difficult to enumerate the facets of his character. Genial, kind, devout, full of chuckling fun, and sometimes shaken with gusts of uproarious laughter at the absurdities and eccentricities of a mad world, his own sane centrality was unshaken.

I do not quite believe all the stories about the sword-stick that he carried in Fleet Street, the huge knife that he kept under his pillow, or the revolver that he bought immediately after his wedding ceremony in order to protect his bride, or the way he would stop the traffic by standing in the middle of the thoroughfare in a rainstorm to argue about St Thomas. The stories have been built up by friends who, knowing that Chesterton was an extraordinary man, seize too eagerly

THE CENTRALITY OF CHESTERTON

on signs and wonders to prove it. There is some danger of obscuring the most extraordinary thing of all, his real self.

The greatest of all his discoveries was that in the Nicene Creed he had a synthesis which superseded all the philosophical systems and transfigured the universe, touching every object in it with beauty and wonder, and giving a supernatural significance to the most obscure and apparently insignificant individual. The 'Father Brown' stories are almost too fantastic and phantasmagorical, but they do make the reader realize that postmen and milkmen, though they come and go like clockwork, have immortal souls.

It was because the Nicene Creed was the firm framework of all his thought that he was able to write a book which experts on St Thomas Aquinas actually believe to be a book about that philosopher. It was of course really an outpouring of Chesterton's own thoughts on God and man. They illuminated the thought of St Thomas because Chesterton himself (like St Thomas) understood the implications of his own creed.

'He began', says his biographer, 'by rapidly dictating to Dorothy [his secretary] about half the book. So far he had consulted no authorities, but at this stage he said to her:

'"I want you to go to London and get me some books."

'"What books?" asked Dorothy.

'"*I* don't know", said GK.

'She wrote therefore to Father O'Connor and from him got a list of classic and more recent books on St Thomas. GK "flipped them rapidly through", which is, says Dorothy, the only way she ever saw him read, and then dictated to her the rest of his own book without referring to them again.'

Gilson, one of the greatest living experts on St Thomas, said that Chesterton's book was the best ever written on that philosopher. But this is surely an expression of graceful courtesy rather than a critical judgement. It is perhaps the only point where the biographer, who thinks the book on St Thomas to be Chesterton's best work, has gone a little astray. Her appeal to authority here, however generous that authority may be, is not an appeal to reason, for the book is not about St Thomas at all. The best page in it is one in which he once more expounds, not St Thomas, but Lewis Carroll in a series of perfectly sound philosophical remarks that he might have made in any other of his own works, except for one clause in which he makes the perfectly true statement that for St Thomas eggs were eggs:

Against all this the philosophy of St Thomas stands founded on the universal common conviction that eggs are eggs.... The Berkeleian may hold that poached eggs only exist as a dream exists; since it is quite as easy to call the dream the cause of the eggs as the eggs the cause of the dream; the Pragmatist may believe that we get the best out of scrambled eggs by forgetting that they ever were eggs, and only remembering the scramble. But no pupil of St Thomas needs to addle his brains in order adequately to addle his eggs; to put his head at any particular angle in looking at eggs, or squinting at eggs, or winking the other eye in order to see a new simplification of eggs.

The truth it proclaims is profound, and St Thomas would have agreed, just as he would have agreed with Alice when she refuses to be only a dream in the mind of the Red King.

The *Summa* is mentioned only at a distance in a fine metaphor as a labyrinthine city. We are told, with sympathetic insight, that St Thomas was a fat bull of a man of whom it would be quite impossible to make an outline, though Chesterton hopes that his book may distil the essence of St Thomas into a kind of beef-tea. The book, in fact, is exactly what it was bound to be from the method of its composition; but when all this is said it must be repeated that, with the Creed of Christendom as his foundation, Chesterton could expand in any direction and still be at one with St Thomas. Indeed, Chesterton's book entitled *Orthodoxy*, if a few sentences had been added about the Aristotelian genesis of St Thomas, his fatness and his general sanity, would have just as sound a claim to be called a study of that philosopher.

In his real self Chesterton was essentially a poet. His 'Lepanto' and *The Ballad of the White Horse* will live, but, fine as these are, there are still finer things in some of his shorter poems. One of the noblest of these is the poem he entitled 'The Convert'. In the first eight lines he describes how a thousand tongues of the pagan world, debating old riddles and new creeds, had deplored his return to the perennial philosophy of Christendom, speaking of him not unkindly, but 'softly, as men smile about the dead'. His answer rang out like a clarion:

> The sages have a hundred maps to give
>   That trace their crawling cosmos like a tree.
> They rattle reason out through many a sieve
>   That holds the sand and lets the gold go free;
>   And all these things are less than dust to me,
> *Because my name is Lazarus and I live.*

RONALD KNOX
Chesterton's Father Brown

WHEN you met Chesterton in life, the physical bigness of the man made him seem out of scale; he overflowed his surroundings. And the same thing is true, in a curious way, of his literary output; he never really found his medium, because every medium he tried—and how many he tried!—was too small a receptacle for the amount of himself he put into it. He stood alone in the remarkable generation to which he belonged in being perfectly integrated; he had a philosophy of life, and not of this life only, which was all of a piece, and it so possessed him that he could not achieve, in any particular form of writing, mere literary perfection. His life of Dickens is an admirable performance, but it is really the Chestertonian philosophy as illustrated by the life of Dickens; his *History of England* is a brilliant résumé, but it is a history of Chesterton rather than of England. Shaw kept on urging him to write plays, but when *Magic* was produced it was too good for the stage; an after-dinner audience was not capable of the intellectual effort demanded of it. Even *The Ballad of the White Horse*, one of his certainly immortal works, cannot be graded among English epics because it is so much more than an epic. And the same fate pursued him in that fortunate moment when he took to writing detective stories. When we founded the Detection Club, he was appointed, without a dissentient voice, as its first president; who else could have presided over Bentley and Dorothy Sayers and Agatha Christie and those others? Yet the Father Brown stories cannot really be graded among mystery stories; they are mystery stories with a difference. As usual, the box has been so tightly packed that the clasps will not fasten; there is too much meat in the sandwich.

When you take to writing detective stories, the measure of your success depends on the amount of personality you can build up round your favourite detective. Why this should be so, is not immediately obvious; it might have been supposed that this kind of fiction had a merely mathematical appeal. But, whether because Sherlock Holmes has set the standard for all time, or because the public does not like to see plots unravelled by a mere thinking-machine, it is personality that counts. You are not bound to make your public *like* the Great Detective; many readers have found Lord Peter Wimsey too much of a good thing, and I have even heard of people who were unable to appreciate the flavours of Poirot. But he must be real; he must have idiosyncrasies, eccentricities; even if he is a professional policeman, like Hanaud, he must smoke those appalling cigarettes, and get his English idioms wrong. And if possible—perhaps that is where Lord Peter fails—he must appeal to us through weakness; when he appears on the scene of the tragedy, the general reaction must be 'A man like that will never be able to get at the truth.' It is because he drops his parcels and cannot roll his umbrella, because he blinks at us and has fits of absent-mindedness, that Father Brown is such a good publisher's detective. He is a Daniel come to judgement.

He was 'based', as we say, on Monsignor John O'Connor of Bradford, whose gracious memory is still fresh among us; it was he who later received Chesterton into the Church. The occasion on which Father Brown came into being is well documented, both in Chesterton's autobiography and in Monsignor O'Connor's memoir of him; and it should serve for a specimen of what is meant when we are told that such and such a character in a book was 'based' on such and such a figure in real life. Two young friends of Chesterton's, having been introduced by him to his new clerical friend, expressed surprise afterwards that a man trained in the seminary should possess such knowledge of the world, especially of the criminal world. Chesterton was delighted with their *naïveté*; was it not to be expected (he said to himself) that a man who spent three hours every Saturday listening to the tale of other people's sins should have some acquaintance with the byways of human depravity? And this reflection was incorporated bodily in the first of the Father Brown stories, *The Blue Cross*:

'How in Tartarus', cried Flambeau, 'did you ever hear of the spiked bracelet?'

'Oh, one's little flock, you know', said Father Brown.

That was all, really; nobody who had met Monsignor O'Connor would have put him down as 'a clerical simpleton'. He may have had difficulties about folding his umbrella; but instinctively you felt that this priest was a shrewd judge of men, with a reading of history and literature beyond the common. The owlish eyes blinking at you, the wooden indifference to appearances, the prosaic trudge in pursuit of his day-to-day tasks—all that was not Monsignor O'Connor as Chesterton saw him, but Father Brown as Chesterton invented him. He simply decided that for his own purposes—if I may put it in that way—he wanted a detective as unlike Lord Peter Wimsey as possible.

There was to be nothing of the expert about Father Brown; he should have no knowledge of obscure poisons, or of the time required to let the rigor mortis set in; he was not to be the author of any treatise about the different kinds of cigarette ashes. All his knowledge was of the human heart; he explains, in *The Secret of Flambeau*, that he is only capable of detecting murder mysteries because he was the murderer himself—only, as it were *in petto*. 'What I mean is that, when I tried to imagine the state of mind in which such a thing would be done, I always realized that I might have done it myself under certain mental conditions, and not under others; and not generally the obvious ones. And then, of course, I knew who really had done it; and he was not generally the obvious person.' He could put himself inside the other man's skin. He could even put himself inside an animal's skin—no, the dog did not know the murderer by instinct and spring at him, that was sentimental mythology. The important thing about the dog was that it howled when the sword-stick was thrown into the sea—howled because the sword-stick didn't float.

The real secret of Father Brown is that there is nothing of the mystic about him. When he falls into a reverie—I had almost said, a brown study—the other people in the story think that he must be having an ecstasy, because he is a Catholic priest, and will proceed to solve the mystery by some kind of heaven-sent intuition. And the reader, if he is not careful, will get carried away by the same miscalculation; here, surely, is Chesterton preparing to show the Protestants where they get off. Unconsciously, this adds to the feeling of suspense; you never imagine that Poirot will have an ecstasy, or that Albert Campion will receive enlightenment from the supernatural world. And all the time, Father Brown is doing just what Poirot does; he is using his little grey cells. He is noticing something which the reader hasn't noticed, and will kick himself later for not having noticed. The lawyer who asks 'Where was the body found?' when he is told about the Admiral's drowning has given himself away as knowing too much, already, about the duck-pond; if he had been an honest man, he would have assumed that the Admiral was drowned at sea. The prophet who goes on chanting his litany from the balcony, when the crowd beneath is rushing to the aid of the murdered woman, gives himself away as the murderer; he was expecting it. We had all the data to go upon, only Father Brown saw the point and we didn't.

What is the right length for a mystery story? Anybody who has tried to write one will tell you, I think, that it should be about a third of the length of a novel. Conan Doyle uses that formula in *A Study in Scarlet*, and in *The Valley of Fear*, filling up the rest of the book with a long story which does not really affect the plot. The modern publisher expects a full-length novel (which demands either a second murder or a great deal of padding), or else a short story (in which it is difficult for the author to give us the full conditions of the problem). Father Brown began life as short stories in the *Saturday Evening Post*, and short stories he remained; for an author so fertile in ideas, perhaps it was the simplest arrangement. But it must be confessed that this enforced brevity produces a rather breathless atmosphere; the more so, because Chesterton was an artist before he became an author, and occupies a good deal of his space with scene-painting. And the scene-painting takes up room—valuable room, the pedantic reader would tell us.

What scene-painting it is! The Norfolk Broads, and the house full of mirrors standing on its lonely island; or that other island on the Cornish estuary, with its wooden tower—you would expect the second of these passages to be little more than a repetition of the first, but in fact it is nothing of the kind; in the one case you have the feeling of being in Norfolk, in the other you have the feeling of being in Cornwall. The atmosphere of that dreadful hotel in *The Queer Feet*; the atmosphere of a winter-bound summer resort in *The God of the Gongs*; the (quite irrelevant) effect of bitter cold in *The Sign of the Broken Sword*—what a setting they give to the story! Flambeau explains, at the beginning of *The Flying Stars*, that in his criminal days he was something of an artist; 'I had always attempted to provide crimes suitable to the special seasons or landscapes in which I found myself, choosing this or that terrace or garden for a catastrophe'; and if the criminal, so limited in his choice of means, can be expected to provide a suitable décor, how much more the writer of stories! Yet it is only Chesterton who gives us these effects, the 'topsy-turvydom of stone in mid-air' as two men look down from the tower of a Gothic church; the 'seas beyond seas of pines, now all aslope one way under the wind' on the hill-side of Glengyle; the 'green velvet pocket in the long, green, trailing garments of the hills' on to which Mr Harrogate's coach overturns, ready for the coming of the brigands. Did Chesterton pick out these landscapes with his artist's eye, and then, like Flambeau, invent crimes to suit them?

But it does take up room. And, if only because the canvas is so overcrowded, you must not expect in these stories the mass of details which you would expect of Freeman Wills Crofts; the extracts from Bradshaw, the plan of the study with a cross to show where the body was found. Hence the severely orthodox readers of detective stories, who love to check and to challenge every detail, must be prepared for a disappointment; Chesterton will not be at pains to tell us whether the windows were fastened; how many housemaids were kept (in defiance of modern probabilities), and which of them dusted the room last; whether a shot in the gun-room would be audible in the butler's pantry, and so on. Even the unities of time and place are neglected; you can never be quite sure whether it is next morning, or a week later, or what. Consequently, you never quite feel 'Here am I, with all the same data at my disposal as Father Brown had; why is it that his little grey cells work, and mine don't?' Not that there is any deliberate concealment of clues, but the whole picture is blurred; the very wealth of detail confuses you. All you can do is to set about eliminating the impossible characters in the hope of finding, by a process of exhaustion, the villain. Women can be ruled out; there is only one female villain in the whole series—it is part of Chesterton's obstinate chivalry that he hardly ever introduces you to a woman you are meant to dislike. People with Irish names (how unlike Sherlock Holmes!) are fairly certain to be innocent. But, even so, the characters of the story elude you; you do not feel certain that you have been told quite enough about them.

For Chesterton (as for Father Brown) the characters were the really important thing. The little priest could see not as a psychologist, but as a moralist, into the dark places of the human heart; could guess, therefore, at what point envy, or fear, or resentment would pass the bounds of the normal, and the cords of convention would snap, so that a man was hurried into crime. Into crime, not necessarily into murder: the Father Brown stories are not bloodthirsty, as detective stories go; a full third of them deal neither with murder nor with attempted murder, which is an unusual average nowadays; most readers demand a corpse. The motives which made it necessary for Hypatia Hard to elope with her husband, the motives which induced the Master of the Mountain to pretend that he had stolen the ruby when he hadn't—the reader may find them unimpressive, because there is no black cap and no drop at the end of them. But, unless he is a man of unusual perspicacity, he will have to admit that he also found them unexpected.

The truth is that what we demand of a detective story is neither sensations, nor horrors, but ingenuity. And Chesterton was a man of limitless ingenuity. What really contents us is when we see at last, and kick ourselves for not having seen before, that the man who was murdered in the Turkish bath without any trace of a weapon was stabbed with an icicle; that the poisoner did drink the tea which accounted for her victim, but took a stiff emetic immediately afterwards; that the time of a particular incident was given wrongly, not because the witness was in bad faith, but because she saw, not the clock, but the reflection of the clock in a looking-glass. All those brilliant twists which a Mason and an Agatha Christie give to their stories, Chesterton, when he was in the mood for it, could give to his.

How to dispose of the body? If it was only for a short time, you could hang it up on the hat-stand in a dark passage; if you wanted to get rid of it altogether, you could bury it in the concrete floor of a new set of flats. A ship could be lured to its doom by lighting a bonfire which would confuse the appearance of the lights in the tideway; you could gag a ruler so securely that he would be unable to answer the challenge of his own sentries, and would be shot. They are all ideas we might have thought of and didn't.

Whether such expedients would be likely to be adopted in real life is perhaps more questionable. But then, how far is the writer of mystery stories bound by the laws of probability? Nothing could be more improbable than Father Brown's habit of always being on the spot when a crime is committed; but he shares this curious trick of ubiquity with Hercule Poirot. The thing is a literary convention; it may not be a good one, but it is well worn. No, when we open a detective story we leave the world of strict probability behind us; we must be prepared for three or four quite independent pieces of shady business happening to happen in the same country house on the same evening. And Chesterton's imagination was flamboyant; he was like a schoolboy on holiday, and could sit as light to realism as P. G. Wodehouse. If you meet him on his own ground—that is, halfway to fairyland—you will have to admit that for sheer ingenuity he can rival Miss Sayers herself. Cast your mind back to your first reading of the Father Brown stories, and ask yourself whether you saw what was the missing factor which linked all the various exhibits in Glengyle Castle, or why *The Insoluble Problem* was insoluble.

No, if we are to judge the Father Brown cycle by the canons of its own art, we shall not be disposed to complain that these are something less than detective stories; rather, that they are something more. Like everything else Chesterton wrote, they are a Chestertonian manifesto. And it may be reasonably maintained that a detective story is meant to be read in bed, by way of courting sleep; it ought not to make us think—or rather, it ought to be a kind of catharsis, taking our minds off the ethical, political, theological problems which exercise our waking hours by giving us artificial problems to solve instead. If this is so, have we not good reason to complain of an author who smuggles into our minds, under the disguise of a police mystery, the very solicitudes he was under contract to banish?

I am inclined to think that the complaint, for what it is worth, lies against a good many of the Father Brown stories, but not all, and perhaps not the best. Where the moral which Chesterton introduces is vital to the narrative, belongs to the very stuff of the problem, the author has a right, if he will, to mystify us on this higher level. In the overcivilized world we live in, there are certain anomalies which we take for granted; and he may be excused if he gently mocks at us for being unable, because we took them for granted, to read his riddle. There is something artificial in a convention which allows us to say that nobody has entered a house when in fact a postman has entered it, as if the postman, being a State official, were not a man. There is something top-heavy about a society in which a fellow guest is indistinguishable from a waiter if he cares to walk in a particular way. And there is something lacking in the scientific investigator who can be taken in when his own secretary disguises himself in a false beard, simply because he has sat opposite his secretary day after day without noticing what he looked like. But it must be confessed that in some of the stories, especially the later ones, the didactic purpose tends to overshadow, and even to crowd out, the detective interest; such stories as *The Arrow of Heaven*, and *The Chief Mourner of Marne*. If we read these with interest, it is not because they are good detective stories, but because they are good Chesterton.

When he wrote *The Incredulity* and *The Secret (of Father Brown)*, Chesterton had perhaps rather written himself out, and publishers pressed him for copy faster than even he could supply it. At the end of his life, he seemed to get a second wind, and *The Scandal of Father Brown* contains some of his most ingenious plots. But how seldom does an author manage to spin out a formula indefinitely; how signally Conan Doyle failed to do it!

SIR ARTHUR BRYANT
Chesterton

GILBERT KEITH CHESTERTON spent his whole life in teaching others how to live. Even today the sound of his name is like a trumpet-call. To him the world was a field in which one went about doing battle with evil in order that good might endure. If from his generation one had to select the name of one man who might have stood as a type of Don Quixote or St George who slew the dragon, it was he. If any literary name of our age becomes a legend transcending letters, it will, I believe, be his. In his lifetime he was often likened to Dr Johnson, and it was an analogy that did not only depend on his giant girth and splendid conversation. For like Johnson he never penned a line or uttered a sentence that harboured a mean or ignoble thought, nor did he ever miss an opportunity of striking a blow for what he felt to be right. In all that he did and stood for there was neither fear nor calculation. He was the kind of man of whom Bunyan was thinking when he drew the picture of Mr Greatheart. His sword was at the service of pilgrims.

And what a sword it was! It is nearly half a century now since, in his dedication to *The Man Who Was Thursday*, he roused the heart of a new generation to challenge the cold decadence of an unbelieving and selfish intellectualism:

> A cloud was on the mind of men, and wailing went the weather,
> Yea, a sick cloud upon the soul when we were boys together.
> Science announced nonentity and art admired decay;
> The world was old and ended; but you and I were gay.
> Round us in antic order their crippled vices came—
> Lust that had lost its laughter, fear that had lost its shame.

Looking back on it, we can still feel the fire of that protest, even as he felt it when, as a boy, he first opened Stevenson's book, and

> ... cool and clear and sudden as a bird sings in the grey,
> Dunedin to Samoa spoke, and darkness unto day,
> But we were young; we lived to see God break their bitter charms,
> God and the good Republic come riding back in arms:
> We have seen the city of Mansoul, even as it rocked, relieved—
> Blessed are they who did not see, but being blind, believed.

It was the greatness of Chesterton's creed that the salvation he preached was the salvation not of the elect, but of the many. His concern was always with the common man. He did not confine his sympathies to the well-behaved and refined: to the recluse in the cloister or the scholar in the study. He was not one of those who thought that only the best were to be saved; his catholicism was an all-comprehending democracy. And it was one that was founded on a deep understanding of all that humanity needs; not only of its sufferings but of its joys. He did not only wish to shelter the oppressed from the clouds that threatened them, but wished also to see them rejoicing in the sunshine. He was the champion of all those things that make common men happy—of laughter and marriage, of home and beer. Like old Samuel Johnson, he loved the poor, not with the perfunctory pity of the professional philanthropist, but with an earnest desire to make them as uproariously happy as he himself was happy. That was why he so admired Dickens, the poet *in excelsis* of the joys and humours of plain, suffering, unvarnished humanity. Yet, though his inferior in sheer creative genius, he had nothing of the egotism and self-pity of Dickens. He was too good for that. I never met a more generous man, and I never saw a happier. I do not believe there is anyone who had the inestimable privilege of knowing Gilbert Keith Chesterton who would not say the same.

We live in a Protestant country and are a Protestant people, and Chesterton, in the latter years of his life, was a Roman Catholic. Having begun his career in a blaze of early triumph and popularity such as comes to few men, he put it by him and set his face against the stream of contemporary thought. Whether one agrees or not with his choice of Faith and Dogma, there is not the least question that in doing so he singled out for attack many elements in our modern civilization which are both powerful and intolerant of criticism, even when the attack was as good-humoured and full of healthy laughter as his always was. He did not thereby make his task any the easier. For ephemeral reasons, though for no others, his influence in his lifetime was therefore limited. Yet by virtue of this very fact, I believe his influence on unborn generations will be the greater. For though his public at the end of his career was probably smaller than at its triumphant beginning, it was a public that counted for more. It was the kind of public that read his books not merely for pleasure and amusement but because it found in them a faith and an inspiration. That faith and inspiration is not diminished now that he is dead, and it will continue to be transmitted to others. And his books will continue to be read, and, I think, increasingly read.

For ninety-nine people out of a hundred, perhaps for nine hundred and ninety-nine out of a thousand, life tends to be a dull and uninspired affair, a round of prosaic duties which are got through for some ulterior end and without joy or relish in the performance. It was Chesterton's supreme merit that he never saw life or presented it to others as anything but a flaming and glorious romance. He did so, without any attempt to overlook its material aspects; on the contrary, he emphasized these and saw in them the complete justification of his creed. The very weakness of man was to him something to be rejoiced over and turned to good account. 'When Christ', he once wrote,

at a symbolic moment was establishing His great society, He chose for its corner-stone neither the brilliant Paul nor the mystic John, but a shuffler, a snob, a coward—in a word, a man. And upon this rock He has built His church, and the gates of Hell have not prevailed against it. All the empires and the kingdoms have failed, because of this inherent and continual weakness, that they were founded by strong men and upon strong men. But this one thing, the historic Christian Church, was founded on a weak man, and for that reason it is indestructible.

Chesterton followed a Master who was born in a peasant's manger and died on a rough-hewn cross. And like Him he knew that in these plain and unadorned instruments of common life were the chains of Hell and the keys of Heaven, angels ascending and descending, and the son of man glorified.

HESKETH PEARSON
G. K. Chesterton

IN my young days there were four outstanding personalities in the literary world: Bernard Shaw, H. G. Wells, G. K. Chesterton, and Hilaire Belloc. We of the new generation admired them all, but though we recognized that Shaw was the most remarkable of the lot, both as man and writer, I think we were secretly fondest of Chesterton, who was not only lovable but had all the human frailties from which Shaw seemed to be so inhumanly free. Chesterton loved company, enjoyed a rousing song, relished a vulgar joke, drank good beer when he could get it, and preferred bad beer to no beer at all. He would talk with anybody in a pub, either sense or nonsense according to the mood of the moment or the conversational atmosphere of the place. His laughter could be heard several houses away. He was a sort of Falstaff in bulk and wit, and like that notable character he laughed at his more serious followers. My first glimpse of him was in a Fleet Street pub in the summer of 1913, and I soon perceived that he was pulling the leg of his youthful companion, who looked so earnest that Chesterton's absurdities were probably uttered for the good of his soul.

'One should always drink port from a tankard', said GKC. 'Because one does not like to see that one is coming to the end of it. Also it takes on a richer hue. Also it has a mellower taste. Besides, one can *grasp* a tankard: and *drain* it. Now one can't drain a glass. One can only *sip* a glass.' To each of these statements the young man made affirmatory sounds, but as he was drinking from a glass he felt a little guilty and nodded his head dubiously when charged with sipping. This gave Chesterton a new idea. 'Glasses', he went on, 'are made to be smashed. It is said that those who live in glass houses should not throw stones. But what man, living in a glass house, would do anything else?

It is the simplest way of getting out of a glass house. Indeed, if he throws a sufficient quantity of stones, it ceases to be a glass house.' He then ordered another tankard of port for himself and a glass of port for the young man.

The following year I was introduced to Chesterton and found him extremely amiable. His way of putting things appealed to me greatly in those days, and he chuckled with amusement when I said, parodying his own style, that Shaw's most serious limitation was that he preferred potatoes to potations. Some years later I published an imaginary conversation between him and Shaw, giving GKC the phrase he had appreciated on that early occasion, and he wrote to say that as I could imitate his mannerisms so well perhaps I would like to do his next book for him.

The unique thing about Chesterton both as talker and writer was the way in which he could ridicule ideas by playing with words. My friend, Edward Fordham, who was at St Paul's School with him, tells me that they were once arguing as to whether some policy or other was good or bad. 'The word "good"', said Chesterton, 'has many meanings. For example, if a man were to shoot his grandmother at a range of 500 yards I should call him a good shot but not necessarily a good man.' He was never stumped by an unexpected question. Following a debate on racial characteristics he was chatting with Edward Fordham when an elderly lady whom neither of them knew came up and asked in a rather affected manner: 'Mr Chesterton, I wonder if you could tell me what race I belong to?' Adjusting his glasses, he replied: 'I should certainly say, madam, one of the conquering races.'

His ability to adapt himself instantly to an unforeseen circumstance and make fun out of it was shown when his hat blew off one day in Fleet Street. A friend of mine witnessed the occurrence, saw him chasing the hat, and to save him trouble dodged the traffic and rescued it almost from under the wheels of a motor bus. Returning to the pavement, he handed it to its owner who was puffing and panting from his exertion. 'That's very kind of you, very kind indeed', said GKC. 'But you shouldn't have taken the trouble. My wife has bought me a new hat, and she will be most disappointed when she hears that the old one has only just been saved from well-merited destruction.' 'In that case', protested my friend, a little vexed that he had uselessly imperilled his life, 'why on earth did *you* run after it?' 'It's an old friend', replied Chesterton with some emotion; 'I am fond of it, and I wanted to be with it at the end.'

Though he wrote a number of detective stories, GKC struck me as the most absent-minded and unobservant person I had ever met. Yet he must have noticed people and things in a quick-glancing way, because when I met him again after a lapse of several years and mentioned my name in case he had forgotten me, he said: 'I remember you well. In fact you appear in one of my Father Brown stories.' Naturally I wanted to know which. 'Ah!' said he, 'It is a detective story and the least you can do is to detect yourself.' His absent-mindedness never worried him, though it caused much inconvenience to others, especially when he had promised to give a lecture at some distant spot. His wife once received a wire from him: 'Am at Wolverhampton. Where ought I to be?' On another occasion he arrived over an hour late at a meeting and apologized by saying: 'My wife always sees me into a train, but she is not there to see me out of it, and I left this one at a station further down the line.'

That aspect of his character displayed his other-worldliness. The acquisition of money meant nothing to him, and when he found money in his pocket he spent it or gave it away. He sold his books for a song. For one of the best, *The Napoleon of Notting Hill*, he received £100. His tastes were simple, and so were his pleasures. Just before going to a tea-party at his house a small girl was told by her mother that she would learn a lot from Mr Chesterton, who was a very clever man. Afterwards the child revealed the nature of his lessons: 'He taught me how to throw buns in the air and catch them in my mouth.'

Even his pugnacity in print and on the platform was unworldly. He loved disputation and enjoyed a fight for its own sake, not for any kudos to be got out of it nor for the pleasure of scoring off an opponent. He probably never wanted to win, because that would have meant the end of the argument. He was once asked to stand for parliament, and replied that he would be delighted to do so if it were absolutely certain that he would be defeated.

D. B. WYNDHAM LEWIS
Diamonds of the Gayest

'AM in Market Harborough where ought I to be—Gilbert.' The anguished wire received by Mrs Chesterton is sufficiently celebrated, and sufficiently misquoted—a dozen towns usurp the primacy of Market Harborough in current British folklore—to have given Chesterton immense satisfaction, he himself being no fanatic for checking quotations. Its psychological significance is considerable. Apart from bearing all the hallmarks of happy wedlock, this urgent scribble expresses the desolation of a man whose waking hours were, like those of Aquinas, filled with exigent creative thought, on colliding suddenly with the mundane. Chesterton's body was, very visibly, in Market Harborough. Recalled with a jerk by some trifling circumstance (say a ticket-collector's raised eyebrows) from Heaven, Hell, or Clapham Junction, his mind had just rejoined it; quite possibly with some of those spiritual 'strivings and yearnings and retchings' endured by Kim's Lama.

It was a situation regularly recurrent in some form or other. Chesterton could, as he said, start off on a mental Odyssey from anywhere at any time. He was a true contemplative; a pebble, a straw, a soap-advertisement landed him swiftly in the Infinite. That devotee of Swift's who cried that the Dean could write powerfully about a broomstick might, had he lived some two hundred years more, have seen this feat brought to perfection at least once a week.

Habitual and concentrated thinking is not a characteristic of the modern English, with whom the absent-minded professor who leaves his trousers at home is a stock figure of fun rather than awe; yet a man more English in every fibre than Gilbert Keith Chesterton can hardly be imagined. The intensity of his Englishness, enriched a trifle, mentally speaking, by a slight strain of Scots on his mother's side, would in fact almost recall Professor Raleigh's petulant libel on the Welsh—'so damned Welsh that it all looks like affectation'—were it not that Chesterton was so obviously, as French peasants say, *bon comme le pain*; utterly good, sincere, single-minded in his integrity, wholemeal through and through. Two great Englishmen before him have blended these qualities with extreme intellectual vigour. With Dr Samuel Johnson he had often been compared, apart from his sharing the same physique; but the Doctor had a rough and bullying provincial streak which London life never eradicated. I think Chesterton's true exemplar is St Thomas More. With goodness, gaiety, wit, and an unquenchable passion for truth More radiated that same chivalrous courtesy which was so marked a trait in Chesterton. They could both hate ideas; never men. Chesterton's almost eccentric love of humanity in gross and detail, and the poor above all, continually disarmed despite themselves the bitterest opponents of everything he fought for, with perhaps one notable exception. Dean Inge's exasperated cry of 'Obese mountebank!' was not in the best tradition of duelling.

Chesterton was born on 29 May 1874, on Campden Hill in Kensington, where the long-established estate agency of Chesterton and Sons flourishes still. His autobiography sufficiently describes the solid middle-class background of his childhood, his amiable father's many hobbies, the standards upheld in that stratum of British Victorian society, the happiness of that environment. At St Paul's School he did not distinguish himself, though the budding of a literary talent was discernible in contributions to a school magazine. From St Paul's he went on to the Slade, and might well have made a name as a satiric artist of the first order, judging by his many drawings, and particularly his illustrations to some of Belloc's light novels in later years. It was the era of decadence and *fin-de-siècle* nihilo-pessimism, so well depicted by Holbrook Jackson in *The Eighteen-Nineties*; a slough out of which young Chesterton at length, after some floundering—his home atmosphere had been Unitarian–Liberal—heaved himself unaided. 'I invented a rudimentary and makeshift mystical theory of my own . . . I thanked whatever gods might be, not, like Swinburne, because no life lived for ever, but because any life lived at all.' Though he was still a thousand miles distant from Christianity, this early revolt against the Zeitgeist was to lead Chesterton to a destination of whose very existence he was, at this time, only dimly aware.

Quitting the Slade abruptly at the beck of a more imperious Muse, he went for a few months into a publishing office, fell ardently to writing, published a slim, well-received volume of verse called *The Wild Knight*, and made his entry as a freelance into Fleet Street, his harridan true-love (years later, with an international literary reputation, he still described himself as 'a jolly journalist') all his life. From this point his career is a straight unbroken line to celebrity. In 1900 began that friendship with Hilaire Belloc, his fellow swordsman, compared with which the legendary bond between Damon and Pythias seems merely a nodding acquaintance. In the same year began that long series of Saturday essays in the Liberal *Daily News* which speedily established him as a portent and won him an eager public, its enthusiasm duly leavened by the customary resentment of what Barbey d'Aurevilly calls the antlered dull. In June 1901 he married Miss Frances Blogg, whom he had met in the 'fantastic Suburb' of Bedford Park, at that time the ivory tower of the Intelligentsia. Of his first eleven books three (*Browning, Dickens, Heretics*) confirmed the arrival of a new star in letters. A notable landmark is the publication of his twelfth, *Orthodoxy*, in 1908; a swashing challenge to sceptics revealing that Chesterton had discarded his vague liberal optimism and accepted Christianity on intellectual grounds. This decision was final. Thirteen years later it took him to Rome.

Lecturing, debating, and pouring out a stream of essays, books, and poems now began to fill a busy life and exalt a reputation steadily soaring. By 1909, when the Chestertons moved from Battersea to Beaconsfield for the rest of their days, he was already a public figure, counting among his friends men of mark like Belloc, Shaw, Wells, Max Beerbohm, Maurice Baring, E. V. Lucas, Masterman, and a dozen more, and had acquired the Chestertonian Shape, for so many years the joy of caricaturists and gossip-writers. The huge cloaked untidy form in the wide black hat had become one of the touristic sights of Fleet Street, and the high tenor voice, *clara vox*, he shared with Belloc and Charlemagne had been heard, at least by Press gossips, laughing at its own jokes in every tavern from Temple Bar to Ludgate. Stories about, for example, his giving up his seat in a bus to three ladies and turning the White City lake into a stormy sea by falling out of a row-boat became current coin among the wits. The note of affection in all these playful libels is significant. Save for stirring the bile of some supercilious organ like *The Academy*, which detested him at one period on principle, Chesterton never roused in his brethren of the pen that tireless malice which is the reward of distinction, and he passed through the roaring literary jungle like Dante through Hell, unmauled. He conducted no feuds and knew no cliques. As the prickly Wells once said, almost in despair, it was impossible to quarrel with him. His secret is conveyed in the last stanza of the *Divine Comedy*, 'L'amor che move...'

By the outbreak of the First World War Chesterton's position as a serious thinker was recognized by most men of intelligence, barring those who could not, and cannot, perceive the profundity of the Atlantic wave owing to the froth on top. High spirits and an incurable passion for paradox continued to baffle and annoy prigs of every hue, and he certainly indulged himself in paradox and puns not infrequently to the point of sheer naughtiness. The realization that hilarious fantasias like *Manalive*, and *The Flying Inn*, and *Tales of the Long Bow*, are built on a strong coherent philosophy, not to speak of the Father

Brown sequence and, in fact, everything Chesterton wrote after 1908, is possibly still hidden from many who esteem what might be called his vaudevilles for their content of pure rollicking. But the note is there; a constant. In the earliest whodunnits it rings clear to the discerning. By disguising himself as a waiter the master criminal Flambeau renders himself, socially speaking, invisible to a company of gentlemen in evening dress, a situation still sartorially possible in the 1900s. By adopting postman's uniform a murderer is able to enter and leave a block of Hampstead flats unnoticed. Such immunity derives in each case from the fact that in refined Edwardian society waiters and postmen and servants in general had no faces, were not 'persons': an affront to the dignity of man, and especially the poor man, which seems to Chesterton ignoble and grotesque, though no doubt 90 per cent of his original readers, unacquainted with the dogma at issue, assumed him to be standing on his head as usual. There was no possibility, on the other hand, of missing the message of a poem like *The Ballad of the White Horse* (1911), one of the outstanding narrative-ballads in our literature, with a single verse of which *The Times* was inspired to conclude a brief and sombre news-bulletin, printed as the 'first leader' and headed 'Sursum Corda', on one of the blackest days of the Second World War:

> I tell you naught for your comfort,
>   Yea, naught for your desire,
> Save that the sky grows darker yet,
>   And the sea rises higher.

It is the Mother of God speaking to King Alfred in a dream during the Danish terror, and the call to endure must have braced thousands of Alfred's posterity equally, as did a second *Times* quotation some months later:

> 'The high tide!' King Alfred cried,
>   'The high tide and the turn!'

The same tonic effect on a large scale was achieved by the splendid clangour of 'Lepanto' (1915), as probably not a few old soldiers can testify. 'The other day in the trenches we shouted your *Lepanto*', John Buchan wrote to Chesterton in the summer of that year. It gave this humblest of men enormous pleasure.

After the war-to-end-war came the diversion of a great part of Chesterton's energy, on recovery from long and severe illness, into a fresh channel; a step which many, including his wife, did not cease to deplore as a waste of his gifts. He added to the burdens of a life of hard work by returning to the rough-and-tumble of Fleet Street and becoming an editor. He deemed it a sacred obligation to his soldier-brother Cecil, dead in France at the age of 39.

Cecil Chesterton, founder and, also with Belloc, editor of *The New Witness*, that formidable little weekly celebrated for, among other things, its onslaughts on the politicians involved in the Marconi shares scandal, certainly knew something about the technique of editing. Gilbert, however supreme a literary journalist, certainly did not. The drain of Cecil's paper, relaunched as *G.K.'s Weekly* in 1925, on his vitality and his purse alike was scarcely compensated, even with Belloc's continued assistance, by the fact that it once more gave him a Fleet Street pulpit—a far smaller one than in the old *Daily News* days, since Fleet Street's most powerful rulers blacklisted *G.K.'s Weekly* with the same cordial unanimity as *The New Witness*, and its circulation was relatively minute—with the renewed opportunity once a week to flail the dragons he had been fighting all his life, from Big Business monopolies and the Prussian Spirit down. Literary work and incessant lecturing meanwhile continued, and it is some measure of his stature that during these years of toil and worry he produced one of his two greatest books, with half a dozen others. *The Everlasting Man* (1925) proved him a master of Christian apologetic; the first-fruits of his reception into the Catholic Church—a step delayed for years chiefly by his determination, like Newman, to be ruled on such a momentous issue by the voice of reason alone—the book is, so to speak, a richer, deeper, more matured and splendid *Orthodoxy*. In some ways its principal successor, *St Thomas Aquinas* (1933), is the more striking cerebral feat. Chesterton was not a professional philosopher. The mind of the Angelic Doctor, like the Himalayas, is no playground for the dilettante. Chesterton's success was warmly proclaimed by leaders of the Thomist Revival in Europe like Etienne Gilson ('I have been studying St Thomas all my life, and I could never have written such a book'), and Father Gillot, Master-General of the Dominican Order. If any doubts remained of the 'obese mountebank's' intellectual powers this book disposed of them, for any unprejudiced critic, once and for all.

He found time, while shouldering his manifold burdens, to make two lecture tours (1919, 1930) in the United States, where he became a public portent immediately on landing and routed the redoubtable atheist lawyer Clarence Darrow in a memorable public debate. Debating invariably called forth his most joyous powers, lecturing was almost always a 'chore'. Between 1922 and 1929 he visited Holland, Palestine, Poland, Italy, and Spain. In Rome he had a private audience with Pius XI, and found his doubts of Fascism much increased by a long talk with Mussolini. At Dublin in 1932 he recognized in the International Eucharistic Congress a true league of nations and Democracy incarnate. Later that year the BBC asked him to attempt a broadcast series. In this vast new field he triumphed so effortlessly from the beginning ('The building rings with your praises. . .') that his death undoubtedly deprived Broadcasting House of a fixed star of the first magnitude. The secret, as ever, was love. In the summer of 1936 he fell seriously ill of overwork and cardiac trouble, and died on 14 June, aged 62. His heart was too small, apparently, for that vast frame; a paradox, as has well been remarked by his biographer, Miss Maisie Ward, in his own manner. On his tombstone—and, not long afterwards, his wife's—Eric Gill carved two familiar Latin lines of exile and longing which are, so to speak, the crystallization of all Chesterton's thought:

> Qui vitam sine termino
> Nobis donet in patria.

Whether his name will continue to live depends, Belloc remarks in his brief monograph *On the Place of Gilbert Chesterton in English Letters*, not upon him but upon his country. Belloc was writing in 1940. In 1956 that Christian ethos which was the source and mainspring of Chesterton's achievement is holding its own in this country still, against increasingly heavy odds. If it could conceivably be destroyed or driven underground, temporarily, by some such violence as that predicted by Orwell in *Nineteen Eighty-Four*, Chesterton's name would vanish with it. If it prevails, his name will be 'among the first of English names'. 'Thatt', to echo Kipling's Babu, seems axiomatic. Those of the irreconcilable who have dismissed Chesterton as a buffoon will be found almost invariably to be declared enemies of his faith and philosophy, using a formula consecrated by experts of the same blend like Voltaire and Gibbon; a formula highly soothing to the uninstructed and saving much brainwork; doubly effective, naturally, if its object is known to the majority chiefly as a practitioner in comic fantasy. The issue, called nowadays 'ideological', has been more precisely defined by Manning, discerning it at the roots of all human conflict, as theological. Fundamental humility prevented Chesterton ever being wounded by cries of 'Buffoon!' He was likely, indeed, to agree cheerfully and invoke, with immeasurable deference and apology, God's clown of Assisi.

Survival, then, to finish with Belloc's forecast, will be due on Chesterton's part to six major characteristics. He is intensely national, of extreme precision in thought, unique in his genius for illumining and explaining realities by vivid comparison and parallel, 'literary' in the highest and deepest sense, infused with true charity—which occasionally weakens his sword-arm—and, finally, rooted in the Faith to which we owe our civilization. Add to these qualities a perpetual 'spouting well of joy within', and it would seem we have sufficient to ensure Chesterton as long a survival, other things being equal, as any of the great in English literature. Like most of them in their day he may be at this moment suffering, to some extent, that resistive eclipse which is the common lot of great men of letters for some little time after death. If so, it can be only temporary. Sir Arthur Bryant's recently expressed conviction that Chesterton's influence on generations yet unborn will be greater than it was on his contemporaries will seem to many to be firmly based, even with Belloc's condition in mind. Braced by deep draughts of Chestertonian wine, the last survivors of Christian civilization hunted down in future catacombs can go singing to their fate.

Like all the great also, he has his lapses, as mentioned already. What the morose have labelled the Chesterton Formula might at times, usually when the teeming brain was tired, decline into a mere verbal parlour-game recalling Lord Henry Wotton's luncheon-table virtuosities in *Dorian Gray*; which, of course, does not affect the essential value of the Paradox. As Chesterton more than once reminded the world, Christianity is founded on the sublimest of all paradoxes, which inspired a poem of St John of the Cross: God incarnate crying in a manger. Fecundity, again, was sometimes too strong for minor pedantries; occasionally Pegasus scorned the snaffle, and a lifelong addiction to quoting (chiefly poetry) from memory, inaccurately, anguished more than one sorrowing silver-haired publisher. But such carelessness is well compensated by the generosity with which Chesterton lavishes beauty everywhere in his pages, *manibus lilia plenis*. Few writers of detective fiction—to take once more a literary exercise which he regarded as pot-boiling and developed at high pressure, almost entirely to assist the tottering finances of *G.K.'s Weekly*—break suddenly into passages like, for instance, the one in a Father Brown story comparing the intense cold in the forest to 'that silver sword of pure pain that once pierced the very heart of purity'. To Art, his earliest love, he owed that characteristic feeling for shape and colour in a cloud, or a pool, or a jewel, or the eyes of a girl, and his fondness for linking the hue of purple, in a sky or a wig, with strong evil is in the best tradition of the medieval colourists. Of the tangible power of evil in the world he was always acutely aware; it could equally evoke an offhand phrase in a short story ('It seemed as though the trees were filling slowly with devils'), and the tense drama of the second act of *Magic* (1913), a play, described by its author as 'a fantastic comedy', which stirred, of all people, George Moore. And in *Magic*, incidentally, Chesterton enriched the British stage with a new type, superbly played by Fred Lewis during a *succès d'estime* at the Little Theatre—that of the Duke, a courteous, exquisite old ass whom Molière might not have been ashamed to acknowledge.

The distance from this kind of writing to the discussion of the sophisms of Siger of Brabant in *St Thomas Aquinas* might well be taken to illustrate the range of Chesterton's genius, ... a glimpse, so to speak, of the total man: the Chesterton whose finest jewels are often enclosed in caskets rarely visited by the majority. One such is the slim volume, which scared its publishers into a foreword disclaiming all responsibility, called *The Victorian Age in Literature*, published in 1913; a feast of brilliant fun and brilliant criticism and appreciation alike (what could be better than the description of Matthew Arnold's general manner—'a smile of heart-broken forbearance, as of the teacher in an idiot school, that was enormously insulting'?) In *Orthodoxy* and *The Everlasting Man*, again, are magnificent prose-passages unknown to those who instinctively avoid books with this special theme. The travel books are another rich mine, and in that immense mass of ephemeral journalism Chesterton turned out in forty years of a working life are to be found diamonds of the gayest and purest, too long buried in dusty files.

LANCE SIEVEKING
'Mr Tame Lion'. Reminiscences of G. K. Chesterton

WHEN I first saw G. K. Chesterton I was speechless: not, as you might reasonably suppose, with astonishment and awe, but merely because I had not yet learnt to speak. I was only a few weeks old and I accepted him without comment. The occasion was that of my baptism at Harrow-on-the-Hill. It was not until a few years after this, in about 1899, that I consciously made friends with him.

He was an enormous man, nearly six feet six tall and weighing something round twenty stone, and his huge head was covered with a tousled mane of pale yellow hair. His gigantic figure was made more gigantic by the loose voluminous black clothes he usually wore, a sort of cloak, or ulster, flapping like a sail as he surged forward. Every now and then the incredible protruding slope of his immense stomach could be seen, tightly encased in an enormous black waistcoat with innumerable buttons, and he usually wore a vast black crumpled sombrero with a wide brim that bent up and down in unexpected places.

But the first thing you noticed were the pince-nez: a pair of spectacles with a straight bar between on which there was a finely coiled spring. The oval glasses had little wash-leather pads and the spring pressed these tightly on either side of his nose, and when he took them off there were two bright red marks on his nose. In those days everyone wore glasses like that. Chesterton's had a cord or wide silk ribbon which used to catch in things. They were always wildly crooked, one glass up and the other down. In the middle of his huge face was a drooping yellow moustache, from beneath which used to come the most surprising thing of all—his voice. It was very high, at times almost squeaky, and what he said was often preceded by a long, high, infectious, neighing chuckle.

When I was a tiny boy I nicknamed GK 'Mr Tame Lion', because that is what he looked like when he played with me in our back garden, and though he did not make deep growly noises like the lions at the Zoo, he roared as sweet and high as any Kokooburra bird. And 'Tame Lion' he remained for the rest of his life.

We lived at number four, Lyon Road, Harrow. One afternoon he drew a little caricature of himself and we cut it out and stuck it on a piece of cardboard and in his miniature condition I cast him as the 'Tame Lion of Lyon Road' in a play I was going to produce in my toy theatre. It was one of those little theatres made by the celebrated Mr Pollock, and I thought the highly coloured scenery of *The Daughter of the Regiment* was very suitable for the new play. One backdrop contained a representation of the Sphinx surrounded by plenty of bright-yellow Egyptian desert. Kneeling down behind the theatre out of sight I announced the title *The Lion and Swinx*. GKC was delighted by my version of the word and insisted on it being 'kept in'. After that 'The Swinx' became our favourite character and crept into every play in the most improbable contexts. That little Pollock theatre gave us endless delight. My audience usually consisted of my mother, my little sister Elinor, and the latest governess from 'The House of Education' at Ambleside. Sometimes the cook and the housemaid were pressganged to swell the audience.

I have often wondered what difference there would have been in his writing and his life if he had not been a gigantically fat man. One thing at any rate is certain: he could not have made that joke about his size and weight which he trotted out in one form or another on so many occasions. He used to make remarks such as 'I am conscious of being rather a heavy object in every sense, to be attached to so airy a subject'; and he invented imaginary newspaper headlines—'The Breadth of Politeness: Mr Chesterton gives up his seat in a bus to four ladies.'

You know how well you know a person with whom you have been intimately associated when you were a child. Owing to your inexperience of the world you cannot judge them in terms of law, or social custom. But you know their character at bottom: you know whether they are 'good' or 'bad', 'cruel' or 'kind', to be trusted or not to be trusted, generous or mean. You know these things in a way you never can in later life. I knew Mr and Mrs 'Tame Lion' as a small child and a boy, and though I was 40 when they died I still knew them as a child knows people. They were both truly good people and full of love. GK died in 1936 and Frances, his wife, wrote to me:

> Dear Tame Lion was only really ill for a week, though he had not been well the last two years. He died as he would have wished, with his brain and sword and pen all at their best. I say to myself over and over again:
> 'Life that is only mean to the mean
> And only brave to the brave.'
> It was brave to him, and please God, shall be to me.
> Yours always,
Mrs Tame Lion

A few weeks later she, too, was dead. She was 66. Tame Lion was 62.

Mrs Chesterton—Frances—usually travelled with him when he went about the country lecturing. My mother, who was her best friend, said that he owed much of his success to Frances, but treated her with a certain lack of consideration. My mother used to be very indignant and give instances of his alleged ungraciousness: but I think she exaggerated. I remember one of her stories of how Frances had sat up half the night collating twenty pages of notes for him and preparing refreshments, spare handkerchiefs, a change of shoes, and so on. I imagined Tame Lion looking through the notes in a desultory fashion, every now and then giving his little high-pitched grunt. When they got there, wherever it was, and were walking out of the station, Frances suddenly exclaimed: 'Oh, Gilbert, where are the notes? You've left them in the train—oh what shall we do!'

'What does it matter?' he said a little ungraciously. 'Don't make such a fuss, Frances.'

My mother was indignant, because if somebody had taken so much trouble on her behalf she would have gone through the motions of using the notes during the lecture, even if she had only pretended to read them. Those voluminous notes of Frances' might have put GK off. One has to remember that Tame Lion was sorely tried by the mountain of flesh he had to carry about with him. If he was occasionally irritable it was not a sign of anything unlovable in his character. I am sure he loved Frances devotedly; I could tell by the way he looked at her. Whether or no he needed the notes she used to make, he certainly ought never to have travelled without her. She was not in the least surprised when, one day, she got a telegram which said, 'Am at Market Harborough. Where ought I to be?'

He often used to write me little letters on scraps of paper. I found one the other day, a picture of himself looking very bewildered, with St Paul's School in the background, underneath which he had written: 'Twenty years late for school.' When he gave me books, he always wrote in the beginning and usually put a drawing as well. Once when I got home I found I had taken the book, *A Miscellany of Men*, before he had written in it. I took it back and we both forgot all about it. Months later he sent it to me with a splendid drawing of us both, he as a lion crouching beneath the feet of an enormously tall Viking with a winged helmet. I was then 16 and nearly as tall as he was. The picture was headed 'The Viking and the very Tame Lion', and this was the inscription:

> The lion before Daniel's feet
> Was meek as any spaniel,
> Because he'd kept for half a year,
> A borrowed Book of Daniel.

He used to recite verse in an inimitable way. I remember sitting with him and Frances in the Oriental Café on the sea front at Hastings: Frances, as usual, was quietly observant and appreciative, not missing t' slightest gesture of the huge pudgy hand, or a note of the high squeaky voice. I have no idea who wrote the verse, perhaps he did—perhaps he made it up on the spot—but King's Lynn had somehow come into the conversation and Mr Tame Lion turned to me and said, his stomach heaving with the long squeaky chuckle:

A friend of mine once composed a poem about King's Lynn. It went like this:

> The King's Lynn moon
> Has an extra touch of yellow;
> And a fellow
>   (*Noise of neigh*)
> Wants to BELLOW
> At the yellow that is extra
> In the King's Lynn moon.

Then the cord of his pince-nez caught in his teacup and there was a moment of confusion.

Our family were what is called 'High Church', and when Mr Tame Lion was received into the Roman Church it caused no flutter at all in our household. My mother's poet cousin, Gerard Manley Hopkins, had done the same thing and so had my mother's aunt, Maria Giberne, who had become a Catholic nun under the influence of Cardinal Newman, and several of our vicars had 'gone over to Rome' as we said. We looked upon Christ's Church Militant here on earth as one army and, by exact analogy, all the different sects in it merely as different regiments.

I wonder how many people there are still alive who have been present, as I was in my godfather's house, when those four giants, Wells, Shaw, Belloc, and Chesterton, were shouting, interrupting each other, arguing and laughing? A stray epigram or two survives, a beautifully turned witticism, or the expression of a passionately held conviction, but scarcely an echo can be recalled of the uproarious zest with which those four men talked, or the astonishing spontaneity with which glittering strings of words sprang out of their mouths. Ideas, splendid, original ideas—always ideas and more ideas: brand-new ideas and ideas so old that they had been forgotten. What did they talk about? All four of them wrote millions of words in books, pamphlets, and plays about their ideas. All I can do now is to recall a general sense of their effect upon me, as one might try to give an account of having listened to the finest orchestra in the world giving a staggering performance of some gigantic and sublime work of music. I still retain the flavour and impact of their four extremely different personalities: Shaw and Chesterton urbane and chuckling; Belloc and Wells combative and scornful.

Once, I remember, they argued about the desirability, or the reverse, of personal immortality. Chesterton remarked: 'HG suffers from the disadvantage that, when he dies, if he's right, he'll never know; he'll only know if he's wrong.'

Wells gave an exasperated exclamation, whereupon Belloc said: 'There is something sublimely futile about discussing the desirability, or undesirability, of the inevitable.' At which Shaw accused Belloc of habitually begging the question. Then he trounced Chesterton for consistent evasion of all points at issue in any argument on any subject, and, without letting the other three get a word in, he proceeded to hold forth on immortality, personal, impersonal, metaphorical, mythological, and so on, and so on. At last, after twenty minutes, he paused, and Mr Tame Lion observed to the ceiling: 'That, I suppose, is what is known as putting the whole thing in a nutshell.'

JOHN RAYMOND
Jeekaycee

G. K. CHESTERTON was the first grown-up writer that I ever discovered for myself. At the age of 12 I purchased, for two shillings, the selection of his stories, essays and poems that had just appeared in the Everyman Library. Then, as now, I had a fierce sense of property (more essential in a private school than anywhere in later life) and the inscription on the fly-leaf—'this book belongs to J. Raymond, 20th June, 1935'—marks the evening after prep when I first made informal acquaintance with Valentin ('head of the Paris police and the most famous investigator in the world'), Flambeau ('this colossus of crime') and the dumpy Essex priest with a face 'as round and dull as a Norfolk dumpling'. After which I tore through the book, only understanding one-fifth of it. For the next two years—until I crossed the floor and adopted Socialism, the Life-Force, the World State, and the Shape of Things to Come generally, and urged Arms for Spain on the junior debating society of my public school—I was Chesterton-sold completely. I read all Father Brown, could recite whole stanzas of the 'Ballad of the White Horse', and much of the other poetry and revelled in *The Man*

Who Was Thursday. Naturally, when I took up with Wells and Shaw, Chesterton lost a devoted reader. Though he never overwhelmed me in the same way a second time, I have since read all Chesterton's better-known work and a good deal of his lesser writings. And lately I have been re-exploring him in Mr D. B. Wyndham Lewis's new World's Classics anthology.

Though the selection itself is admirable—it aims 'to take as wide a sweep as possible, discarding for the most part pieces in prose and verse already over-anthologized and familiar to all'—Mr Wyndham Lewis's introduction is quite the reverse. Its rumbustious, pint-pot, claret-stoup, have-with-thee-to-Saffron-Walden! manner is just the tone calculated to put the contemporary common reader off Chesterton for life. There is a certain kind of professionally Catholic writer who is so steeped in Original Jeekayceeism that he cannot see the wood for the paradoxes. Mr Wyndham Lewis is obsessed by the GKC myth. He begins his essay on the accustomed Jeekaycee note:

> 'Am in Market Harborough, where ought I to be—Gilbert'—The anguished wire received by Mrs Chesterton is sufficiently celebrated, and sufficiently misquoted—a dozen towns usurp the primacy of Market Harborough in current British folklore . . . etc., etc.

Though I enjoyed this anthology immensely, I do not recommend it to any adult reader who is embarking upon Chesterton. Instead, I would recommend him to begin with the *Autobiography* (1936)—not just for the jokes, the stories, the brilliant cameos of James, Belloc, and the rest, but for the picture it gives us of an England viewed from the original perspective of the men who fought vainly against the Boer War, the Marconi Scandal and the sale of Campbell-Bannerman peerages, a body of journalists and writers whose campaign for the English political decencies has left a permanent mark on our democracy. They failed in each of their specific causes; they sacrificed their time, their money and their energy as artists in the various struggles they undertook, the while men like Sir Rufus ('Gehazi') Isaacs and Sir Alfred Mond were climbing step by step—or, rather, in leaps and bounds—up the rungs of the Georgian ladder. In many cases—in Belloc's particularly—the struggle soured and embittered their lives and, like all good Christian self-diagnosticians, they were too intelligent not to realize this. But in the long run they made their point and their reward has been posthumous.

Perhaps an even better introduction to Chesterton is Miss Maisie Ward's *Life* (1945)—one of those voluminous fighting biographies in which our literature is fortunately so well-stocked. Miss Ward does for

GKC what Mr Robert Speaight has just done so admirably for Belloc. (His book is the first exuberant indication of the strong tide now flowing in Belloc's favour, a tide that will eventually sweep him into harbour among the supreme masters of English prose, alongside Swift and Gibbon.) Like Mr Speaight, Miss Ward quotes copiously and to the point. Her book is quite un-Jeekayceean in tone: she gives us the essential Chesterton, the man behind the flagons and the tavern staves. She describes his solid middle-class background (Kensington with a dash of Scotch), his boyhood at St Paul's, and under the 'great grey water-tower' of Campden Hill, his spell at the Slade, his life as a journalist, his greatly blessed though tragically childless marriage. Chesterton is too often represented as a kind of TV Dr Johnson. (It is unfortunate that Mr Gilbert Harding publicly voiced his admiration of Chesterton, since these three wildly disparate figures are now firmly welded in the illiterate public imagination as a trinity of sane, truculent good humour and we only need to sandwich the late James Agate between Chesterton and Harding to complete the Bluff Old Sage succession.) The vast figure in the sombrero and billowing cloak, the cigars and the Beaujolais, the messengers waiting all day for copy, the taxis waiting all day in Fleet Street—all this has long ago been written to death. The great merit of Miss Ward's book is that it shows us the serious Chesterton—the man who waited fifteen years before becoming a convert, the writer who slaved for five years as a journalistic drudge in order to carry on his dead brother's newspaper and impose Cecil Chesterton's ideals upon the indifferent and sceptical 1920s.

In her book Miss Ward suggests that Chesterton was a great social prophet. I do not believe this to be true. Like all good publicists, he was a great simplifier. No one who reads the effective and moving poem about birth control which Mr Wyndham Lewis includes in his anthology, can doubt that he had never seriously begun to think about the subject. His whole cult of the poor or 'The Secret People' ('Our wrath come after Russia's, and our wrath be the worst') seems, even in the context of its time, a piece of splendid, but unhistorical rodomontade. And it is the same with his laments for the Middle Ages, the guildy-livery nonsense of *Chaucer* and *A Short History of England*. Yet he had not read his Cobbett, Carlyle and Morris for nothing. Take this passage from *G.K.'s Weekly*, written after the Wembley exhibition of 1924 when the shadow of unemployment was already stalking the country. It might well be addressed to the men of Suez:

... That was the real weakness of Wembley: that it so completely mistook the English temperament as to appeal to a stale mood. It appealed to a stale mood of success; when we need to appeal to a new and more noble mood of failure, or at least of peril. The English ... no longer care to be told of an Empire on which the sun never sets. Tell them the sun is setting and they will fight though the battle go against them to the going down of the sun: if they do not stay it, like Joshua. ...

This is the kind of peacetime fighting talk that this country needs if it is to survive at all. Chesterton may not have seen England steadily and whole, but his vision of it was as morally clear as Orwell's—if no more so.

His permanent literary value is far harder to assess. Shaw unwittingly put his finger on Chesterton's shortcomings as a writer in a letter written in 1923, discussing the prospects of the revived *New Witness*. Almost all that he says about the paper applies to Chesterton himself:

> Thus there is absolutely no public for your policy: and though there is a select one for yourself one and indivisible, it is largely composed of people to whom your oddly assorted antipathies and pseudo-racial feuds are uncongenial. Besides, on these fancies of yours you have by this time said all you have to say so many thousand times over, that your most faithful admirers finally (and always suddenly) discover that they are fed up with the *New Witness* and cannot go on with it. This last danger becomes greater as you become older, because when we are young we can tell ourselves a new story every night between our prayers and our sleep; but later on we find ourselves repeating the same story with intensifications and improvements night after night until we are tired of it ...

Those paradoxes—repeated and repeated, intensified and improved, in essay after essay—how they charmed us when we were young, and how infuriating we find them today! Belloc's strength lay in his versatility; Chesterton by contrast seems a distressingly homogeneous writer. Yet, he, too, can ring his changes, as in the unusual and little-known poem *To the Jesuits*:

> Flower-wreathed with all unfading calumnies
> Scarlet and splendid with eternal slander
> How should you hope where'er the world may wander,
> To lose the long laudation of its lies?

This has the counter-attacking *panache* of Claudel (one of Chesterton's greatest admirers); it echoes his 'Ces vastes piliers de Saragosse, ces bruyantes racines de Manrèse.'

Certainly, Chesterton's verse will be read as long as Kipling's—and that is saying a good deal. Men will continue to shout 'Lepanto' as Buchan told the poet they shouted it in the trenches in 1915. 'Chuck it,

Smith!' and that wonderful Bronx cheer for Stratford-upon-Avon ('Lord Lilac thought it rather rotten, that Shakespeare should be quite forgotten') have already passed into the tag lines of English literature. Alfred will be making his last stand at Ethandune long after the sun has gone down on Notting Hill and the Flying Inn. Adolescents will continue to take fire at *Orthodoxy* and their elders be stirred to contemplation by *The Everlasting Man* (here, much more than in Belloc's *Objections*, is the real answer to Wells's *Outline of World History*). Personally, I think that his literary criticism—*Robert Browning, Robert Louis Stevenson*, and *The Victorian Age in Literature* (my own favourite of all Chesterton's books)— is certain to survive. At his best, he was a superb *jongleur* of ideas and *St Thomas Aquinas* is his greatest single intellectual achievement.

I sometimes think that Chesterton will be remembered chiefly through his character, as one of the few genuinely good human beings in the annals of literature. A lesser writer than Shaw, or Belloc or Wells, he had a larger heart than any of them—and all three men cared more for him than they did for one another. As Wells remarked in exasperation, he was an impossible man to quarrel with. Like all journalists, like Belloc himself, he wrote thousands and thousands of words, far, far too many. Belloc wrote dozens of bad books, but you have to look hard to find a bad sentence in any of them; similarly, in Chesterton you have to look hard to find a bad sentiment, a hint of malice or meanness or ungenerosity, and this, in a man who spent most of his adult life as a controversialist, is remarkable. For Belloc, the Faith was the supreme *donnée*, the composition of time and place into which a man was born; for Shaw it was a tragicomic fantasy, a sidechapel in the vast temple of Evolution; for Wells it was an evil, deluding rigmarole, a huge piece of mumbo-jumbo invented by celibate sadists to subdue the free Cockney spirit of man; for Chesterton it was a quest, the great adventure. Having taken up these positions, all four grasped their rackets firmly and set to work, 'united in the strife that divided them'. What a grand, sloshing men's doubles they played for a quarter-century on the centre court of English ideas! Today, alas, the centre court is empty and all that is left of that unresolved tournament are the ball-boys of the TV panels lobbing at one another in a vain pretence that they are still playing the great game.

WILFRED SHEED
On Chesterton

THE turn of the century was a happy time for the light essay. Life was just serious enough, without being too serious. People of the reading persuasion had the time and patience to listen to abstract discussion on every level of importance; and the popular essayist, the man who put things well, was expected to write with equal vivacity about chasing his hat, votes for women, and whither are we drifting. Naturally, he was expected to deal with each of these in the same well-known and beloved style.

After two wars, etc., much of this light-hearted discussion appears pretty thin and beside the point. Nowadays we like things to be either very funny or very serious, and that special blend of the two moods seems as a consequence curiously outdated. The humorist sticks firmly to his business (monotonously stamping his products 'joke', 'joke', 'joke'), and leaving the heavy stuff to the expert and the prophet. We are much quicker than our immediate fathers were to question an essayist's qualifications: and God help the contemporary light author who strays without the appropriate degrees into, say, economics, as Mr Belloc did, or into history, *à la* Mr Wells. Intelligence is no longer enough.

In short, the atmosphere has grown intolerably cramped and tense for essay-writing and essay-reading. Most of the essayists of the spacious days are either forgotten completely, or at best embalmed in bedside readers—and thoroughly embalmed they look. Those who are too good to be ignored remain to perplex us: and strenuous efforts are made from time to time to prove that Bernard Shaw was wholly serious (in the modern manner) and that G. K. Chesterton was entirely frivolous, a specialist in humour.

Chesterton lived long enough to see the Age of Specialization preparing its epitaphs for the carefree Edwardians, and there is perhaps just a suggestion in his later work that he had reached the decision that, if one had to make a choice, he would rather be taken seriously than not. His last essays seem therefore a little more intense and didactic than his early ones; and although they establish his sincerity beyond a reasonable doubt, they are not so excellent purely as essays. The growing urgency of his convictions, together with a certain amount of literary weariness (very honestly earned), often makes for unevenness and strain.

The newspapers and magazines for which Chesterton wrote regularly allowed him a flexibility in subject-matter and mood, which is rarely allowed to modern journalists (especially when they are still in their middle-twenties). Chesterton took abundant advantage of this licence, and in the middle of his wildest flights, he was always likely to become disconcertingly serious—even in fields where he was not an expert. Sometimes he actually switches without warning from one attitude to the other and back, in the course of a single sentence. To the modern reader, this freakish dexterity poses something of a problem. Even in his own day, Chesterton sometimes seemed a little too clever to be true. Nowadays it is tantalizingly difficult to assess his true value.

One value is obvious immediately. Like Bernard Shaw and very few others, he had a genius simply for having original ideas, even as a few people have had a gift for composing original tunes. It is hardly possible to read a page of Chesterton without finding an unexpected idea, at best wise, at worst fiendishly ingenious. On some pages the ideas come in a thick and tangled profusion, a formidable verbal jungle. Of course, the ideas vary considerably in significance and validity. Chesterton was a chronic debater from childhood, and a few of his arguments look suspiciously like debating-points. I don't think that even the warmest Shavians will deny that their hero was sometimes guilty of much the same thing. In each case, the brittle parts can be pulled off from the main body of the man's philosophy, without disfiguring it. And, left where they are, they still have considerable entertainment value.

Along with this rare fertility of mind went an almost unbelievable range of self-expression. As a boy, Chesterton had been a promising enough artist to be sent to the Slade. And what he couldn't say with paints and crayons he poured into notebooks full of poetry and prose. Later, the emphasis of his life shifted, and while he kept up his sketching mildly, he wrote torrents of essays, stories, and verse, light and not so light. He had, if anything, too much to say and too many ways to say it, so that he wrote quickly and carelessly at times, with a terrible creative impatience. The bursts of pure inspiration which visit other artists only occasionally seemed to come to him again and again, so that at times he seems to treat his muse almost cavalierly. When he complains in his essays about the aesthetes and the 'art-for-art's sakeists', it is not their posturing so much as their sterility that seems to have got him down. The artistic temperament which labours endlessly over its mice was incomprehensible to him. And because he found artistic creation so easy and amusing himself, he was uncharacteristically insensitive to those who got nothing out of it but a 'green pain', or the equivalent. (Of course they have since had their revenge by rejecting *him*.)

The pressures of creation (and overwork) kept him from reaching complete perfection in most literary forms. His novels and stories depend more on repeated injections of ingenuity and imagery, than on the steadier pleasures of form and style. In his essays, there is plenty of evidence that he could write excellent prose when he had the time and inclination: but so often we get the unmistakable signs of hurry—'nay indeed', 'in fact I would say', 'mere' this, 'jolly' that, and the other stock phrases. His writing is never ugly, because he had an ear for words: but much of it has an unfinished, first-draft quality.

His artistry was instinctive, a capacity, among other things, for not having to take pains. His light essays and verse have an almost unconscious elegance and sureness of touch. In fact it is in these casual forms that his genius seems to be most at home. Wherever instinct could replace the work of time and discipline, Chesterton was a superb artist. His more abstract thinking indicates an almost mystical absorption in shapes and designs; and it is not entirely surprising to find that even after his many severe denunciations of art for its own sake, he should have wound up by proclaiming that art and not reason is man's real signature on nature—art, that is, for the sake of anything but itself. His essays on literary topics illustrate again and again his sensitiveness to form—as well as his preference for matter. But perhaps, all the same, an inconsistency can be argued; for in his best humorous work (see 'The Tomato') he comes close to pure self-justifying art.

The trouble with a man who possesses such rare individual gifts as Chesterton did is that he is likely to take an eccentric, personal view of the universe, which is not of much use to anybody else. And some critics, while admitting Chesterton's value as an intellectual gymnast, have claimed that he had this limitation in a crippling degree. It is a question that must ultimately be settled according to taste (one man's blinding common sense is another man's raving nonsense) but I think that at least one aspect of it can be examined more clearly if we compare Chesterton with his three great contemporaries, Wells, Shaw, and Belloc.

There has always been a tendency to lump these four men together, simply because they were all writers who kept hammering away (entertainingly) at large views of human society. I think the tendency has done less than justice to all four of them; but my concern here is with a possible injustice to Chesterton.

Shaw, Wells, and Belloc were each to some extent eccentrics, or, to use the current vogue word, 'outsiders'. Shaw in particular gained much power from the impression he gave of being outside everything. Not content with shedding family, nationality, religion, and other such large realities, he even went so far at times as to shed species itself. His main dramatic joke was the clash between the disembodied, 'angelic' intelligence (Caesar, Undershaft, the Superman) and the rest of humanity. The fairly obvious point of it all was that people without human passions, loyalties, and appetites could undoubtedly handle the world's problems with laughable ease: a restatement of the view held by Aristotle (and your grandfather) that human nature is our chief problem.

I realize that this hardly does justice to Bernard Shaw's full position. But I don't think anybody can deny a measure of fundamental eccentricity in a vegetarian teetotaller who was fussy about his underwear and who left all his money for the propagation of a phonetic alphabet. These are not the instincts of the average man, as Shaw was well aware.

Belloc also illustrates the literary uses of perversity. Casual readers may reasonably wonder whether Belloc was prouder of being a Sussex farmer from Balliol or of being a French artilleryman from the Pyrenees. That great man also had an interestingly ambivalent attitude towards the very rich, whom he despised whole-heartedly in print, but, I believe, only spasmodically in private.

The case of Wells is not quite so clear-cut—nor so relevant to Chesterton, who was so palpably different. For myself, I feel a sense of a man uprooted, if only from his past and his class: Bernard Shaw's new man, Henry Straker, the boy from the Polytechnic. It is hard to account otherwise for a man so consistently in tune with his own times.

I am not mentioning these eccentricities simply to deride them. The frictions and tensions which arose from them were sources of tremendous artistic strength, and without them, these men might never have written at all. Bernard Shaw in particular was admirably candid about the foibles of his strange, lonely character. He admits that as a young man he once waited half an hour outside the door of some fashionable people, trying to get up the nerve to go in. Later on, he got his own back by tweaking the noses of the same sort of people, and creating a string of dramatic heroes to do so for him. His eccentricity was converted into superb and valuable entertainment for the rest of us. And whatever the psychological basis of the proceeding, the noses undoubtedly were the better for tweaking.

But valuable as these personal eccentricities may have been as triggering mechanisms, they must have offered serious disadvantages whenever their owners set out to tell other people how to live. There is no space here to discuss how much the political and social fancies of these men were dictated (as so many of our own are, of course) by whim and prejudice. My point is that Chesterton, although a man of strong inherent tastes and opinions, was not driven by personal discontent, goads in the flesh or splits in the personality. He was, almost uniquely among creative writers, adjusted to his surroundings and to himself. He loved his country with an unsentimental passion, and he saw nothing wrong with his family or his class. People who knew him well say that he lacked even the normal minimum of snobbery. He hated what he thought the rich were doing to his country, and he admired the poor for putting up with it so patiently. But it is inconceivable that he would ever have waited, racked in doubt, outside any man's door. He was, in his own word, 'centric', socially and psychologically; and for this reason, his view of the *cosmos*, right or wrong, was never bitter or cantankerous.

This is worth bearing in mind when we read his blistering criticisms of his own times. Many satirists have had a kind of built-in itch which would almost certainly have made them irritable in any circumstances. Chesterton on the other hand had to draw his inspiration from a great fundamental store of contentment, more suitable for pastoral ramblings or nature notes. The result is that his anger seems more muscular than nervous; and his reasons for it are generally calm and deep. He does not just jump suddenly and swear, as many of our more biting satirists have done.

The massive Chestertonian self-control has unfortunately made him seem less sincere rather than more so. The total absence of bile and malice enabled him to heap rounds of good-natured abuse on people's ideas, without, apparently, making any enemies at all. (Even Lord Cadbury and the cocoa-magnates never seem to have been stung into retort by his inspired badgering.) At the same time, the absence of any trace of vindictiveness added to the ever-growing notion (which we have inherited today) that he could not really mean what he was saying. His towering irony was forever breaking down into pure humour; and he had a further disarming habit of making sure that a disproportionate share of the mockery was reserved for himself. In art, as in life, he made the tactical error (as far as reputation is concerned) of taking himself much too lightly.

Nowadays, his unruffled serenity is a great source of comfort to some, but it seems to get on the nerves of others, who consider it a trifle smug and unrealistic. There is (after our two wars, etc.) a strong inclination to feel that anybody who enjoyed life as unflaggingly as that must have been fundamentally childish; and the tag 'Peter Pan' has been used more than once. According to this view, Chesterton is much too innocent to have anything relevant to say about our own naughty problems. His apparent psychological adjustment is thus put down to a quaint detachment from the unpleasant side of life.

Some of Chesterton's warmest admirers must take a share of the blame for this. They have tended, I think, to overestimate his innocence and simplicity. For, while Chesterton undoubtedly had those qualities to a marked extent, he did not come by them without something of an inner struggle. His humorous sketches show a surprising sombreness of character, a starkness and almost a grotesqueness of line, which is curiously reminiscent (to me, at least) of Goya's grisly cartoons. As an adolescent, he toyed with black magic, and while we cannot tell for sure how far he went with it, we know from his own statements that he went far enough to be horrified. Throughout his life, he was powerfully aware of the fact of evil, and when he talks of man's primary duty of gratitude to God for 'things as they are', there is always a suggestion that he has glimpsed other possibilities. Once, in reference to those who don't believe in the existence of positive evil, he says mournfully, 'I can only say that they must be much better people than I.'

In fact, the vividness and wildness of his inner-vision were of the kind that have been more often associated with writers like Blake, Poe, or Rimbaud than with the famous humorists; and it was probably only by a supreme exertion both of the will and the intellect that Chesterton was able to escape morbidity by so wide a margin, that he was actually accused by many of being too incorrigibly cheerful. His awareness and pictorial sensitivity were of the kind that could disintegrate a man or at least reduce him to habitual melancholy. This indeed may help to explain Chesterton's profound understanding of Dickens, another writer who had gone far into the darkest corners of the imagination and come out cheerful and sane. Chesterton always made a point of defending Dickens against charges of 'mere geniality'; now perhaps he needs a similar defence himself.

To a man who sets out to improve his times, a strong sense of evil can, of course, be as much of a handicap as social maladjustment. He may be inclined to see the power of evil in everything he does not like (like one very intelligent person I know who divines the presence of evil in the architecture of the United Nations building). But once again, we find a calmness in Chesterton, a superb absence of the jitters, which forestalls this difficulty. Having looked steadily upon the face of evil itself, he was not inclined to imagine its presence in muddling politicians or corrupt institutions. He was competely without superstition (even of the most scientific kind); the sorrows of his own life (which he experienced with cruel intensity) he never ascribed for a moment to anything but the ordinary rhythms of life, which he shared with the rest of humanity. He had, in his writing and his life, a rare blend of hardheadedness and compassion; which qualities save him time and again from what he would have called mere brilliance.

The point hardly requires further labouring. The essays probably provide enough additional evidence that if Chesterton did make occasional mistakes of judgement, it was not so much through excessive and erratic fancifulness as through haste and high spirits. In view of the fact that he was banging out essays at a furious pace about practically every subject under the sun, it is a tribute to his indomitable sanity that he so seldom talked nonsense. Even where his knowledge of a subject was obviously limited, he usually managed to produce some penetrating and relevant ideas about it.

Chesterton had neither the time nor the temperament for the laborious gathering of specialized information. This at least is a valid criticism, and nowadays it is rated an extremely serious one. He liked to preside over ideas 'wrestling naked', but hardly over facts wrestling naked. Even in matters concerning English literature—in which he was genuinely expert—he was capable of absent-minded misquotation. In matters of history and economics, he leaned very heavily on Hilaire Belloc. And because Belloc also possessed tremendous artistic gifts, as well as a quite formidable air of conviction, he gave Chesterton a good deal more than the plain facts. He managed in the course of time to infect his friend with many of his own assorted enthusiasms, which included Napoleon, the Middle Ages, and practically everything to do with France. As usual Chesterton found fresh and interesting reasons for admiring all these things; but I have never felt convinced that he would have admired them quite so much if he had studied his history on his own. His historical tastes are sometimes surprisingly out of character.

This is the only instance I know of in which Chesterton ever yielded an inch of his massive independence of mind; and I think it can largely be put down to the quasi-hypnotic powers of Mr Belloc which were, so I am told, next thing to irresistible. And if Belloc did lure him into certain uncharacteristic attitudes concerning things, he never altered his basic stance in relation to ideas. In matters of pure intellectual judgement, Chesterton had absolute integrity and permanence.

This is a distinction which many of his critics have failed to recognize. For instance, they accuse him of glamorizing the Middle Ages; and perhaps he did. Possibly he was over-sanguine about the working hours and living conditions of the average medieval peasant. But of the 'idea' which he saw behind the Middle Ages, and which he presented to the reader with all clarity and fairness, there was no such vagueness. Perhaps the idea as he perceived it was never properly realized in the Middle Ages: look at it anyway, Chesterton says in effect, think of it as abstract idea, and as a possible concept of human society. Similarly in economics: Chesterton was perhaps the last man to help us all to make the actual move to the country which his Distributist philosophy dictated. He may have had fairly imprecise notions as to how the factory and commercial systems could be effectively dismantled first. But he had a most penetrating understanding of what was good for human beings, and of how a whole society could grow sick and die. And if his economic facts were not quite equal to Shaw's, his ignorance was neither wilful nor half as great as it has been painted.

Ideally, Chesterton's thinking should have been implemented by numerous trained economists and sociologists. But in this respect he has been poorly served by his admirers, many of whom (like myself) know far fewer facts than he did, and who prefer, on the whole, to take a large view of things. So there he must remain: still by far the best exponent of his own cause. Has the cause got any value for us today?

Perhaps there is a tendency now to underrate the type of theorist who comes to us out of the past without a blueprint in his hand. (If he *has* a blueprint, we can patronize his floundering attempts to be practical.) We seem to be so far removed from first principles and causes that nothing can be served by discussing them further. Everybody knows that we *have* to have two-party democracy, that we *have* to have socialization, that we *have* to have compulsory education, and so on. There is no turning back from anything, the treadmill only turns in one direction, and there is really nothing to talk about any more except how quickly we can do what we have to do. It is hard to believe that we still have anything at all to learn from the examination of naked ideas.

In Chesterton's time, this feeling of inevitability was not yet so compelling, and it was still possible for a man to think freely without escaping into fantasy. Chesterton used this freedom to magnificent effect, and he practised a scepticism of breath-taking scope, which even encompassed the other sceptics, and scepticism itself. He took absolutely nothing for granted, and many of his apparent paradoxes are simply the results of a restless, sceptical tinkering with our least-questioned beliefs. For instance, take the saying: 'You can't turn back the clock.' Well, can't I? says Chesterton—and we think at once that he is asking a contrived and trivial question. But follow him a bit farther. Does the saying mean that society can only move forward to more and more new fashions, rigidly predestined, or can we in fact make an effort of choice, as easy as putting back the clock really is, to revive certain good things from the past? If we are going to live by catch-phrases and slogans, Chesterton insists on examining them seriously, first on the verbal level (what are we actually saying?) and then on an intellectual level (what kind of thinking do the slogans represent? And what do we really think we mean by them?). This sort of examination can be called paradoxical only if we assume that our commonplaces are beyond serious questioning.

Earlier, I have suggested some of the disadvantages that beset a man of artistic sensibilities when he undertakes intellectual work. But consider one tremendous advantage which follows if the artist's intellect is sufficiently powerful to keep order in its own house. After Chesterton has examined an idea verbally with a lawyer's thoroughness, a door seems to open into a new, wider place where we can see the idea in pictures, as an artist might see it, as well as in words. Take for example the essay 'Oxford from Without'. Chesterton is discussing the suggestion that the Oxford of his day might be a place where rich boys could have their edges smoothed by contact with boys from other milieus. Ah yes, says Chesterton, by contact with navvies, coalheavers, shipping clerks—and he need say no more. A picture of something ludicrously unlike Edwardian Oxford is already sketched in our minds. He goes on to say that, while Oxford is undoubtedly a playground for the governing classes, it is just as well that the governing classes should have such a playground, an Arcadian spot in which to romp good-naturedly, under the amused eye of the general public. This is excellent pictorial writing (and remember that Chesterton was a painter of some skill). He is asking his persistent question, 'What does this idea *look* like?'

In another essay, Chesterton discusses the case of a destitute mother who has been sent to gaol for not keeping her children clean, although she has unquestionably kept them healthy and cheerful. Chesterton quickly paints a picture of his disapproval: a courtroom in ancient China, silent under a hot sun, the emperor bored and cruel: how shall we tease the prisoners today? This, he says, is the proper setting for that kind of justice. It is a shocking method, and one which must sometimes miss the mark, but it has unique value as a way of showing people what is being done to them. Chesterton's central position, as a genuine man of the people (a man who had the nerves to enjoy Bank-holiday crowds and music-halls) and as a devout Englishman, gave additional impact to this method. If the nation's rulers showed any tendency to alien tyranny or mystagoguery, there was Chesterton with his verbal cartoons, massively representing the people of England, 'who have not spoken yet'. It may have been sleight of hand, a kind of sham ventriloquism, but they really seemed to speak through Chesterton, as if he had a cross-section of them wrapt up in his vast frame. As Ezra Pound said despairingly, 'Chesterton *is* the mob': certainly the music-hall mob, and perhaps the mob of the cinema too.

Are the Chestertonian methods of examination still relevant today? Some of the problems which he discusses in his startling manner are still with us, and his opinions about them have a refreshing vitality, as coming from someone who was looking at them with fresh, open eyes, and not with our own sleepy, leave-it-to-the-experts resignation. But it is in respect to pure method, quite detached from conclusions, that Chesterton makes the greatest demands on our own vitality. Can we still look at the world of things completely spontaneously, and with intense concentration, as for the first time, or does that already mean too much effort for too little reward? He talked of a story he meant some day to write, about an intrepid explorer who discovered Brighton by accident, and who thus saw that drearily familiar spot with the wild, curious gaze of a stranger. It takes a supreme effort to discover our own country over and over again in this way, with appropriate alertness and wonder, and to decide each time what we want to do with our discovery (among other things a strong belief that it is not too *late* to do something with it). And this perhaps is the most significant challenge that Chesterton throws at us. To meet it calls for a type of reflection which can easily be dismissed as meaningless, not to say time-consuming.

Take another quick look at this method. Suppose Chesterton touches on the subject of eating. First he sees a man filling himself like a furnace through a small hole in his head. He sees other men urging this fuel patiently out of the ground, animals being deprived of life to supply it; but beyond that he sees the philosophical value of man, and why he is worth feeding at all; why it is a crime to deprive him of food, or to feed him against his will. Before long, we are at the hub of his philosophy, discussing larger issues of law and psychology, and the purpose of life.

In a sense the method is Socratic, except that Chesterton is still talking about eating at the end of the essay and not about something more abstract. In his mind, ideas take and keep firm shape (remember his fascination with shapes); the word has been made flesh for the sake of human understanding. He was hardly a materialist, but he had no fastidious desire to escape from matter either. In fact he belonged even more closely to this earth than most of the materialists with whom he did battle. He accepted both spirit and flesh, and all the facts of human life, with love and courage, and with open eyes. Only thus, he seemed to say, can one spare oneself a thousand small slaveries of the mind and body, and perhaps a few major ones as well. (Incidentally, and apropos of this, the Chestertonian type of total scrutiny might in fact prove the safest answer to 'double-think', the rewriting of history, and the other horrors of an Orwellian future.)

Behind this intense activity of the brain and heart is the massive Chestertonian assumption that life is worth all this trouble: that in gratitude for the gift of living, no price is too high to pay in love and understanding. And here perhaps the lines can be drawn. This assumption on this scale was either great wisdom or great folly, a blazing vision or a tipsy hallucination. Whatever it was, it was spectacular: one of the loudest, truest voices for sanity, or absurdity, in the whole of literature. If his intuitive appraisal of life was right, then his work must surely stand as a great masterpiece of human wisdom; if it was wrong, then he is at least a giant among clowns, dancing wildly and beautifully in the empty moonlight. In either case, the effect is magnificent and unforgettable; and quite certainly unique in literature.

R. G. G. PRICE
A Check-up on Chesterton's Detective

WHEN Monsignor Ronald Knox satirized the Higher Criticism by applying the methods used on the Bible and Greek and Latin texts to the Sherlock Holmes stories, he started something that has begun to grow dull. The game has got too elaborate. Questions like the date of *The Norwood Builder* have been upgraded from Sunday morning crossword status to postgraduate complexity. There is no place for the novice in the game, for the man who would never survive the first round of the qualifying examination for the entrance tests to the Baker Street Irregulars. The only thing for the young student is to begin somewhere else, somewhere that has fewer fixed facts and hardly any dates. Here are some notes on G. K. Chesterton's Father Brown stories:

A FEW CRUMBS OF PERSONAL INFORMATION

Father Brown's initial was J. (*The Eye of Apollo*). He was born in Essex (*The Vampire of the Village*). His sister had married into a race of refined but impoverished squires and had a daughter Elizabeth Fane (*The Worst Crime in the World*). His eyes were large and grey. He had stubbly brown hair and a round face (*The Hammer of God* and elsewhere). Repeated comparisons with turnips prove he was brachycephalic. He carried a stumpy umbrella with a knob, not a crook (*The Fairy Tale of Father Brown*). He smoked a big pipe, which once dropped out of his mouth and broke in three pieces (*The Honour of Israel Gow*). He took red pepper, lemons and brown bread-and-butter with his whitebait (*The Duel of Dr Hirsch*). He was not unsociable: Frank Harrogate, the great financier's son, vaguely remembered seeing him at the social crushes of some of his Catholic friends (*The Paradise of Thieves*). No doubt it was at these that he learned that Lord Amber, who had gone into wild society in a sort of chivalry, was now paying blackmail to the lowest vultures in London (*The Flying Stars*). He had seen wicked things in a Turkish carpet (*The Wrong Shape*).

SOME NOTES ON FATHER BROWN'S EDUCATIONAL BACKGROUND

The doctor at Mandeville College asked him curiously, 'But you haven't anything particular to do with Oxford?' to which he replied, 'I have to do with England' (*The Crime of the Communist*). There is no other reference to formal instruction or the lack of it. He had read Keats (*The Pursuit of Mr Blue*) and he knew enough about the Encyclical *Rerum Navarum* to lecture on it (*The Oracle of the Dog*). He had once been fairly good at thinking and could paraphrase any page in Aquinas (*The Secret Garden*). He was familiar enough with the *salvo managio suo* and the *servi regis* to risk hinting that his companions were not (*The Curse of the Golden Cross*). He talked natural history with much unexpected information (*The Perishing of the Pendragons*). He knew the Donkey's Whistle, a thing so foul that Flambeau, though an ex-thief, had never heard of it (*The Blue Cross*). He knew how a wax nose spots in wet weather (*The Head of Caesar*). In the middle of the night in a Mexican road-house of rather loose repute, he sat in the hall and read *Economics of Usury* (*The Scandal of Father Brown*).

A CHECK-LIST OF FATHER BROWN'S ECCLESIASTICAL APPOINTMENTS

He was a curate at Hartlepool (*The Blue Cross*). There were a number of unnamed posts, including one in or very near Belgravia (*The Queer Feet*). Cobhole, Essex, is the most frequently referred to, though he seems to have kept more restrained there and it was not the site of any of his adventures. His work tended to lie in the north rather than the south of places—North Scarborough (*The Absence of Mr Glass*). North of South America (*The Resurrection of Father Brown*), North London, St Mungo's (*The Wrong Shape*). It is probable but not quite certain that the chapel the blacksmith's wife attended in *The Hammer of God* was in Bohun Beacon itself. He once went south to the extent of being attached to St Francis Xavier's Church, Camberwell (*The Eye of Apollo*). St Dominic's, in *The Mirror of the Magistrate*, might have been a North London church. It was near enough in for a judge to go home after a legal dinner and he had several distinguished neighbours.

Where was the provincial church often referred to? It seems unlikely to have been Scarborough, first because, illogically, 'provincial' always sounds inland, and second because it was near the hamlet in *The Song of the Flying Fish*, which was close enough to London to make it worth while to take the six-forty-five in the evening and return the next morning. Unless he were on the move to an unprecedented extent, the number of Father Brown's parishes must be limited, and it is reasonable to assume that this provincial church was situated in the place where his rooms were next to the huge block of flats (*The Point of a Pin*). It was probably a small town as it had fringes and a forest and park beyond it. The period was that of the building boom after the First World War—Sir Hubert Sand had served in France. This would fit in with the fact that the hamlet in *The Song of the Flying Fish* was obviously on the point of expanding too: though there were under a dozen inhabitants, there was a bank with a resident manager.

THE DATE OF *THE MISTAKE OF THE MACHINE*

The action took place twenty years before the time Father Brown recounted it to his friend Flambeau in the Temple Gardens. At the time he had been chaplain in a Chicago prison. Now, *The Arrow of Heaven* took place immediately on Father Brown's first arrival in America and therefore before he took the chaplaincy. During this episode he had reason to suspect that his young friend Wain had been engaged, 'with only too conspicuous success', in evading the last

Amendment to the Constitution. This cannot be the Volstead Act, as Chesterton, the recorder of these events, died in 1936, and twenty years, even if the Chicago case had been recorded at the very end of his work, would bring it back to 1916. To bring it back much further would over-run the Amendments of 1913, which concerned Federal Income Tax and the method of election to the Senate, and there is no stopping place before the much less probable Amendment of 1870 about Negro Voting. If the first arrival was 1913, the chaplaincy could have been held at the very end of the year or the beginning of 1914 and just give time for Father Brown to tell Flambeau about it, but it is a tight fit.

FATHER BROWN AND THE POLICE

Relations were easier than between Sherlock Holmes and Lestrade. For example, a private detective who approached them about the probable murder of his client was referred to Father Brown, whom they described as 'an able amateur' (*The Pursuit of Mr Blue*). Perhaps they relied on using him because they were themselves going through a bad patch: this is shown by *The Secret of Flambeau*—they were still searching for him after he had retired, not only from being a thief but from being a semi-official crime investigator, with offices by the door of Westminster Abbey (*The Eye of Apollo*).

Their goodwill was demonstrated not merely in passing cases on to him but by ready co-operation with his suggestions in cases of their own. In the episode of *The Quick One* he expressed a desire for a certain Scotsman to be found. The police and, for some reason, the postal services worked all night. Traffic was stopped and correspondence intercepted. Finally, the rattle and rumble of heavy vehicles was heard outside the bar in which the police had their headquarters. The Scotsman was outside and it had taken five men to handcuff him. Yet, though Father Brown said he was not a murderer but a witness, the police continued to fawn on him. They even put up with his habit of waiting to produce a solution until the trial of the wrong man was half-way through. In *The Man in the Passage* Father Brown broke the case against the prisoner wide open, although he had been called for the prosecution. He may, perhaps, have been piqued at being treated as a hostile witness. In *The Mirror of the Magistrate* it was only during a break in the proceedings that he mentioned to the police he was going to base the defence on the guilt of the prosecuting counsel.

The advantage that Father Brown had over the police was that he found it much easier to get to the scene of a crime than they did. Their lethargy is amply documented. When a woman fell down the lift-shaft of the block of offices in which Flambeau worked there was time for some minor crime with a will, preliminary investigations by Flambeau and a good-deal of theological improvization by Father Brown. Although a small mob had gathered in the street, it was possible to be leisurely because, as Father Brown said, it would be half an hour before the police arrived, and this although the disturbance must have been visible across Parliament Square (*The Eye of Apollo*). When a squire disappeared from a small hamlet, causing increasing alarm during the day, his ward's fiancé wired to the police in the evening; the next morning they wired back that they were sending a man down (*The Vanishing of Vaudrey*).

One of the few occasions, if not the only one, when they arrived before Father Brown was the episode of *The Three Tools of Death*; but this was at a house beside a railway line in Hampstead. Summoned by a message sent by a passing train, the police hung about until Father Brown had been fetched to solve the case. Their only attempt at any kind of action was to send off in pursuit of the wrong man, who turned out to be making for the police station. They chartered a special train to bring him back. Father Brown worked quickly, gave them what they wanted and made off. As he left, 'an acquaintance from Highgate' stopped him to say that the coroner had just arrived; but he refused to wait for the inquiry. That could safely be left to the police.

Probably the clearest evidence of the police's lack of mobility is to be found in *The Insoluble Problem*. There was an inn, 'The Green Dragon', forty-five miles away on the Casterbury road and the cathedral town of Casterbury itself was a sixty-mile motor drive from Father Brown. It seems to follow that the inn was at the most fifteen miles from the cathedral. When Father Brown and Flambeau drove up to the inn they found a man sitting amid some lumber that consisted mainly of seventeenth-century pamphlets. Remarking 'Curious circumstances, I may say, curious circumstances indeed', he led them into the garden, where his sister-in-law's grandfather was not only hanging from a tree but had a sword stuck through him. In reply to questions, the man said the old gentleman had been there about half an hour and that he had rung up the police. He added, 'But they can hardly be here for several hours. This road-house stands so very remote.'

Carry on from here, if you want to; but don't write to me with your findings: I've had my fun.

BERNARD LEVIN
Pantomime Horse

THAT strange, galumphing beast, the Chesterbelloc, is neighing again. Rereading them has been a weird experience; every night since I started I have dreamed that I was back at school. For I, like many others who have saddled this curious animal, discovered it in adolescence, and I can still feel the intellectual shock that the discovery occasioned. Here were writers many of whose ideas I found preposterous or loathsome; yet for the first time in such circumstances I found myself enthralled, stirred and delighted by their works. And in this schizoid state I believe I was very far from unique, for the Chesterbelloc spoke in a voice that was immediately intelligible to a schoolboy; it spoke of rebellion, and non-conformity, and romanticism, and above all (particularly the Chesterlegs) it spoke in paradox, picturesque and exciting. Even Shaw was temporarily forgotten as I drank up *The Man Who Was Thursday* and *The Napoleon of Notting Hill, Mr Petre* and *The Path to Rome*.

Yes, but I am no longer a schoolboy, and the Chesterbelloc does not prance and sport in the imagination any longer. Thirteen post-war years afterwards it appears as a pantomime horse rather than a dashing white charger, a pantomime horse, moreover, that has come apart in the middle. And the King Charles's Head (to weight the metaphor still further) that it wore at both ends—booze behind and the Jews before—can no longer be overlooked so easily; too many people have died of anti-Semitism, and not a few of drink.

But the first thing one notices is that the animal has been classified the wrong way round by the literary zoologists; on closer examination it turns out to be a Bellocerton. The fact is, Belloc was a very much better writer, thinker and poet than Chesterton, a fact obscured by the greater noise made by Chesterton and Belloc's own willingness to play the back legs. (The 'Lines to a Don' are the work of a disciple, not of an equal—still less a superior.) Yet the two are still thought of as one and interchangeable, with the result that they have been written up and written down together—the very coinage 'Chesterbelloc' is witness of that.

And if we could separate the two, and evaluate each on his own, what should we find? To begin with we would find that the purpleness of Chesterton's prose is wearisome today, that the thread has worn threadbare. Before I reread the Father Brown stories, I thought I remembered the writing as being admirably evocative, with just a hint of the exotic in the priest's summings-up. How wrong I was!

Is there no connexion between the idea of a winged weapon and the mystery by which Philip was struck dead on his own lawn without the lightest touch of any footprint having disturbed the dust or grass? Is there no connexion between the plumed poignard flying like a feather arrow and the figure which hung on the far top of the toppling chimney, clad in a cloak for pinions?

The alliterations are too obvious, the dashing words like 'poignard' and 'pinions' too out of place to be anything but in the way. If this sort of stuff were confined to the fiction it would detract less from Chesterton's talent, but it is not; the love of striding about in a verbal cloak affects almost everything he wrote, even the great 'Lepanto'. Of course, Belloc had it too, but not so badly; or rather he could work it more skilfully into its surroundings. And an exaggerated style, like a garden in which everything is allowed to grow as it will, soon fades and rots. The verbal control and discipline of a Beerbohm, say, was beyond Chesterton, which is why Beerbohm is more readable today. And it was this discipline, which he entirely lacked, that Belloc to a far greater extent had.

And then Belloc was cleverer, and a much better scholar, than Chesterton. If we cannot on this occasion compare their novels, we can most instructively compare their essays. Chesterton on *Charles II* sinks without trace under the weight of paradox ('Reason is always a kind of brute force; those who appeal to the head rather than the heart, however pallid and polite, are necessarily men of violence') and the search for effect ('He could not keep the Ten Commandments, but he kept the ten thousand commandments'). Belloc's 'On the Method of History' is the work of an historian, and a good one; what is more, it might almost be a criticism of his friend's work.

Thus a man will have a just appreciation of the thirteenth century in England, he will perhaps admire or will perhaps be repelled by its whole spirit, according to his temperament or his acquired philosophy; but in either case, though his general impression was once just, he will, if he considers it apart from reading, tend to add to it excrescences of judgement which, as the process continues, will at last destroy the true image.... Does he admire the thirteenth century? Then he will tend to make it more national than it was because our time is national.... He will tend to lend the thirteenth century a science it did not possess, because physical science is in our own time an accompaniment of greatness.

As for the poetry, neither of them was of the first rank, but Belloc was of the second and Chesterton was not. 'Ha'nacker Mill' is not a great poem: but in its simple prettiness it can stand, say, with Housman, and not a line of it jars. Chesterton, on the other hand, in 'Love's Trappist' is a man standing on tiptoe, trying to do by artifice what can only be done by art. Only in his satirical and light verse (apart from the huge setpieces, all flawed, like 'Lepanto' and 'The White Horse') was he really in his element, and even here he never did anything to match the bolo-punch brevity of Belloc's

> Good morning, Algernon: Good morning, Percy.
> Good morning, Mrs Roebuck. (Christ have mercy!)

Still, the Chesterbelloc (or Bellocerton) exists; it is not a mythical beast, and the two halves were not associated simply because they were friends or because they both happened to be fat men. They did have some things in common, chiefly three. They were both militant Roman Catholics and they were both obsessed with drink and the Jews. (Not, of course, that either was a drunkard, and some of their best friends were Jews.) There is a supreme arrogance about their religious beliefs. They were something to do battle for, valiantly and continuously (indeed, *The Flying Inn* is the story of the defeat of Mohammedanism in an attempted conquest of England). They got it mixed up with all sorts of things; drinking, and storytelling, and patriotism (there are times when one is almost persuaded that Britain was a Catholic country when they were writing), but they battled for their creed in fair weather and foul and command one's respect alike by their skill and valour in the battle and the gay certainty of their convictions.

On the Jews, however, they do not command respect; quite the contrary. True, a good deal of water has flowed under this particular bridge in the last quarter of a century and we should beware of retrospective judgements. (Though, on the other hand, we might say that from the scurrilities of the *New Witness* during the Great War sprang many a seed which was to flower horribly later.) But it is not the *fact* of their anti-Semitism that appals, nor even the sliminess of many of their *obiter dicta*, in prose and verse, on the subject, nor yet the stupefying ignorance and inaccuracy displayed by both on the subject. (Belloc's *The Jews* is surely the only serious rival to the Webbs' *Soviet Communism* for the title of the silliest book in the English language.) What distresses one who, like myself, owes them an immense debt for their stimulating and provocative influence, is the obsessional nature of their attitude. There are times when they seem to be literally mad on the subject; there are, for instance, passages in Belloc's book *The*

Contrast (in which he even manages to discuss the Dreyfus case at length without ever admitting that Dreyfus was innocent!) which really might be the work of some deranged Mosleyite. They managed to get even this stuff, at times, mixed up with their Catholicism, which must be unstomachable to more than unbelievers, and it is practically impossible to get away from it for long anywhere in their works. Still, they were both great warts-and-all men, and however strongly we condemn their vice we should not allow ourselves to be blind to their considerable virtues.

And it is not, after all, on account of its anti-Semitism that this curious pantomime horse cuts so odd a figure today. Colour and ingenuity for their own sakes have gone out of fashion in writing and thinking. That, indeed, is what is wrong with front legs and back; they have become unfashionable. The whirligig of fashion, it is true, brings in even more changes than that of time; but somehow I doubt if this particular fashion will come in again. The truth is, they were romantics, and we are not, and there is an end of it.

BERNARD BERGONZI
Chesterton and/or Belloc

THE papist critic trying to reassess these two dead and illustrious figures had better put his cards on the table at once: the opposing dangers are sufficiently obvious, I think. On the one hand it is temptingly easy to genuflect before them in their established position in the front rank of English Catholic literary sainthood; to remember gratefully the spell they exercised in boyhood, to contrast their triumphant upholding of reason and right-thinking with the pestilential errors of our own day, and then, without too indecent haste, to pass on to other matters. This is implicitly to remove them from the province of criticism altogether. On the other hand, there is an equal danger of leaning over backward in a fit of unbiased (or 'undeflected') critical rigour, and dismissing them as posturing adolescents whose reputation was the result of temporary fashion and largely extra-literary considerations. However, a re-reading of some of their work . . . suggests that the truth is both duller and harder to formulate exactly than either of these two extremes would imply.

The first thing to be done, of course, is to reduce the compound monster, the Chesterbelloc, to its two components and look at them separately. The persistent coupling of their names may have had some point fifty years ago, when they seemed to their contemporaries a well-drilled propagandist musical comedy team, moving across the intellectual stage with supreme timing and precision. But in any longer view it does no good to either of them. Certain basic attitudes—apart from their Catholicism—they did, of course, have in common, and notably the one which seems to us most obvious and deplorable; namely, their anti-semitism. To some extent this was a product of their cultural environment rather than anything unique to Chesterton and Belloc; in the years before the First World War anti-semitic attitudes seem to have been extraordinarily common, both in literary circles and out of them. As Orwell once remarked, there have been very few English writers who were not in some degree anti-semitic; and this was certainly true, for instance, of the socialist and 'progressive' H. G. Wells. Chesterton's attitude to the Jews does not seem to have developed much beyond a kind of rowdy schoolboy prejudice, but Belloc's opinions had a more sinister tinge, for they linked up with those of the French anti-Dreyfusards, and thus had a remote cousinship with those subsequently associated with fascism in Europe. He was, for instance, advocating in the twenties the kind of measures adopted by Hitler in the thirties. And it is saddening to recall that he lived on in silent old age right through the era of the gas chambers and the 'final solution'. One can only regret that the Engish Catholic writers who have been most assiduous in perpetuating the Chesterbelloc cult should have been so reticent about admitting—let alone condemning —their heroes' racialism.

But it would be giving way to prejudice of another kind to pretend that this basic flaw, once admitted, totally negates their value as writers. It does not; or so I assume. Certainly, the impression that Belloc gives of a greater intolerance and potential degree of violence makes Chesterton seem, in comparison, a decidedly more genial figure. The personalities of both men dominate their work; and Belloc appears as arrogant, egotistical and—what is more—a fundamentally unhappy man. Chesterton, by contrast, has a basic serenity and assurance which blends with his exterior flamboyance to produce the popular cult-image: the lovable and convivial 'toby jug' figure, with neo-Johnsonian overtones. Yet, when one comes to discriminations of a specifically literary kind, he becomes increasingly hard to place, for all his great output of poems, novels, stories, and essays. Much of his writing was originally journalism, for both Chesterton and Belloc grew up in an age when it was still just possible to combine serious literary intentions with journalist practice. Chesterton, in fact, was proud of his newspaper work and disliked the aestheticism of the self-conscious artist. The daily newspaper, he wrote somewhere, was the greatest body of anonymous work to have been turned out since the cathedrals of the Middle Ages. But in the era of Northcliffe this generous attitude was already something of an anachronism.

It is true, however, that Chesterton's occasional essays are usually more interesting than his attempts at wholly imaginative writing, whether in fiction or verse. His novels are perhaps even more neglected than they deserve, for one of them at least—*The Man who was Thursday*, which is not a novel at all but a symbolic romance—deals with a potentially major theme: the whole problem of personal identity, so intriguing to our own age. But we need only compare it with, say, Kafka's symbolic fables on comparable themes, to see that Chesterton's story never rises above the level of a charade, or at least a prolonged and ingenious joke. His poetry, it must be admitted, is even less worthy of serious consideration. Reading it again, one is struck most of all by the predominance of the Pre-Raphaelite mode—a literary tradition which had become extremely tired by the time Chesterton began to write in it, *c.*1900. And yet, at the same time, it must have had a certain obstinate viability to have lasted so long, for Chesterton seems never to have questioned its suitability for his poetic purposes. Indeed, one of the mildly disconcerting things about Chesterton is that though his Catholicism and medievalism represented extremely serious religious and social ideals, he was perfectly prepared to give them imaginative expression in well-worn trappings borrowed from the secondhand wardrobe of William Morris and others. . . .

So we are forced to return to Chesterton's discursive prose if we are to find whatever is perdurable in his massive *oeuvre*. It is in the essays that we most often find the characteristic Chestertonian vehicle and weapon, the paradox, which has been variously described as 'truth standing on her head to attract attention' and 'truth cutting her throat to attract attention'. Hugh Kenner, in his typically ingenious little book on Chesterton, has admitted his deficiencies as an artist, but has claimed that he was a 'metaphysical moralist' who employed the paradox as a mode of perceiving and understanding the nature of reality. There is something in this: it is certainly true that Chesterton's use of paradox frequently degenerates into a mannerism, but it is equally true that he used it as earlier nineteenth-century thinkers had used other stylistic devices—in the way described in John Holloway's *Victorian Sage*—as a means of mediating a total view of life. It is well known that Chesterton 'invented' Christian orthodoxy for himself before discovering that it really existed, and he similarly arrived at a *Weltanschauung* that was basically Thomist before he had even heard of Aquinas. (His book on Aquinas is recognized by modern Thomists as a classic of its kind, and it was written with only a casual knowledge of the relevant texts.) Whatever the limitations of his imagination, it would be a great mistake to underestimate his intelligence, which was remarkable. But for the modern reader it is probably too strictly rationalistic. Every event in man's life, and everything in the universe, had for Chesterton an ultimate ontological purpose. Secure within an all-embracing and rationalistic world-picture, he could make adroit semantic analyses of his opponents' statements and wittily point out the errors and false assumptions contained in them. And this is the essence of the 'paradoxical' method. Engaged by the wit and precision of the observation, and the urbanity of the tone, one is forced to concede that some kind of point *is* being made, even if one does not agree with it: to that extent one has had, momentarily, to accept the Chestertonian *Weltanschauung* and abandon one's own. In this way Chesterton's paradoxes have a rhetorical function, comparable to the stylistic tricks of Arnold. But they worked a good deal better in that pre-1914 Age of Reason, the period of the great debate between Chesterton and Belloc and Shaw and Wells, when men believed in the power of reason and were prepared to use it vigorously and combatively to prove that they were right and their opponents wrong, than they do now, when we are suspicious of *any* discussion of pure ideas, and a logic which deals in such clear-cut categories as 'right' or 'wrong' seems pretty crude. Whether this is a good thing or not is not at the moment my concern, but it explains why Chesterton—and the rest of his generation, Catholic or otherwise—seem such strange and remote figures nowadays. We are more concerned with a subtle and discriminating attention to particular facts than with fitting events into a total world-picture (of course, we will still have one, though it will possess the dubious advantages of being unconscious and unexamined). And here, admittedly, Chesterton is often deficient. His overriding concentration on a two-valued dialectic sometimes led him into some very unsatisfactory literary valuations. Thus, he complains in an early essay of the contemporary neglect of Scott:

The ground of this neglect, in so far as it exists, must be found, I suppose, in the general sentiment that, like the beard of Polonius, he is too long. Yet it is surely a peculiar thing that in literature alone a house should be despised because it is too large, or a host impugned because he is too generous. If romance be really a pleasure, it is difficult to understand the modern reader's consuming desire to get it over, and if it is not a pleasure, it is difficult to understand his desire to have it at all.

This is plainly nonsense, with an 'either-or' being worked to death, since there is absolutely no reason why one should not enjoy romance or anything else in fairly small amounts and get bored by large quantities. The underlying notion is that good things partake of Being, that Being is good *en soi*, and the ultimate fullness of Being is God himself. Chesterton would have found it blasphemous to suggest that one *can* have too much of a good thing. It is this mistrust of subtlety, of the possibly infinite gradations of Being, that gives Chesterton's prose at its best a fine dialectic sweep and at its worst a tiresome obtuseness. Nor is this ultimately very much to do with Chesterton's Catholicism: the typically twentieth-century reader is likely to feel more at home with the poetry of Hopkins or the prose of Newman, where finer discriminations do operate.

Yet having said so much, I must hasten to add that Chesterton could be a very good literary critic. If it were a question of deciding which of his works were really worth preserving, I think I would opt for the two books on Dickens, which, though they may seem a little dated now, contain—along with Gissing's—some of the first respectable criticism to be written about Dickens. Some of the other writings on Victorian authors, too, would be well worth preserving. Indeed, instead of the present *ad hoc* collection of essays on a variety of topics that Mr Sheed has given us, I would much rather have seen a straightforward reprint of the *Criticism and Appreciations of Charles Dickens*, embodying as it does some of the best of Chesterton's criticism. In that book he adopted some surprisingly modern (and, I think, sound) attitudes. Thus, in the essay on *The Old Curiosity Shop* we find him taking a sternly anti-Intentionalist line:

> The function of criticism, if it has a legitimate function at all, can only be one function—that of dealing with the subconscious part of the author's mind which only the critic can express and not with the conscious part of the author's mind which the author himself can express. Either criticism is no good at all (a very defensible position) or else criticism means saying about an author the very things that would have made him jump out of his boots.

A fit rejoinder to what Northrop Frye has called 'the oh-come-now school of criticism'. And again, in the essay on *Great Expectations* there is a remarkable anticipation of some modern critical approaches:

> A great man of letters or any great artist is symbolic without knowing it. The things he describes are types, because they are truths. Shakespeare may, or may not, have ever put it to himself that Richard the Second was a philosophical symbol; but all good criticism must necessarily see him so. It may be a reasonable question whether the artist should be allegorical. There can be no doubt among sane men that the critic should be allegorical. Spenser may have lost by being less realistic than Fielding. But any good criticism of *Tom Jones* must be as mystical as the *Faery Queen*.

Reading the best of Chesterton's literary criticism, one can only regret that he dissipated his energy into so many other forms of writing for which he was less or not at all suited. But he was a humble man and sure of his readers, and there is no reason to think he would have regretted it himself.

If Chesterton was at his best when writing about books, Belloc was certainly at his best away from them. One need only glance at the brief essay on Jane Austen in these *Selected Essays* to see how little of a critic he was, for his essential egotism prevented him from removing his attention from himself to the work in front of him for very long. But, on the other hand, he was very much more of an artist than Chesterton. Re-reading his poems for the first time in several years, I was agreeably surprised, for Belloc appears in them as a good minor poet who deserves at least the same reputation as that now enjoyed by Housman: his wish to be remembered primarily as a poet seems eminently reasonable. The epigrams, of course, are deservedly famous, and one wishes we had now a political satirist who could achieve the lethal compression of 'Epitaph on the Politician':

> Here richly, with ridiculous display,
> The Politician's corpse was laid away.
> While all of his acquaintance sneered and slanged
> I wept: for I had longed to see him hanged.

There are few places outside Pope where precisely this note has been struck, and struck so well. Belloc's sonnets suggest that he was one of that remarkably small company of English poets who could genuinely think and feel and write—without apparent effort—in sonnet form. The language is almost as derivative and literary as Chesterton's, but the best of Belloc's poetry does give one the traditional romantic pleasure of hearing the voice of a vigorous and authentic personality coming through the somewhat threadbare diction and conventional rhythms. To this extent, Belloc is not merely bookish. And a similar quality redeems quite a lot of his bellelettrist prose, which is written according to prescriptions not very likely to recommend it to the modern reader, for Belloc came to literary maturity when the classically inspired canons of 'beautiful English' and the 'fine style', largely divorced from content, were still the norm. Nowadays we see prose style as much more intimately a function of content, feeling, and attitude, and are readily reminded of Max Beerbohm's comment on Pater—'that sedulous ritual wherewith he laid out every sentence as in a shroud'. But the strange thing is that though Belloc wrote prose according to these somewhat external prescriptions, much of it is still good when judged by other standards. One might mention, for instance, the set-piece called 'The Relic' in the present book of essays, which conveys extremely well the intensely personal quality of Belloc's experience in a Spanish church.

In fact, Belloc's apprehension of the world seems to have needed the measured and calculated quality of his prose, or the formality of his verse, in order to be coherently conveyed at all. For despite his aggressively dogmatic and assertive manner, I feel that his inner life often existed on the edge of chaos and near-despair. The tension is certainly apparent in his writing. Though, like Chesterton, he wholeheartedly accepted the Catholic world-picture, Belloc did so as an act of disciplined intellectual assent, whereas Chesterton believed because the whole nature of his mind was constituted to do so. It is, perhaps, these sustaining tensions that give the sense that Belloc was both a greater artist and, for all his disagreeable attitudes, a great man. One sees, for instance, the inner isolation and unhappiness appearing for a moment beneath the rigid mask in the conclusion of his essay, 'On Unknown People':

How often have I not come upon a corbel of stone carved into the shape of a face, and that face had upon it either horror or laughter or great sweetness or vision, and I have looked at it as I might have looked upon a living face, save that it was more wonderful than most living faces. It carried in it the soul and the mind of the man who made it. But he has been dead these hundreds of years. That corbel cannot be in communion with me, for it is of stone; it is dumb and will not speak to me, though it compels me continually to ask it questions. Its author also is dumb, for he has been dead so long, and I can know nothing about him whatsoever.

Now so it is with any two human minds, not only when they are separated by centuries and by silence, but when they have their being side by side under one roof and are companions all their years.

In fact, Belloc was a good deal closer to the characteristic masters of modern literature than we may at first imagine. He, too, was a *déraciné* figure with a bewildering variety of *personae*: ex-scholar of Balliol, ex-French artilleryman, Sussex farmer, Liberal politician, anti-Dreyfusard, London man of letters, sailor. All these figures in turn inspired different aspects of his writing but never gave him anything like the conviction that a genuine set of cultural roots would have done. It is certainly true that the English connection, and in particular his friendship with Chesterton, were beneficial to him as a man, and modified the potentially sinister elements on his Gallic side. Had he remained wholly a Frenchman it is only too easy to imagine him as a supporter of Maurras, professing a purely 'political' Catholicism, and subsequently a man of Vichy. As it happens, Belloc's devotion to the culture of Western Europe as a whole, and to Catholicism as the incarnation and guardian of that culture, were clearly a form of compensation for his lack of more intimate roots. This devotion, too, was not without its unfortunate side, for it could lead him into such dangerous half-truths as the pronouncement that 'the Faith is Europe, and Europe is the Faith'. Yet it is impossible not to be moved by the extent of Belloc's knowledge and love of England and France and Italy and Spain and the Catholic parts of Germany. He knew these countries and their people and buildings intimately because he had been over most of them on foot. *The Path to Rome* is as much the record of a love-affair as a travel-book. Belloc's concept of 'Western culture' was something much more personal and existential than the purely literary and eclectic kind of 'tradition' compiled by Pound or Eliot, in their rather Adam Verver-ish fashion.

But in a final judgement it is Belloc's lack of interest in a specific literary tradition, and his tendency to oppose flatly the deeper tendencies of his age rather than to interpret and explore them, which makes him remote and inaccessible to present-day criticism. And much the same is true of Chesterton. Together with their non-Christian contemporaries they lived and argued in a world that seems almost as strange and distant as the Paris of Aquinas. Nowadays we tend to agree with Russell that 'it is better to doubt than to believe', and the conceptual apparatus of our literary criticism is made up of hints from Arnold and Richards about the 'free play of ideas' and the 'organization of impulses'. So we go in fear of the stock response and the pre-existing *Weltanschauung*: thus far has criticism become itself ideological, with its own built-in 'deflections'. Yet this is clearly another and larger matter. Where Chesterton and Belloc are concerned, the Christian humanist is as likely as the agnostic to find them fallen idols rather than living gods in the heaven of literature. But their sleeping features deserve, at the very least, a long and respectful stare, before they are finally eroded by the winds of time.

GARRY WILLS
Rhyme and Reason

CHESTERTON'S favourite reading from childhood was the poets—Isaias and Job and the psalms, Shakespeare and Browning and Swinburne. His taste was catholic, including Pope as well as Shelley, though he was always faintly irritated by Milton's inhuman epic. Very soon he had begun his own versifying, interspersing ballads modelled on Scott with his first romances and fairy tales. The next two stages of his work [were] the sprawling monologues in *The Debater* and the attempt to put aphorisms into rhymed form. The poems of his first volume of importance, *The Wild Knight*, arose from that effort, and they mark the end of his specifically poetic ambitions. After this first volume of such promise, he became a jester in verse as in prose.

The Wild Knight includes poems like 'The Fish', which Chesterton had been working on since the Slade days. 'By a Babe Unborn', which the *Autobiography* relates to his struggle to burst out of his own mind, is a concentration into verse of the story he had variously recast, then published, as 'A Crazy Tale'. The binding theme of the volume is Chesterton's universal theme of praise for existence, but the nightmare-separation from the world of reality is still vividly remembered. In eight poems we find the fate of the outcast depicted, becoming at its most intense moments the final loneliness of an unregarded death; while in two poems solipsism is directly described. To Kenner's thesis it might, therefore, be objected that Chesterton *was* writing from his own experience, and of things he had intensely felt. But this response is too pat. The threat Chesterton felt was of madness, and the response is intellectually defiant, that is, argumentative. Argument rarely leads to good poetry; it would finally divert Chesterton from the course of the poet. But in this first volume, the intention and the workmanship are a poet's.

Diffusion, a Swinburnian laxity and facility, were the faults Chesterton had inherited from his early work; but we can see here his conscious opposition to this drift. He strives for a tight and pregnant simplicity, and often achieves it:

THE SKELETON

Chattering finch and water-fly
Are not merrier than I;
Here among the flowers I lie
Laughing everlastingly.
No; I may not tell the best;
Surely, friends, I might have guessed
Death was but the good King's jest,
   It was hid so carefully.

The entire volume shows an attempt at verbal asceticism which did not last. Swinburne's influence is still here, but in a very short flower of the sea; Browning's, too, but in a very short monologue. The greatest resemblance, strangely enough, did not arise from direct limitation. These poems, like Blake's, are very fierce auguries of innocence, their rhetoric patterned after that of the prophets, as in the poem whose first two lines are a paraphrase of Isaias' opening distich:

To teach the grey earth like a child,
To bid the heavens repent.

If the lion does not exactly lie down with the lamb, 'The Donkey' moves surprisingly like the Tyger, and 'The Fish' and 'The Skeleton' are also of this company. Another brief poem traces the world's vitality into the narrow chamber of a seed, and discovers there

God almighty, and with him
Cherubim and Seraphim,
Filling all eternity—
Adonai Elohim.

Even more important than these stylistic considerations is the fact that many of these poems achieve insight without argument—an attainment difficult enough to one of his critical cast of mind, and one which he never made his own during the many years after this first book appeared. The poem to St Joseph is an example:

If the stars fell; night's nameless dreams
   Of bliss and blasphemy came true,
If skies were green and snow were gold,
   And you loved me as I love you;

> O long light hands and curled brown hair
>   And eyes where sits a naked soul;
> Dare I even then draw near and burn
>   My fingers in the aureole?
>
> Yes, in the one wise foolish hour
>   God gives this strange strength to a man.
> He can demand, though not deserve,
>   Where ask he cannot, seize he can.
>
> But once the blood's wild wedding o'er,
>   Were not dread his, half dark desire,
> To see the Christ-child in the cot,
>   The Virgin Mary by the fire?

The poem does not hinge on intellectual paradox, but on a real emotional ambivalence: Chesterton sees rather than understands man's conflicting instincts—that sex is fruition and expenditure, that virginity is barren yet has in itself some holiness. The poem's structure is perfect, rising to one climax, then reaching a deeper climax and mystery. The symbols are theological, as in Blake's work, but the reality is analogously physical and moral and mystical and mad.

METAPHYSICAL MINSTRELSY

But other poems in the book, especially the exercises in political rhetoric, show that paradox was driving out the autonomy of direct emotional vision. This is not to say, as Kenner does, that philosophy was driving out poetry. It may have been poetry that was receding, but the new element was not philosophy. 'Paradox' is constructed of the same symbolic stuff in Chesterton's verse as in his novels, and it must be judged by the same norms. Chesterton ceased to be a poet, in the conventional sense, but he became a rhyming jester. His later volumes, except for one poem which was in itself a book, were mere collections of his occasional verse written for the newspaper or for his friends. They are haphazard collections, and include things which are merely topical and things which are worthless. Three things Chesterton could write, those three which Shaw could not—a long song, a war song, a drinking song. They are songs rather than poems, spontaneous as some feudal bard's 'journalism'. It is extraordinary how completely Chesterton's talents did express themselves in the manner of a jongleur. His drinking songs have less argument in them than any of his later verses; they arose from a deep enjoyment of his role as a truly festive Feste; for only the metaphysical jester can escape the melancholy of all other clowns. If we add two more genres to the list, both entirely in keeping with the minstrel's repertoire, it will be complete—satire and carols.

Haste and extravagance are marks of the jester's style—for it is not a mere lack of style. Even in *The Wild Knight* we can see this new mode of song coming to birth. One of the solipsist poems, in which the poet tries to escape a horrible world wherein everyone bears his own recurrent countenance, ends with these lines:

> Then my dream snapped; and with a heart that leapt
> I saw across the tavern where I slept,
> The sight of all my life most full of grace,
> A gin-damned drunkard's wan half-witted face.

There is a jolting bathos in the revelation of the poem's scene—as if the nightmare had been caused only by whisky fumes; but Chesterton had to admit this setting in order to make a drunkard the vision of grace, as Orm had been in another context. There is in the relishing of this extreme contrast a humour that is answered by the vivid, even gaudy, violence of the last verse. This is not poetry, but it is artistic in its aim; it is a metaphysical joke. The minstrel is laughing, and especially laughing at himself.

Chesterton's ability to play with pictures brightly picked out was simply another aspect of his central rhetoric of jest. Yet if it be granted that Chesterton was that kind of wild and logical symbolist which I have called a metaphysical jester, many objections to this kind of writing remain: for instance, the claim that it is merely philosophy in disguise. . . . A more serious charge is that such a rhetoric of ideas is 'propaganda' in the modern sense of intellectual seduction.

Propaganda is not art; neither is it education of an honest sort. Propaganda is that bastard form of art and instruction which sugars a doctrine with colours and forms not integral to it or to the artefact used for this indoctrination. But Chesterton did not build an argument, then stick it all over with figures of speech. The symbols came first to him, as his real form of expression. These are always apt symbols, carrying their own meaning; there is no previously formulated expression of the ideas to which they must be fitted. They are the expression. We see this in the case of colours, which had a symbolic urgency which was activated in many ways, but always from within. Chesterton felt the symbol's vitality before he knew what it signified. Even persons and events—Joan of Arc, Lepanto—were symbols to him before he understood what they might symbolize. The figure of the white horse, too, had haunted him in various guises long before he thought of it primarily as Alfred's horse. If anyone thinks he is a mere propagandist, let him say what explicit concept or course of argument that white shape is meant to promote. An honest mind will soon disappear into the depths of that richest of Chesterton's symbols.

Chesterton's verses are those of the court fool, who does not pretend to make things as sane men do. The characters in his novels are not symbols of men, like Hamlet, but of ideas, like Zarathustra. Such a jester stands outside propaganda, the illegitimate use of art *ab extra*, in the same way that the satirist does. Oblique comment is the jester's mode, as syllogisms are the technique of the logician. A propagandist can use attractive but fallacious syllogisms, as he uses appealing but inept symbols. But this does not mean that the real logician or satirist or jester is a propagandist.

The proof that Chesterton was not a propagandist is that he ceased trying to write poetry and fiction of the conventional type. He forswore the quiet stories and self-contained nonsense verses which he had first excelled in. His first slim publication, *Greybeards at Play*, distorted the form of nonsense verses by inserting satire on decadent artists and philosophers. He had the good sense not to attempt this again, and I think he grew to dislike his one essay into 'pure' nonsense. His work from that time on has a uniformity of texture which Chesterton sustained so well that men forgot it was artificial. It is reasonable to say that his artefact is worthless, but it is simply absurd to claim that it is not an artefact.

But if the verse be called a metaphysical minstrelsy—neither philosophy nor propaganda—other criticisms can still be adduced. For instance, an *a priori* argument against such verse is that its blatant and over-vivid colouring forbids all subtlety and makes delicate insight impossible. But jugglery is entirely a matter of delicate balance; and, as Chesterton frequently insisted in his comments on pantomime, even an exaggerated phrase or gesture can 'act out' what is shadowy or deep-hidden. The poem in which Christ's double role is described— he is Zeus; he is Prometheus—is projected as a whirling pantomime of the eagle and the vulture, cornfields and chasms, daylight and fires in the night. Yet the concept is a delicate one, nowhere else in Chesterton, nor in any other writer, given this uniquely illuminating expression. One cannot make the eagle and vulture mere trappings for the statement that God is both order and act. Such abstract words, or any multiplication of them, do not express what Chesterton says. There is no other way of saying it: God is Zeus, God is Prometheus. Similarly, there is no other way of saying what he says of the Virgin in

'The White Witch'—that Mary is an anti-witch, who drives out all the evil phantasms which the image of Hecate—of that deepest of human perversions, the woman of evil—has spread in the imagination of men since the dawn of time.

Even single lines have this ability to say much in a simple and direct stroke. The entire map of Europe is sketched in 'Lepanto', so that a single brush-stroke fixes, in his heraldry, France:

> The shadow of the Valois was yawning at the Mass.

The cobwebby assonance is overdone, as the slumbering heaviness of 'Sloth' is overdone in a medieval carving; but the exaggeration touches depths that careful statement often does not reach.

Medieval art is aptly brought in here—heraldry and the windows and the stiff chasubles and the gold everywhere. We have long since overcome the misconception, held for centuries, that only chiaroscuro can achieve subtlety and open man's soul. The density and richness of a window are complicated by the very brilliance and multiplicity of colours. Every Gothic cathedral wore motley. That dazzling multiplicity is what we notice in Chesterton's tales—the jostling suspension of all colours in the scheme. The same technique is used in the poems. All the issues of liberty and Europe's fate are traced in the whirl of 'Lepanto's' colours, and then the whole thing is set in a comic frame which deepens the significance and interrelation of the historical sections of the poem. Chesterton looks backward from Quixote, using the comic knight as a means of understanding the hopeless, victorious knight of Austria.

Another objection, tangential to those just considered, is that such obvious and crude rhetoric as Chesterton used is easily reeled off by anyone with a modicum of talent. The fallacy at work here should be recognized. The juggler's art may not be worthy of exalted attention, but it is not easy. No one ever juggled by accident. It is an art, and it takes practice. The same is true of Chesterton's symbolic juggling of words and colours. Even in its simplest form this is a difficult thing. It may be a waste of time to write drinking songs; but when good ones are written, they are written only by skilled artificers, like Horace and Shakespeare. Where are the swarms of men succeeding in this easy pastime?

Chesterton certainly considered facility a part of his task as a minstrel; not simply a temptation, as it is for most poets. It is a mistake to think there is no asceticism in the immediate volley, as in the six years' chiselling. The asceticism lies in the difficulties undertaken, the willingness to accept any challenge. Look through the collected poems and notice the variety of forms Chesterton used: sonnets, triolets, ballades, odes, blank verse, couplets. His ballad forms are often deceptively simple in appearance. He tried one difficult metre in his first volume, with little success. It was used in a poem devoted to the praise of Woman. Many years later he used the rhythm again in a fine tribute of the Virgin:

> And a dwarfed and dwindled race in the dark red deserts
> Stumbled and strayed,
> While one in the mortal shape that was once for immortals
> Made, was remade.

His early ballad of Gibeon, otherwise worthless, is a panting obstacle course of feminine rhymes. He frequently used the rhyme-scheme of *In Memoriam*. All this is *tour de force*, admittedly; but *tour de force* is by definition that which is not easy. And, like another master of *tour de force*, Chesterton as he ends the refrain thrusts home, not only in symbolic poems like 'Lepanto' and 'The Monster', but in the acerb[ic] verses of F. E. Smith and the Bishop who called St Francis flea-bitten, in the brilliant translations of Dante, Du Bellay, Guérin.

THE BALLAD OF THE WHITE HORSE

Because Chesterton was a true balladeer, he could use certain traditional forms with a spontaneity and sense of the form's genius which is denied most poets by their very acuteness and personal accent. This was true not only of the drinking song but of the Christmas carol. Because of this, it was possible that Chesterton could write one kind of poem which would not be only jest or *tour de force*. He could retain his loose and rapid spontaneity, yet work to a larger plan, polishing and reshaping its parts. He could stitch together the ballad stanzas as the original singers had done at the dawn of epic, when the stories of Robin Hood and Roland were fashioned from the old, sporadic material to a new and larger pattern.

Chesterton's instincts led him to this form, and they did not fail him in the choice or execution. This was his most serious artistic endeavour, and he made the attempt but once. For at least four years that we know of he worked on *The Ballad of the White Horse*, the only example in his career of such extended labour and delayed publication. More than this, he used symbols, stanza forms, and words which he had been collecting in his mind for years. The white horse had been his private symbol of chivalry since the time when he owned a hobbyhorse of that colour; an inn sign, a canvassing trip in Wiltshire, a honeymoon memory—these and many other experiences had been stored in the symbol's energy. The first two lines of the finished poem were the inspired opening of a boyhood poem on Moses. One of Alfred's prayers had come to Chesterton in his sleep, and was copied down, in the first person, many years before the ballad appeared. We can see the changes made in one section of the ballad by comparing it to the 'Fragment of a Ballad of Alfred' published in the *Albany* for 1907, four years before publication of *The White Horse*.

Chesterton's fascination with the ballad form dated from his school days, from the time when first he read Scott and Macaulay. A paper he delivered to the J.D.C. shows that he soon went to the original English ballads. Discussing them in this paper, he praises especially their vigour of epithet and 'signature phrase' which can bring a character to life at one stroke. Many of the figures in *The White Horse* are picked out in this vivid manner: the deserted king, for instance, is 'Alfred of the lonely spear'. One of his own first ballads, sung by the minstrel in a story written at the St Paul's School, resembles the strain of the melancholy minstrel, Elf:

> Softly and silently
> Sail the fair Valkyrs,
> Spirit-receiving ones
> Whispering to warsmiths . . .
> Bending their bright course
> Laden with thane-spirits
> Up to high halls
> Where with the wise Woden
> Baldur the beautiful
> Reigneth for Right.

Whenever Chesterton's ballad is brought up, the inevitable comparison is with Coleridge's 'Rime'. Maurice Baring, in his review of *The White Horse*, remarked that Chesterton told a vivid story as well as Coleridge but did not let the tale alone carry the theme. He thought Chesterton's ballad was, for this reason, less 'authentic'. But this is the one point on which the more recent poem has an unquestionable superiority. Coleridge's theme is not more complex and exalted than Chesterton's, but it is less 'popular'. Chesterton works from popular sentiment, as the ballad must; his poem is full of patriotism and the spirit of a single landscape:

> He sang of war in the warm wet shires
>   Where rain nor fruitage fails,

> Where England of the motley states
> Deepens like a garden to the gates
> In the purple walls of Wales.

Coleridge filled the English ballad with Oriental horrors, but *The White Horse*

> Seems like the tales a whole tribe feigns,
> Too English to be true.

Chesterton recaptures, moreover, that moment when the primitive ballads were woven together to become national epic. His poem is the record of a war from the heroic age; epic boasts and similes, a national hero, the hushed eve of battle and the screaming day that follows, make *The White Horse* echo the tales of Roland and Henry V as well as of Robin Hood. Coleridge's ballad, on the other hand, is a weird voyage into the self, its introversion making the 'authentic' heroic note impossible.

The local scene and fiery patriotism do not limit *The White Horse*. The vale of Alfred is England, and England is Christendom, in the poem. The 'triple symbol' mentioned in the prose introduction is an interpretation of the elements in the English greatness—Saxon, Celtic, Roman. But they are also symbols of that complexus which Chesterton described in *Blake* as the formula of Western man—pagan poetry, Roman order, and Christian religion. Chesterton always insisted that Christianity did not drive out pagan things but subsumed them even in growing from them. There is a real conflict between poetry and reason, faith and doubt, the supernatural and nature, but this dialectic is life-giving, not destructive. This is the meaning of Colan, Eldred, and Marcus, united by Alfred, in the van of the English armies. In *The Everlasting Man* Chesterton would again describe a triple dialectic, considering three stages of history—not chronological stages, simply, but co-existent levels of human reality. First there is 'minimal man'—child, savage, artist, god among the beasts. Then there is civilized man, giving law to the nations and searching the heavens. Finally, completing and ordinating the former elements, without banishing them, there is Christian man, exiled citizen of the City of God.

Against this Christian balance and complexity come the barbarian forces of simplification and destruction. The Danish heroes are merely the Christian thanes simplified, isolated from the balancing discipline which Christianity imposes on man's nature. Saxon Eldred typifies the love of life, of wine, of 'slow moons and certain things'; his farm has become, by vow and charity, a haven of the poor and storehouse of earth's good things. Danish Harold is the same type of man, but one whose vigour and blood have flamed into a destructive sensualism. Gaelic Colan is a mystic whose gods must be harnessed by the new God from Rome; but Danish Elf is the poet whose gods are free, and who spread their beautiful barrenness everywhere. Marcus is the lover of order for its own sake, of force and process subjected to a constructive asceticism; whereas 'Ogier of the stone and sling' loves destruction for its own sake, and expresses the nihilist mystique in clear-etched lines which reveal the depths Chesterton's verse could reach:

> There lives one moment for a man
>   When the door at his shoulder shakes,
> When the taut rope parts under the pull,
> And the barest branch is beautiful
>   One moment, while it breaks.

Guthrum rises over his warriors as Alfred towers among the thanes. The pagan leader is a 'clerk', whose weary sentences have the beauty of lyrics from some Greek tragedy:

> Do we not know, have we not heard,
> The soul is like a lost bird,
>   The body a broken shell?
>
> And a man hopes, being ignorant,
> Till in white woods apart
> He finds at last the lost bird dead;
> And a man may still lift up his head,
>   But never more his heart.

This love of life is Horatian, strong within narrow limits, but desolate when the man lifts his eyes to the horizon and the encircling dark:

> The little brooks are very sweet,
>   Like a girl's ribbons curled,
> But the great sea is bitter
>   That washes all the world.

Even Guthrum, therefore, must try to forget his philosophy in the drunkenness of war. All the Danes' joy and wisdom leads to sterility and destruction. Marcus builds as a pagan, but he slowly changes into Ogier unless Christianity intervenes, 'because it is only Christian men Guard even heathen things'. Creation is loved with permanence only by men who believe in the Creator. The simple love of the Danes, uncomplicated by the mysteries of faith and humility, shifts with every mood, and brings ruin with its shiftings—fire on Ely fen, molten lead on Glastonbury, Roman colonnades left like 'the spectre of a street', and the great hieroglyphic Horse fading, unkempt, to an undecipherable smudge on the hills. The Christians, on the other hand, keep arch and book, sing a hymn of the crafts at the height of battle, and tend the strange white sign stamped on their native earth. Alfred sees God as a labourer who fills the vines and tends the fields, and Marcus calls him 'God that is a craftsman good'.

'LIKE A GOOD CHILD AT PLAY'

Alfred becomes the central Christian warrior by undergoing defeat, by 'hardening his heart with hope' when there is no earthly hope. Only then is the vision given to him:

> In the river island of Athelney
>   With the river running past
> In colours of such simple creed
> All things sprang at him, sun and weed,
> Till the grass grew to be grass indeed,
>   And the tree was a tree at last.

He is granted that carelessness about the world which alone reveals the world. The pagan must cling to the tree finally—making of it a god, a portent, an end—or destroy it in his rage. But the Christian recognizes the endangered and divine world as a picture wrought by the Artist who 'saw that it was good'.

Yet Alfred still asks for a pagan wisdom, for prevision and some power over destiny: will he win, or again be sent reeling? The Virgin answers that the Christian clarity of wisdom goes with an ignorance of such evil reckoning of the odds. The tree is a tree at last because it cannot be twisted into a magic instrument of power. It is a simple tool in the toy world of nature which reflects, in time, the City of God. Faith, like Christian hope, is based on a certain ignorance; it recognizes simple facts like trees. On this ignorance true knowledge can be built. Only Christianity has made men content with such knowledge, justifying the wisdom which grows from ignorance by making it correspond to the created reality which came from nothingness. The pagan does not believe in this radical creation of the world from nothing, nor in the formation of wisdom from innocence. He seeks always to know why he knows what he knows. He cannot begin at the beginning—his own beginning and the world's—because this involves a preliminary self-annihilation through humility, a humility which re-enacts one's passage out of the abyss of nonexistence. But Alfred, stripped of all pretensions in defeat, accepts that state of ignorance in which man can be taught, as by one's mother:

> And he saw in a little picture,
>   Tiny and far away,
> His mother sitting in Egbert's hall,
> And a book she showed him, very small,
> Where a sapphire Mary sat in stall
>   With a golden Christ at play.

This is the argument of *Orthodoxy* carried to its completion. In *Orthodoxy* Chesterton traced the folly of self-sufficient wisdom and claimed that the Church alone preserved knowledge—until we pulled the mitre from pontifical man and saw his head come off with it. But the approach was negative in that book; the pure intellect destroyed itself, and poetry was referred to as a sane counterbalance to such intellection. Here Chesterton looks at the *positive* wisdom of Christian innocence, the vision of trees as trees, the childlike hope and love which make Alfred follow a pillar not of fire but of darkness:

> The men of the East may spell the stars,
>   And times and triumphs mark,
> But the men signed of the cross of Christ
>   Go gaily in the dark.

Christianity saves even heathen things, but this 'pragmatic' value of the Faith is only made possible by a deepest state of complete innocence which does not seek to use or to reject the world. The figure of the White Horse on the hill is a sacrament meant to reveal Alfred's love of his native earth in all its autonomy. But he does not fight, ultimately, to save the White Horse, or his kingdom, or the churches. He fights as Joan of Arc did, gravely and in ignorance of the reasons for which heaven bade her go forth, ready for victory or defeat, to defend that strange thing, scrawled on the earth of Europe, which we call France. Alfred is ready to lose all or win all with the same cheerfulness of faith, knowing that his home is here and yet not here.

The specifically Christian 'detachment' and mysticism are articulated with precision in this attitude of Alfred. Christian asceticism does not arise from the simple opposition of matter to spirit, of time to eternity, of this life to some other. This cannot be the Christian's mind on these issues, for God not only made this world. His being supports and pervades and continues to activate it. He is in Orm and in Guthrum.

He 'labours' to

> Build this pavilion of the pines,
> And herd the fowls and fill the vines,
> And labour and pass and leave no signs
> Save mercy and mystery.

Yet God constructs no lasting city here; it is a world of mystery and adventure, where His champions must go gaily in the dark, prepared to see their wisdom and planning collapse as ludicrously as the Council's did in *Thursday*. Around Alfred gather all these reflections on existence as something immeasurably valuable, yet something won by a carelessness and self-forgetfulness which only real faith in God has ever given men.

Of course, Alfred's vision at Athelney is only the beginning of the revelation, in and through him, of his Christian mysticism. The poem builds to a higher vision at its climax, the centre of calm light in which a child plays—an odd, idyllic interlude placed dramatically in a setting of war and slaughter. It is this white light which fires the motley colours of battle and heroism and victory. Every line of the poem leads to that vision or follows from it. Alfred expresses a new facet of it in each episode. In his call to the warriors he repeats the Virgin's summons to the wise ignorance of faith:

> I call the muster of Wessex men
> From grassy hamlet or ditch or den,
> To break and be broken, God knows when,
> But I have seen for whom.

In the Danish camp he scorns the pagan desires, shouting his joy at defeat against their weariness of triumph. In the forest he takes as his ensign of royalty a blow from a peasant woman; and when humility lights him through and through, laughter follows:

> The giant laughter of Christian men
> That roars through a thousand tales,
> Where greed is an ape and pride is an ass,
> And Jack's away with his master's lass,
> And the miser is banged with all his brass,
> The farmer with all his flails.

> Tales that tumble and tales that trick,
> Yet end not all in scorning—
> Of kings and clowns in a merry plight,
> And the clock gone wrong and the world gone right
> That the mummers sing upon Christmas night
> And Christmas Day in the morning.

Laughter is very near that childlike innocence which the Virgin brought to Alfred, as we see in the laughing Madonna and Child of medieval art. Chesterton has fitted a separate 'discovery' of his youth—laughter as a medium of knowledge and sanity—into the total Christian reality.

Before the battle, the pagan Harold delivers the ritual epic boast, but Colan makes real in this context St Paul's 'boast in the Lord':

> Oh, truly we be broken hearts;
>   For that cause, it is said,
> We light our candles to that Lord
>   That broke Himself for bread.

When Colan flings his sword at Harold, Alfred sees this act as a parable of the whole campaign's meaning. Christian men are given victory because they surrender it, they are given the world as a gift only when they recognize that it is *God's* gift and possession: 'Man shall not taste of victory Till he throws his sword away':

> For this is the manner of Christian men,
>   Whether of steel or priestly pen,
> That they cast their hearts out of their ken
>   To get their heart's desire.

Then the champions fight according to their character—Eldred straightforwardly, by mere mass; Marcus in order, stemming the retreat when superstition begins to dissolve the Wessex line; Colan fighting, by some unworldly energy, longer than the others, only to fall like them at last:

> As to the Haut King came at morn
> Dead Roland on a doubtful horn,
> Seemed unto Alfred lightly borne
>   The last cry of the Gael.

These thanes are not only dead; they have been forgotten. They fought as Christians, but without the crystalline innocence of Alfred. 'The spirit of the child' which Christ recommended is seen playing on the shore, along the line which God laid to sunder earth and sea. The child builds a tower, and the sea destroys it, and the child builds it up again. So does Alfred fight, 'gravely, As a good child at play'. Again defeated, his brave allies dead, he continues to obey the Virgin's call to battle with as simple trust as if the day had been his:

> Came ruin and the rain that burns,
> Returning as a wheel returns,
> And crouching in the furze and ferns
>   He began his life once more.

Then comes the last hopeless charge, the vision of Mary over the field, the impossible victory, the baptism of Guthrum, the peace of Wessex instituted under Alfred. But in time the sea washes in on the tower again, and Alfred in his age must buckle on armour and fight, calmly as the child rebuilds in sand. Alfred now disappears from the poem; only echoes and dusty visions return from his campaign. We are left watching the grass grow around the Horse, news returning only at intervals of the grey horse that carries Alfred in his new battles. So the king disappears into the smoke and dust of history which had been stirred up in the first few lines of the poem. The bright vision breaks up and fades, and the White Horse is seen only through a blur of green.

HESKETH PEARSON
Gilbert Keith Chesterton

MOST socially correct men of 40 would perspire at the thought of behaving as they did at the age of 14, and it is refreshing to know that Max Beerbohm, as a middle-aged model of propriety, surrendered to an impulse that shocked him on reflection. Walking through the dark streets of London with William Rothenstein, perhaps a trifle lit-up with liquor, they suddenly decided to revive their schooldays by ringing the front-door bells of two stately houses and running away. They barged straight into the arms of a policeman, who would have taken them into custody if they had not suggested that he needed a drink to appreciate their action. It must have been the lurking child in Max that was instantly attracted to the unabashed child in G. K. Chesterton, who, according to Shaw, 'might be trusted anywhere without a policeman. He might knock at a door and run away—perhaps even lie down across the threshold to trip up the emergent householder; but his crimes would be hyperbolic crimes of imagination and humour, not of malice.'

The child, Gilbert, who was to win the affection of all who knew him, first appeared in Sheffield Terrace, Campden Hill, London on 29 May 1874, but his parents moved to No. 11 Warwick Gardens when he was about 5 and there his youth was spent. His only sister died when he was too young to remember much about her, and his only brother, Cecil, was his junior by five years. An ancestor had squandered what money the family possessed in the time of the Regency; but his son re-established their social standing by selling coal and starting an estate agency, in which Gilbert's father, Edward Chesterton, was a partner, retiring in due course and engaging in various hobbies, one of which, the making of a toy-theatre, enabled Gilbert to enter fairyland. The boy inherited nothing of his father's manual dexterity, but something from his mother, Marie Grosjean, whose Swiss and Scottish ancestry combined to make her practical, commanding, and amusing. As a child Gilbert loved fairy-tales and as a man he wrote them, all his stories being fantastic, all his characters fabulous. He was brought up in a tolerant liberal household and allowed to do much as he liked.

He went to school at Colet Court in Hammersmith, then crossed the road at the age of 12 and entered St Paul's School, where he remained for about five years. As a pupil he failed to give satisfaction. Untidy, absentminded, clumsy, living a world of his own, and unable to take an interest in work or games, he became the butt of masters and boys. He delighted in poetry, the novels of Scott and Dickens, and wandered about in a dreamy fashion, never quite certain of where he was, and conjuring up pictures of Scottish warriors or English oddities. He muttered and laughed to himself, filled his schoolbooks with drawings that obliterated the text, and was quite unable to take the school routine seriously or even to be aware of it. He later referred to his days of pupillage as 'the period during which I was being instructed by somebody I did not know about something I did not want to know'. The High Master of his time, F. W. Walker, had a memorable personality. With a leonine head and a voice like thunder, his roars of rage and mirth echoed through the school and were audible beyond the building. So impressive was he that years later Gilbert pictured him as 'Sunday' in *The Man Who Was Thursday*. The tall and gawky lad even managed to move Walker with a prize poem, and Gilbert, then in the sixth form, was ranked with the eighth by order of the High Master, a promotion that disconcerted the youngster.

Gilbert was seen at his best in the Junior Debating Club, which was started by Lucian Oldershaw, who later became his brother-in-law, and strongly supported by Edmund Clerihew Bentley, who was Gilbert's greatest friend and afterwards achieved fame as the inventor of verses known as 'Clerihews'. There were about a dozen members, one of whom,. Edward Fordham, told me that after a particularly noisy and troublesome session Gilbert, the permanent chairman, took him aside and in a very gentle but serious manner said that the throwing of buns and slices of cake did not create the right atmosphere for intelligent debates. The Club began with the object of discussing Shakespeare, but it soon took in a hundred other subjects, the meetings being held at the houses of the members. A journal was founded, *The Debater*, and Gilbert's first essays appeared in it, mostly on poets, but his later style, thought, and actions were apparent in an essay on 'Dragons', written at the age of 16. After describing the ancient dragon as resembling 'an intoxicated crocodile', he went on to say that the modern dragon had grown prudent:

> He doesn't see the good of going about as a roaring lion, but seeks what he may devour in a quiet and respectable way, behind many illustrious names and many imposing disguises. Behind the scarlet coat and epaulettes, behind the star and mantle of the garter, behind the ermine tippet and the counsellor's robe, behind, alas, the black coat and white tie, behind many a respectable exterior in public and in private life, we fear that the dragon's flaming eyes and grinning jaws, his tyrannous power, and his infernal cruelty, sometimes lurk.
>
> Reader, when you or I meet him, under whatever disguise, may we face him boldly, and perhaps rescue a few captives from his black cavern; may we bear a brave lance and a spotless shield through the crashing mêlée of life's narrow lists, and may our wearied swords have struck fiercely on the painted crests of Imposture and Injustice when the Dark Herald comes to lead us to the pavilion of the King.

A fairly full programme, which he did his best to carry through.

His schooldays began with fighting and ended with friendship. His debating companions were devoted to him and remained his friends for life. He rather enjoyed being stupid in form, doing his best not to learn Latin; he forgot to do his homework and sometimes forgot to attend school, for he was discovered wandering in the playing-field one day when he should have been at his desk and said that he thought it was Saturday, a full holiday at St Paul's. Any other boy would have been frequently punished, but somehow punishment seemed unsuitable in his case. I heard him confess that a master had tried to cane him, 'but the expression on his face and my own ridiculous posture made me laugh, and instead of laying it on thick he suddenly appreciated the absurdity of the situation and desisted'. He wrote a deal of verse in addition to his essays for *The Debater*, and six lines from a poem called 'Adveniat Regnum Tuum' were remarkable for a boy of 16:

> Deep in the heart of every man, where'er his life be spent,
> There is a noble weariness, a holy discontent.
> Where'er to mortal eyes has come, in silence dark and lone,
> Some glimmer of the far-off light the world has never known,
> Some ghostly echoes from a dream of earth's triumphal song,
> Then as the vision fades we cry 'How long, oh Lord, how long?'

He left St Paul's at the age of 17 and spent the next three years at the Slade School for Art, at the same time attending some lectures on English literature at University College. His physiological development having been retarded, he now passed through a somewhat exaggerated period of pubescence, his rather morbid fancy creating dreadful mountains out of common-place molehills: 'I had an overpowering impulse to record or draw horrible ideas and images, plunging deeper and deeper as in a blind spiritual suicide.' This mildest of young men pictured the maddest of crimes, and as his companions were 'a good representative number of blackguards', one of them a diabolist, he became aware of the seamy side of life. He did no work at the Slade, but read and thought a great deal, at last emerging from darkness into light, or, as he put it, 'I was engaged about that time in discovering, to my own extreme and lasting astonishment, that I was not an atheist.' From Shakespeare, Scott, Dickens, and Macaulay he moved on to Swinburne, Whitman, and R. L. Stevenson, the last two lifting his vision out of the twilight in which his soul had been straying, giving him a belief in the redemption of the world by comradeship, and influencing his earlier work. For a short while he called himself a socialist, but he never studied economics and soon drifted into Liberalism. Leaving the Slade he worked for a publisher opposite the British Museum, and then joined another publisher, Fisher Unwin, for whom he slaved away at a salary of twenty-five shillings a week.

In 1896 his friend Oldershaw took him to call on the family of Blogg living in Bedford Park, an artistic quarter of London not unlike the garden-cities of the future. The family consisted of a mother and three daughters, one of whom died shortly after, another was eventually married to Oldershaw, and the third, Frances, was secretary of an educational society. Although Gilbert described Frances as 'a queer card', he fell in love with her at once. She was an Anglo-Catholic, quite remote from the artistic circle in which she lived, and to Gilbert she was a sort of goddess. He was too nervous and diffident to speak of his love, though his whole behaviour expressed it, and Frances wondered if he would ever come to the point. At long last he did, as they were crossing the bridge in St James's Park, and he later confessed that he was thoroughly frightened before putting the dread question. Her answer gave him a foretaste of heaven and he seemed to be floating over the earth. He was so fearful of his good fortune that he wrote a long letter breaking the news to his mother, who was making cocoa for him in the same room. Little wonder that she said: 'I always give money to pavement artists, as I am quite certain that is how Gilbert will end.' Indeeed the momentary difficulty was to establish himself financially before he could marry Frances, and during their long courtship he wrote poetry to her as well as daily love-letters, which were those of a poet in love with prose. His clothes were still so untidy, his general appearance so neglectful, that Frances's mother asked Oldershaw to tackle him on the subject; and Frances too dropped hints which he treated playfully, assuring her when apart that his boots were on his feet, his laces and buttons done up, his tie was round his neck, his hair brushed, cut and shampooed.

By nature a knight-errant, Gilbert had arrived at the conclusion that individuals were right to fight institutions, that minorities were right to fight their oppressors, and that all fighting against odds was noble. He therefore championed the Boers in their war of 1899–1902 against the British imperialists and financiers; but unlike most of the pro-Boers he was not a pacifist. On the contrary he was a violently patriotic Englishman who thought imperialism the reverse of patriotism and believed it his duty to criticize his country's faults. He was pushed into journalism by friends and began writing for *The Speaker*. Utterly unknown in the autumn of 1899, he sprang into fame early in 1900 with the publication of an anonymous volume of poems, *The Wild Knight*, which included his most famous and oft-quoted verses on 'The Donkey'. One critic said that the work was by John Davidson, who repudiated it; whereupon Gilbert acknowledged it. By the following year he was writing regularly for the *Daily News* and other papers, and it seemed as if he could now depend on a regular income. On 28 June 1901 he and Frances were united at Kensington Parish Church, and when he knelt down the sole of a new shoe exhibited the price ticket. He never lost his boyish love of weapons, carrying a sword-stick and a knife that resembled a dagger wherever he went; and after the marriage service, on the way to the station, he bought a revolver and cartridges with which to protect his wife on the Norfolk Broads during their honeymoon, the first night of which was spent at the White Horse Inn, Ipswich.

Mrs Cecil Chesterton states that their married life opened disastrously because Frances was horrified by the sexual act and refused to consummate their union, that Gilbert, full of self-reproach over his lack of consideration, unburdened himself on the subject to his brother Cecil, and that he remained for the rest of his life an enforced celibate. What actually happened may be conjectured. With his invariable clumsiness and no premarital experience of sex, Gilbert rushed the encounter. In his romantic fancy he may have pictured his wife as an imprisoned maiden who had to be carried away recklessly. Whatever the cause he was too impetuous, and instead of feeling his way he assaulted the shrine. But their deep love for one another must soon have taught him restraint and given her comprehension, for the time came when Frances underwent an operation to make her capable of childbirth, which would not have been necessary if they had not proved her incapacity by frequent experiment. Unfortunately the operation was not fruitful, and they made up for their childlessness by taking an endless delight in the children of their friends.

In the years following his marriage Chesterton became a great Fleet Street figure, quite the most notable since Dr Johnson, and the size of his body grew with his fame. Six feet two inches tall, with small feet, delicately shaped hands and a mass of wavy chestnut hair, he became excessively corpulent, weighing some twenty stone before his fortieth year. He loved company, and journalists gathered together at whatever public-house he frequented to hear his discourse on every conceivable topic. His best sayings were repeated from Charing Cross station to St Paul's Cathedral, and his huge person, covered in a large cloak and wide black sombrero, was known to every cabby in the city, for his habit of taking a hansom for a distance of a few yards, keeping it waiting for hours, and then driving in it for another few yards, made him popular with the jarveys. He talked and thought and wrote under every imaginable circumstance, standing under a lamp-post to finish an article, chatting on the pavement to an acquaintance while the rain streamed down, meditating in the middle of the road while the traffic swirled round him. He seemed to be unconscious of time, place or climatic conditions; he was blithely unaware of his personal appearance; and, as his barber once said, he was always 'head over ears in thought'.

In his speech and writing he ridiculed popular fallacies, produced absurd analogies to drive home his point, and burlesqued any argument by carrying it a step further. For instance, he once referred to the type of modern philosopher or journalist who threw off some preposterous assertion in a parenthesis, as though it were too obvious to be discussed. He illustrated this method by quoting someone who had written: 'In two hundred years' time (when the individual consciousness is merged in the communal) . . .', which, said he, was as if himself had written: 'In 1649 King Charles I was condemned to be executed by a Council presided over by Bradshaw (whose mother was a walrus) . . .' or 'My home is in Beaconsfield, which is seven miles from High Wycombe and four miles from Gerrards Cross (where I ate ninety negroes) . . .' He followed these examples with a specimen of the kind of thing then being said by those who opposed Home Rule for Ireland: 'England is bounded on the east by the North Sea, on the south by the English Channel, and on the west by Ireland (whose inhabitants are incapable of governing themselves)...'

His first home with his wife was in Edwardes Square, Kensington, but they soon left that for No. 60 Overstrand Mansions, Battersea, afterwards settling down at Beaconsfield in Buckinghamshire. Soon he was writing a weekly causerie for the *Illustrated London News*, and his articles appeared in all sorts of papers. Every little group or society asked him to lecture or debate, and his evenings were constantly taken up with public disputation. He had a thin high-pitched voice, and his lecturing was not impressive; but the battle of debate roused him, and no one but Shaw could beat him at that game. Humour and wit bubbled out of him and he roared with laughter at his own jokes, whether spoken or written. His Anglo-Catholic wife brought many clergymen into his life, and he enjoyed discussion on theology as much as he revelled in Fleet Street vulgarity. Frances had to keep an eye on his personal appearance and check his absent-mindedness, for he was quite capable of dining at the House of Commons with a boot on one foot and a slipper on the other, and equally capable of sending her a telegram: 'Am in Market Harborough. Where ought I to be?' She wired back 'Home', knowing that she could direct his movements more easily in person. He and his brother Cecil argued about everything interminably, sometimes for twelve hours at a stretch, and not even Frances could tear them apart; though as a rule it pleased him to obey her, for he knew that she alone had his interests at heart.

His high spirits and apparent happiness were infectious. For him the commonplace was always miraculous, and so every hour of the day gave him something to wonder at, or to laugh at, and always to enjoy. His brief comments amused everyone including himself. Here are a few:

Man is a biped, but fifty men are not a centipede.

The word 'good' has many meanings. For example, if a man were to shoot his grandmother at a range of five hundred yards, I should call him a good shot, but not *necessarily* a good man.

I would tell a man who was drinking too much 'Be a man', but I would not tell a crocodile who was eating too many explorers 'Be a crocodile'.

After a meeting at which he had discussed racial characteristics, a woman came up to him and simpered: 'Mr Chesterton, I wonder if you can tell me what race I belong to?' Carefully adjusting his glasses and peering at her, he replied: 'I should certainly say, madam, one of the conquering races.'

Following the success of the female enfranchisement movement, he declared: 'Twenty million young women rose to their feet with the cry *We will not be dictated to*, and promptly became stenographers.'

Hundreds of ideas, flung away by him, were appropriated and used by other journalists. As he sat in a tavern with a bumper of burgundy, his talk was torrential, and when alone he scribbled away unceasingly, whether in tea-shops, public-houses, cabs, omnibuses, trains, or streets. His mind was so full that he found it almost impossible to remember appointments, and he frequently wrote to explain why he had failed to turn up. Once he called on a publisher at the exact hour agreed upon, but he spoilt the effect by handing the man a letter explaining elaborately why he could not keep the appointment.

His stories, like himself, were so full of ideas that they gave the impression of being written by an absentminded man, one whose mind was elsewhere. The first to make him popular with the reading public was *The Napoleon of Notting Hill* (1904), written at the urge of poverty. With ten shillings in his pocket, he left a harassed wife for Fleet Street, where, after a shave, he went to the Cheshire Cheese and ordered a luncheon which included all his favourite dishes and a bottle of red wine. Feeling thoroughly braced he called on John Lane the publisher, gave a synopsis of the story he wished to write, and said that he could not begin it without twenty pounds in his pocket. Lane did not like parting from money, especially at short notice, and promised to send it in a few days. 'If you want the book', said Chesterton, strengthened by wine, 'you will have to give it to me today.' Lane disbursed; and when finished, with his usual indifference to money, Chesterton sold the book outright for a hundred pounds. As in his other fantasies, *The Man Who Was Thursday* (1908), *Manalive* (1911), and *The Flying Inn* (1914), the last of which contains a quantity of rousing verse, Chesterton's Napoleon fights for the liberty of individuals to lead their own lives in their own way against the soul-destroying influence of monopolies and laws made by the few for the control of the many. In his stories we hear for the last time the cry of human beings for the freedom increasingly denied them by cranks, reformers, and financiers. The people in his tales are the embodiments of ideas, for he was interested in opinions, not personalities, and life appeared to him as a conflict between good and evil, not as the clash of actual characters. His pockets were always stuffed with detective stories and tales of bloody combat. Like Don Quixote he pictured himself engaged in an endless battle against diabolic forces, the sword-stick, the revolver, and the bowie-knife which he always carried being the protective armour of an incurably romantic nature. It would have been a dreadful day for him if he had ever drawn a drop of real human blood.

Because he lived solely in a world of ideas, he could not help filling his books with his own thoughts, and the personality of GKC found expression under such titles as *Robert Browning* (1903), *Charles Dickens* (1906), *George Bernard Shaw* (1909), *William Blake* (1910), *William Cobbett* (1925), and *Robert Louis Stevenson* (1927). The ground he held in common with those writers was described perceptively and luminously, but outside that territory he was at sea. Many instances could be given of his failure to understand their peculiarities, but we must content ourselves with one example of misreading in his book on Shaw, and one of misreporting. 'Chesterton's book is a very good one in itself', Shaw said when I touched on the subject. 'It has little to do with me, as GKC has never made any study of my works, and in one place actually illustrates my limitations by telling the world something I should have made one of the characters say in *Major Barbara* if I could have transcended those limitations: the joke being that it is exactly what I did make the character say, as Chesterton might have found had he taken the trouble to open the book and refer to the passage. But if you leave me out of account, you will find, I think, that the book is full of good things, and very generous into the bargain.'

The case of misreporting occurred in GKC's *Autobiography* (1936), wherein he stated that at a male party held in a vast tent in a Westminster garden, the festivities including the boiling of eggs in Beerbohm Tree's top hat, Shaw alone remained sober, at an one point got up, sternly protested, and stalked out. Shaw told me:

Chesterton's memory played him a trick. I usually avoided all social gatherings for men only, as men would not enjoy themselves decently in the absence of women. All that about a tent in Westminster, and boiling eggs in a top hat, if it really took place, is quite new to me. I wasn't there. But there was a male party in the house of a friend in Westminster at which I sat next Chesterton. After dinner, they began throwing bread at one another; and one of them began making smutty speeches. They were actually drunk enough to expect a contribution from me. I got up and went home. You must remember that I am a civilized Irishman, and you cannot civilize an Englishman, nor an English woman either, except superficially. When I first saw an assembly of respectable and sober English ladies and gentlemen going Fantee, and behaving like pirates debauching after a capture, I was astounded. I am used to it now; but it is not possible for me to take part in such orgies.

Against such *gaffes* as these, due to a lack of perception where personalities were concerned, Chesterton could sometimes say in a phrase what other historians would take a chapter to explain, as in his *Short History of England* (1917), where the real cause of the Civil War was shown to be a quarrel, not between the people and the King, but between the squirearchy and the monarchy: 'There was no village Hampden in Hampden village.' Again, in his book on Cobbett, the attitude to the poorer classes of the Whig territorial aristocracy in the eighteenth and early nineteenth centuries was put succinctly: 'They were more interested in pheasants than in peasants: that was an aitch they were most careful not to drop.'

Shaw fully appreciated Chesterton's historical work, and in spite of their dissimilar natures they were very fond of one another, often exchanging visits after GKC's removal from Battersea to Beaconsfield in 1909. Frances was wise to rescue her knight from the Fleet Street tourney, for he drank too much and gave away all his money. He needed a nurse, a valet, an accountant, and a protector, and she now took control of him, giving him a sense of stability. Expenditure was restricted on his visits to London, and he found it impossible to stand a round of drinks on half a crown, but his friends were eager to keep his glass replenished. Like everything else, money was for him an abstract idea, not a concrete counter, and he had a poor opinion of people's acquisitiveness. 'To be clever enough to get all that money, one must be stupid enough to want it', says 'Father Brown' in one of the stories which GKC began to write around that delightful little priest soon after the move to Beaconsfield. Although some of his poems such as 'The Ballad of the White Horse' and 'Lepanto' contained lines that were quoted by all sorts of people, his most popular productions were the 'Father Brown' detective stories, a series of fairy-tales which contain the one sleuth in literature not dimly descended from Dupon or Bucket or Cuff or Holmes. He could splash his colours all over these yarns, and make his paradoxes seem like the deductions of logic.

Frances and Gilbert were happy at Overroads, their first house at Beaconsfield, but happier still when they bought a field and built a studio on it, from which evolved a house called Top Meadow, where they could entertain all their friends, perhaps the most welcome being Father John O'Connor, who provided a few hints for 'Father Brown', Hilaire Belloc, who brought several of his books to be illustrated by GKC, and Maurice Baring, who later helped him to become a Roman Catholic. The paper started by Belloc in 1911 and then edited by Cecil Chesterton received contributions from GKC. It began as *The Eye-Witness*, became *The New Witness* in 1912, and created a sensation over the Marconi affair. Cecil's attack shook the authorities; and when he was found guilty of criminal libel but fined instead of being imprisoned, his friends regarded it as a triumph. Indeed, the whole business smelt a bit fishy, and Cecil's action was justified.

The outbreak of war in 1914 made many people aspire to heroism, and especially the romantic GKC, but a broken right arm that had never recovered prevented him from lifting it above his shoulder, which disqualified him from the infantry, the fact that no horse could carry a man of his weight put the cavalry out of the question, and he mournfully confessed his only value to his country: 'I might possibly form part of a barricade.' Overwork brought on a breakdown in health towards the close of 1914, and for many weeks he hovered between life and death. On recovering he took over the editorship of *The New Witness* from Cecil, who managed to join the army in 1916. It was a dreadful blow when Cecil died in hospital shortly after the conclusion of the war, and Gilbert, with his knight-errant's idealization of action, always declared that Cecil had been killed 'in the trenches' or had 'died fighting in the Great War'. Undoubtedly the conditions of active service had caused Cecil's death, but Edward Fordham assured me that the only time he had ever known GKC anything but open and honest was over his brother's end, insisting that he had perished in battle. As we have seen, he was a little shaky on facts, particularly in regard to persons, and no doubt he honestly believed what his fancy pictured. But he was steady enough on one point in connection with the war. Years before it started Belloc had convinced him that Prussia was the real menace to civilization; and in this they were right, for the financiers and politicians quickly undid all that our soldiers had done, and within a generation Prussian savagery, organized by a monomaniac, was again let loose on the world.

Gilbert and his wife visited Palestine and Italy soon after the war, and then went on a lecture tour of the United States. On Sunday, 30 July 1922, Father John O'Connor received Gilbert into the Roman Catholic Church, the ceremony taking place in the chapel of the Railway Hotel at Beaconsfield. Ever since writing a book on *Orthodoxy* (1908) GKC had been drifting in that direction, and would have reached his destination sooner if he could have taken Frances with him; but her conversion followed in 1926. Just as a realistic sceptical mind like Belloc's needed direction, so did a romantic mystical mind like Chesterton's need anchorage. He wrote:

Catholicism gives us a doctrine, puts logic into our life, ... it is a base which steadies the judgment ... To be a Catholic is to be all at rest! ... The Church of England does not speak strongly. It has no united action. I have no use for a Church which is not a Church militant, which cannot order battle and fall in line and march in the same direction.

The chivalrous warrior and the man who craved for certitude had at last found his spiritual home. Many of his admirers thought that conversion had changed his character. But no one's character ever changes fundamentally, though some circumstances will bring out aspects that are innate, other circumstances other aspects. The most we can say is that GKC's character was diluted, his individuality watered down, by baptism in the Faith.

Frances was out of sympathy with his editorial duties, which indeed wore him out, but loyalty made him carry on the torch lit by his brother. He was not a good editor because the technical side of the business bored him; but he was wonderfully patient, he never got rattled, and everyone on the staff loved him. He hated personal quarrels and animosity, doing his utmost to avoid giving judgement between angry disputants. In 1926 he reluctantly allowed the paper to be renamed *G.K.'s Weekly*, and out of that the Distributist League came into being, its object being to preach the distribution of property, an idea originally formulated by Belloc as an antidote to capitalism and socialism. Chesterton worked hard for it, lecturing and debating in all parts of the country, and incidentally sacrificing a good deal of the money earned from his books to keeping the paper afloat, much to his wife's discomfort. Following his grave illness he was compelled to restrict his wine-drinking, but he never attempted to lessen his labours, and it was a fortunate day for him when Dorothy Collins became his secretary, for she arranged his affairs in every department, from income-tax returns to social and speaking engagements. Frances, too, was overjoyed by the coming of a perfect secretary, and soon Dorothy Collins was an adopted daughter, accompanying them everywhere, driving them about on holidays, and organizing everything. Frances suffered from arthritis of the spine and was not always capable of attending to her husband's wants, though he still shouted for her when he wanted a tie to be tied or a bootlace to be done up.

They again visited America in 1930–1 and GKC lectured all over the States. His physical clumsiness seems to have impressed the Americans as much as his mental alertness and they were concerned over his difficulty in getting into and out of motor cars. Someone suggested that he should do it sideways. 'I have no sideways', said he. Particularly impressed in New York by the advertising signs that flickered on Broadway, he thought of them as fairy-lights, crying: 'What a paradise of beauty this would be for anyone who couldn't read!' Frances was ill at Chattanooga in Tennessee, and Gilbert at once cancelled all his lectures. His agent was in despair and begged Dorothy Collins to accompany him. Frances got better, and he agreed to continue his tour with Dorothy in attendance, but the cancellations had cost him several hundreds of pounds. Incomparably better as a talker than as a lecturer, he suddenly won a new popularity in the early 1930s by proving himself an ideal broadcaster.

The last year of his life was saddened by fierce personal quarrels between members of the paper's staff, and increasingly he sought solace at home, where he loved entertaining children, throwing his dagger about the lawn, making passes at the flowers with his swordstick, practising shots with a bow and arrow, and playing with his dog. One little girl was told by her mother that she was going to visit a very great and clever man, from whom she could learn a lot. After the tea-party the child explained what she had learnt from the great and clever man: 'He taught me how to throw buns in the air and catch them in my mouth.' In much the same casual way he wrote stories and essays, throwing ideas into the air and leaving his secretary to catch them on paper. Whenever Dorothy Collins told him that money was needed he would say: 'Very well. Let us write a "Father Brown" story', and start it at once. Fortunately he never heard that his last 'Father Brown' story was declined by the editor to whom it was sent. It would have distressed him to know that he was written out and that the story was too bad to be published.

Early in 1936 it became clear to Frances and Dorothy that he was failing, and a motor-trip abroad was advisable. They went to Lourdes and Lisieux, and he seemed better. Driving them home from the English coast, Dorothy said 'Oh, sing something!' GKC promptly obliged and sang extracts from Gilbert and Sullivan operas for an hour or so, interspersing them with nonsense and chit-chat. When they got back he tried to work as usual, but his mind had lost its usual clarity and he fell asleep at intervals. He was soon laid up with a weak heart, and had periods of unconsciousness. The thought of death had terrified him in the past, and when he pricked his finger or felt slightly ill his wife and secretary had to console him that all was well. But when he knew that he was dying, he was not in the least afraid. On 13 June 1936 Frances and Dorothy entered his room to see that he was comfortable. He opened his eyes and said: 'Hullo, my darling! Hullo, my dear!' They were his last words, and next day his heart ceased to beat.

I heard from Dorothy Collins that the coffin in which he lay was so big that it could not be got through the door of the bedroom, and that windows were removed for its egress. This reminded me of a talk I once had with him about my favourite character in literature. I said that Shakespeare had been wrong to kill Falstaff, who was deathless, and GKC played with this idea:

> You are quite right. Falstaff was merely shamming death, as he had done at the battle of Shrewsbury. But when they came to fetch his body, they could not

get it down the narrow stairs; so they removed the roof and hoisted him up. When they had got him so far, his spirit took him further, and he soared upwards until he was out of sight; and the gravedigger who had talked to Hamlet stared at the sky muttering: 'Is he to be buried in Christian burial that wilfully seeks his own levitation?'

Nearly a decade before his death the passing of GKC, one of the most lovable personalities in the history of English letters, had been celebrated in a 'Premature Epitaph' by Colin Hurry:

> Place on his hand the jewel, on his brow the diadem,
> Who in an age of miracles dared to believe in them.
>
> > Chesterton companion
> > > His companions mourn.
> >
> > Chesterton Crusader
> > > Leaves a cause forlorn.
> >
> > Chesterton the critic
> > > Pays no further heed.
> >
> > Chesterton the poet
> > > Lives while men shall read.
> >
> > Chesterton the dreamer
> > > Is by sleep beguiled;
> >
> > And there enters heaven
> > > Chesterton . . . the child.

NEVILLE BRAYBROOKE
The Poet of Fleet Street

NUMBER ONE, Leinster Square is a Victorian terraced house standing on the north side of the Park. It has a deep basement area, with steps leading down to it. Yesterday I peered into its lower window, and it was like looking into the past. It was in the early 1920s that 'The Maida Vale and Hampstead Literary Society' held its meetings at this Bayswater address. The topsy-turvydom about the society's name and address had appealed to Chesterton, and gladly he had come to give a talk entitled 'Father Brown and Myself'. I remember my father telling me how our flat had been crammed for the occasion and that some members had been forced to sit on wooden crates. I remember too being told how once GK had sat on a particularly frail chair in my grandfather's home and it had collapsed under his weight. 'You cannot say on my first visit to your house I fell on my feet', he quipped.

Chesterton loved to make play on popular expressions, and he was as much attracted by the topsy-turvy in life as he was delighted in turning sentences upside down and transforming them into paradoxes. To him, paradoxes, like puns, were a kind of poetry; they were a means of saying several different things at one and the same time. His vision was highly personal:

> If a man saw the world upside down, with all the trees and flowers hanging head downwards as in a pool, one effect would be to emphasize the idea of *dependence*. There is a Latin and literal connection; for the very word dependence only means hanging. It would make vivid the Scriptural text which says that God has hanged the world upon nothing.

That passage occurs in his life of *Saint Francis of Assisi* (1923). Five years later he reworked and telescoped this idea in *The Poet and the Lunatics*: 'You remember [Saint Peter] was crucified upside down.... He also saw the landscape as it really is: with stars like flowers, and the clouds like hills, and all men hanging on the mercy of God.' And so the echoes continued throughout his life—sometimes growing repetitive— but more frequently striking with an added clarity.

Chesterton's father was a prosperous London auctioneer who always thought of his son as a poet, and when Gilbert married at the age of 27 he found that he had acquired not only a daughter-in-law but an ally. For Frances Blogg also thought of Gilbert as a poet. But Chesterton preferred to call himself a journalist. To those who exalted the vocation of the poet above the journalist he gave this answer in the *Darlington North Star* (3 February 1901):

> The poet writing his name upon a score of little pages in the silence of his study, may or may not have an intellectual right to despise the journalist: but I greatly doubt whether he would not morally be the better if he saw the light burning on through darkness into dawn, and heard the roar of the printing wheels weaving the destinies of another day. Here at least is a school of labour and of some rough humility, the largest work ever published anonymously since the great Christian cathedrals.

Years later he returned to this theme in his *Autobiography*, which was published posthumously after his death in June 1936. True to paradox there, he claims that his early success in Fleet Street depended upon doing exactly the opposite of what he was advised. To the Nonconformist *Daily News* he would send articles about the cathedrals of France and French café life (which the readers loved 'because they had never heard of them before'), and to the robust, pro-Labour *Clarion* he would write on 'all [those] things which their readers had never heard of' such as Thomist theology or the medieval guild system. I once heard Douglas Woodruff say that Chesterton could dictate as many as 10,000 words in a morning's work.

He himself was fond of the term 'a roaring journalist'. It meant a windmill attack, with words and ideas flying in all directions: 'Dickens was a man whom anybody could hurt but nobody knock down,' or 'when we call a man "manly" or a woman "womanly" we touch the deepest philosophy'. How well such sentences are aimed straight at the heart. But there were, alas, those which landed nowhwere: 'Most people either say that they agree with Bernard Shaw or that they do not understand him. I am the only person who understands him and I do not agree with him.' This is pure silliness, and there were moments when Chesterton allowed himself to become subject to the laws of rhythm rather than reason. By the time he was 50 there were tedious spells when his early originality would be replaced by a technical trickery—always the danger for the popular journalist who must go on writing, writing, writing.

Yet his finest book, *St Thomas Aquinas*, was published three years before his death when he was 57. Half-way through dictating it, he turned to his secretary: 'Now I had better read some Aquinas.' It was no affectation. Etienne Gilson, the greatest Thomist of this century, has pronounced it 'a work of inspiration'. For Chesterton had the optimism that goes with an inspired guesser; where theologians and philosophers had tentatively searched out the way before him, he guessed the right road—and it proved a Roman road that ran straight through all the previous arguments of speculative scholarship and meandering meditation.

'I do not search, I find. I do not believe, I know.' These paradoxes were the texts by which he lived and wrote, just as they remained the texts by which Father Brown carried out his detective ministry (which was spiritual as well as active). Alone with a criminal on a Gothic tower both priest and murderer share the same vision: 'the monstrous foreshortening and disproportion, the dizzy perspectives, the glimpses of great things small and small things great; a topsy-turvydom of stone in the mid-air.'

In the summer of 1922, twenty-one years after his marriage, he was received into the Catholic Church. These two events he considered the most important in his life. Some men lose their identity in religion, some in marriage. But Chesterton discovered his in both. Catholicism offered him a system in which he could develop his ideas (Father Brown, one of his many *alter egos*, was created ten years before his conversion), and in his wife he had already discovered a woman who, in managing his press-dates and lecture tours, allowed his mind the necessary freedom to wander. Wandering was a prelude to wondering, and 'I wonder' was a phrase seldom off his lips.

Before his marriage to Frances, he had told her that trying to love everyone had acted as a preparation for loving her; and in his early notebooks there is an entry that reads: 'My great ambition is to give a party at which everybody should meet everybody and like them very much.'

His own life was an example of that ambition, and his marriage proved to be as happy as Robert Browning's whose life John Morley had commissioned him to recount in 1903 for 'The English Men of Letters' series. Towards the close of his own engagement, he had ended one letter: 'As long as one hasn't by some oversight got too much money there is a great deal of poetry in a penny. Particularly in the one that stamps this [envelope], dear.'

The days of the penny post, like those of the penny press, are over; but the poetry in a penny remains. In Fleet Street, Chesterton was always a poet—even in his prose. That is the final paradox about his achievement. His poems at the best are hymns—like 'The Donkey'—or alternatively good roaring verse like his ballads about 'St Barbara' or 'The White Horse'; 'Lepanto', the famous anthology piece, is full of rousing shouts:

> Don John of Austria has burst the battle-line!

In the early 1920s, on both sides of the Atlantic, undergraduates could be heard yelling his songs late into the night.

Now the perspective has changed. Twenty-five years after his death he is best remembered as an essayist of genius and the author of a handful of inspired biographies. There has been no customary eclipse and there have been numerous reprints of his books. He has kept the bond going between author and journalist at a period when much has happened to discredit the latter term. And that surely would have pleased him more than anything. It would have been his answer to those critics who had argued, and still argue, that the weekly essay cannot be a work of art. Yet to recall those weekly essays is to remember those press-dates which Frances made her husband keep with such diligence. Gallantry calls for another ending—and what better than those lines that he inscribed on the flyleaf of her copy of *The Wild Knight*? They are not included in his *Collected Poems*. They are dated Christmas, 1900, and were first found some years after his death:

> Dearest: what others see
> Herein, it is no mystery—

That I find all the world is good
Since you are all the world to me.

You will not blame my boastful hours,
It is not of such souls as yours
To spew the warmth of sorrow out
Upon the harmless grass and flowers.

Do you fight on for all the press,
Wise as you are, you cannot guess
How I shall flaunt before God's Knights
The triumph of my own princess.

Almost this day of the strange star
We know the bonfire old and far
Whence all the stars as sparks are blown
Piled up to warm us after war.

There when we spread our hands like wings,
And tell good tales of conquered things
The tale that I will tell of you
Shall clash the cup of all the Kings.

I swear it shall be mine alone
To tell your tale before the throne
To tell your tale beside the fire
Eternal. Here I tell my own.

JOHN WAIN
Manalive, *a Good Bad Book*

It is always a pleasure to be given an opportunity of testifying to a past admiration. In my own case there are a good many writers whom I couldn't, with any honesty, praise nowadays, or recommend people to read, but who gave me the greatest enjoyment at one time—more, perhaps, than any writer will ever give me again. And I am always suspicious of teachers, and adults generally, who are for every trying to hurry the young on to the next step. To guide them from the easy to the hard, from the simple to the more satisfyingly complex, is no doubt an excellent thing to do, but if a boy or girl is enjoying a book, getting a lot out of it and looking forward to reading everything written by that author, then hands off, I say: let the appetite cloy itself naturally, but don't implant a sense that what they genuinely like is worthless or they will only turn into insincere snobs who praise what they think they ought to praise. I'm not ashamed to admit that there was a time when I admired the dialogue of Dodie Smith and Esther McCracken, saw genuine merit in Lytton Strachey as a biographer, hung enraptured over the pages of Walpole—and Cronin, and considered Rupert Brooke a mighty poet. Good luck to them! If they hadn't existed I'd have given up reading altogether. I owe them a crucial debt.

My feelings towards Chesterton's *Manalive*, however, go slightly beyond this generalized sense of gratitude for a past benefit. Although *Manalive* fits well enough into the category of books I enjoyed in my teens but don't want to read again, it also transcends it. Strange though it may seem to me now, I know that that book did something to my mind, gave me encouragement and help of a kind that I was beginning to need intensely, without knowing that I needed them.

To explain this I shall have to go back over the book for a moment, since most people will either have forgotten it or never read it. *Manalive*, which Chesterton published in 1912, is about a character called Innocent Smith, whose life is devoted to a kind of one-man crusade against world-weariness and pessimism. He fights these qualities in himself as well as in other people, putting himself through a continual series of difficult and dangerous exploits purely to remind himself that he is still alive and that the ordinary details of existence are valuable. He organizes a cat-burgling expedition which turns out to be on his own house, he deserts his wife and family and travels right round the world so as to approach them again through distance and difficulty, and so on.

All this gives rise to a lot of knockabout paradoxical fun, but it has a more serious side which is anchored to the author's deepest convictions. Chesterton grew up during the *fin de siècle*, when atheism and pessimism were more or less compulsory, and the hallmark of a serious mind was a gloomy sense of futility—and he reacted so strongly against this attitude that it is not too much to say that he made it his life's work to combat it. Even his conversion to Roman Catholicism, was only one more means to this end: he was drawn to it, one feels, because it was a scarlet-and-gold religion with firmly marked-out positives and negatives.

However that may be, the real heart of Chesterton's work is in his hatred of the *fin-de-siècle* spirit. The figure of the typical lumpen-intellectual of the 1890s, sitting with folded arms staring darkly into his glass of absinthe and working out new ways of stating his final disillusion with the universe, remained always in his mind as the image of his enemy. He fought his enemy in everything he wrote; in his poetry, with its noisy insistence on the miraculous nature of everyday existence in the world; in his criticism, with his clever but slanted presentation of Dickens and Browning as propagandists for cheerfulness and his venom against a writer like Hardy ('the village atheist brooding and blaspheming over the village idiot'); in his personal life, with its cult of the zestful, blameless eccentric; and of course in his fiction. Even Father Brown solves crimes which baffle other minds because he is alert to the miraculous nature of the universe and therefore spots the vital clue at once, the clue that leads further out into mystery rather than safely back towards the reasonable.

Chesterton was a writer more notable for energy and audacity than for subtlety, and his vast output has worn badly, though I don't think it is likely to disappear altogether. Certainly he never wrote a book that one doesn't have to make allowances for. There are spots of silliness and prejudice, improbabilities and rickety workmanship generally, everywhere one looks. He never slowed down long enough to write one book, even one short story or one poem, that was as good as his gifts could make it. The result is that a lyric of Housman's or a novel of Bennett's is more effective in presenting the case *for* pessimism than anything Chesterton wrote to present the case *against* it. Art, to be good propaganda, has first to be good art.

All this must be said. And yet the delight with which my sixteen-year-old mind dived into *Manalive* is a testimony that Chesterton's books could work, and work powerfully. On an immature mind, perhaps. On a mind so much in need of a certain kind of solace that it was willing to forgive, or simply not notice, any number of blemishes. (This must be why the young of today can be so genuinely addicted to writers like Jack Kerouac, William S. Burroughs, Gregory Corso, Lawrence Ferlinghetti, and Uncle Ben Beatnik and all.) Given the one vitamin that the mind craves (or the one drug?), a book can simply slide past the defences. Because it had what I needed, I thought *Manalive* the most brilliantly written book I had ever read. Chesterton's language seemed to me more poetic than Shakespeare's. His purple patches, which I should now dismiss as 'fine writing' that he ought to have blue-pencilled out, acted on me like Drambuie. The book opens, you will recall, with a symbolic gale, and its first words are:

A wind sprang high in the west, like a wave of unreasonable happiness, and tore eastward across England, trailing with it the frosty scent of forests and the cold intoxication of the sea. In a million holes and corners it refreshed a man like a flagon, and astonished him like a blow. In the inmost chambers of intricate and embowered houses it woke like a domestic explosion, littering the floor with some professor's papers till they seemed as precious as fugitive, or blowing out the candle by which a boy read *Treasure Island* and wrapping him in roaring dark. But everywhere it bore drama into undramatic lives, and carried the trump of crisis across the world.

I don't say that is bad writing exactly; on a certain level it is very good, and this was precisely the level on which I, at 16, suddenly and breathlessly found myself. (A year or two previously my idea of a stylist was Richmal Crompton, bless her.) As Chesterton's prose trumpeted and roared, I felt like cheering. And now and then—to be fair—I stumbled on a sentence or two that really *was* excellent; or, to put it more precisely, that my maturer judgement still finds excellent. 'With our weak spirits we should grow old in eternity if we were not kept young by death. Providence has to cut immortality into lengths for us, as nurses cut the bread and butter into fingers.' Thus Innocent Smith—and the metaphor still seems to me a genuine poet's image, homely and yet telling.

Not that I distinguished between passages like that on the one hand, and mere rant on the other. The rant is of course much more plentiful: Chesterton's impulse to pull out all the stops overcame him several times in the course of each chapter, so that one is always having to wade through over-emphatic verbiage such as:

We shall have gone deeper than the deeps of heaven and grown older than the oldest angels before we feel, even in its first vibrations, the everlasting violence of that double passion with which God loves and hates the world.

Still, at 16, I swallowed the book whole. The fact is I needed it. My situation, intellectually and emotionally, was such that it might have been written especially for me. Deeply unsure of myself, very much at odds with my surroundings, quite certain that I could never 'get on' in any way that would seem like 'getting on' to those about me, and yet only very dimly aware of another kind of life beyond the horizon, I was naturally a prey to moods of despair. But beneath this despair, like a granite foundation, lay a stubborn optimism about the whole nature of life, a sense that to be allowed to participate at all was a privilege, even if nothing seemed to fit together at the moment. This optimism was instinctive; it was drowned by the voice of experience which pointed out that I had always been a misfit, and also by the waves of self-pity that attack every adolescent; and it was completely without any intellectual foundation. 'Religion', into which I took fitful plunges, seemed mainly to concern itself with punishment and penitence; and I knew nothing of philosophy.

Suddenly the classics master at school, whose name was Bacchus (no kidding), took it into his head to lend me *Manalive*. Whether he knew what he was doing for me I don't know. But he did it.

Chesterton's manipulation of 'ideas', unsubtle though it was, sufficed to give me the feeling that I could justify my instinctively yea-saying attitude, buttress it with some sort of intellectual defence; his strongly affirmative and uninhibited writing made me feel that 'style' was something you did for a purpose and not just to show off. And I did the rest. What harm there was in the book, what over-simplification or bigotry, didn't harm me. No, not even the gratuitous anti-Semitism. There is a Jewish character who seems to have been introduced purely to show the Jews as the eternal enemies of 'innocence'; at a crucial moment, when Smith's case is being pleaded, this character smiles 'a certain smile. It was that smile of the Cynic Triumphant, which has been the tocsin for many a cruel riot in Russian villages or mediæval towns.' But I didn't know what a tocsin was, and didn't even realize that Chesterton was justifying, as nearly as he dared, the persecution of the Jews. I simply raced past it without bothering.

Manalive was the first 'novel of ideas' I had ever come across. I must have been aware that the characters were unconvincing and the action incredible; after all it is part of the charm of the book that Chesterton doesn't even try to make them anything else. And the story has vitality because, for all the artificiality of its people, it is about something real, the struggle between denial and affirmation in which Chesterton believed with all his might. I had already taken sides in the struggle, though I did not even know it had been formally joined. When Innocent Smith declared 'I don't deny that there should be priests to remind men that they will one day die, I only say that at certain strange epochs it is necessary to have another kind of priests, called poets, actually to remind men that they are not dead yet'—the effect on me was of a revelation. All at once I knew that my instinctive, unformulated feelings could be put into words, that they had the dignity of a 'philosophy of life', that other people, too, wanted what I wanted: to push down the blank walls behind which we were hiding and step out boldly to larger feelings, broader perspectives, and life more abundant.

CHRISTOPHER HOLLIS
Chesterton's Paradoxes

CHESTERTON justified his use of paradox by the valid Aristotelian argument that in order to attract the reader's attention it is first necessary to surprise him, and a paradox embedded into a passage of common sense does indeed surprise, but Chesterton's style means that he is not an author well suited for extracts.

A mere collection of paradoxes, extracted from other matter and standing by themselves, ceases to surprise and they reveal—what was indeed inevitable in so prolific a writer—how often Chesterton repeated himself both in his argument and his image. Nor can one quite help noticing how often the argument could be turned just as effectively round the other way, if it had suited the writer's purpose so to turn it.

In one respect Chesterton is enormously up to date. He wrote about the ultimate truths of religion, and there is far more space given in the Press today to that debate than was the case in his time. On the other hand, there is another respect in which he is very much out of date.

To those who grew up before 1914 it was almost impossible not to believe that civilization was a state that had been finally and irrevocably achieved or that there was nothing that man could do that could possibly break the crust of it. Therefore Kipling, when he writes *Kim*, is able in a fashion that seems to us almost intolerably simple-minded to write of the life of a secret society without ever considering whether the inevitable daily mendacity of that life would not have had a corrupting effect on character. In the same way with Chesterton and the effect of violence.

Chesterton himself said that his first serious book was *The Napoleon of Notting Hill*. It is both a witty and a wise book, whose fundamental lesson is, as are all Chesterton's fundamental lessons, well worthy of consideration.

Yet it has about it a maddening *naïveté*. It describes an imaginary civil war in the London of the future without ever a hint that that civil war could possibly in any way interrupt the normal pattern of life. During it men eat at their leisure five-course luncheons in a restaurant and the buses, run according to their normal schedule.

More important, it never occurred to Chesterton that Londoners could kill one another in civil strife without that violence in any way diminishing their natural kindliness or affecting their characters.

The events in Notting Hill are cast in the year 1984 and it was in order to rebuke Chesterton's schoolboy innocence which made him think of a civil war as something like an exchange of views at the St Paul's debating society, which caused Orwell to choose that same year of 1984 for his picture of the future—for his prophecy of what life would really be like in an atmosphere of violence.

MALCOLM MUGGERIDGE
G K C

G. K. CHESTERTON is one of those writers for whom either too much or too little is usually claimed. Unlike some of his eminent contemporaries (Arnold Bennett, for instance), he still has devoted readers. The Father Brown stories continue, I suppose deservedly, to be popular . . . .

It is often contended that Chesterton wrote too much and too hastily; that his financial exigencies and polemical inclinations led him to prefer journalism to literature. This, in my opinion, is highly dubious. In the first place, I question the hard and fast distinction between the two categories of writing. Obviously, there are literary masterpieces like, say, *War and Peace* and *L'Education Sentimentale* on the one hand, and hack journalism like, say, writing editorials for the *Daily Express* or ghosting Christine Keeler's memoirs for the *News of the World* on the other. Between them lies a vast no-man's-land in which most writers, from Johnson to H. G. Wells and Shaw, have been content to forage, if not to reside. Wells, indeed, actually preferred to call himself a journalist, though in point of fact his early novels like *Kipps* and *Mr Polly* have better credentials to be regarded as imaginative literature than many more highly regarded works—*Howard's End*, for instance, or *Mrs Dalloway*.

Then again, I question whether Chesterton had it in him to be much more than the journalist he was. It seems to me that periodical writing exactly suited his talents, as it did Orwell's. They were both atrocious novelists and, on form, superb essayists. Chesterton's most notable talent was for the sudden crystallization of illuminating observations. It would be a good idea, incidentally, to make a collection of these, like La Rochefoucauld's *Maximes*. The following are chosen at random. 'When you break the big laws, you do not get liberty; you do not even get anarchy. You get the small laws.' 'The meanest man is immortal and the mightiest movement is temporal, not to say temporary.' 'In the whole world of things conceivable there is nothing so unmercifully hopeless as an infinity of mere facetiousness, a tyrannical nightmare of jesting.' Let me add two others which have always stuck in my mind. When Chesterton first saw the lights of Broadway he remarked that the spectacle would be marvellous if only one couldn't read. And somewhere or other he writes that when people cease to believe in a deity they do not then believe in nothing, but—what is much more calamitous—in anything. This sort of facility is used to best advantage in occasional journalism. With too portentous a presentation the charm is lost. Chesterton's flashes are like Chinese lanterns—pleasingly spread about a sprawling garden on a warm summer evening; frail and even tawdry, indoors.

Mr Frank Swinnerton has, it seems, suggested 'that it will be at least a hundred years before Chesterton's greatness is fully recognized.' Chesterton himself, with commendable modesty, was convinced that what he wrote was for the moment only.

I would rather live now and die, from an artistic point of view, than keep aloof and write things that will remain in the world hundreds of years after my death ... It so happens that I couldn't be immortal; but if I could, I shouldn't want to be.

One is always a little hesitant about accepting assurances of this kind, but all the same Chesterton's own estimate of his work would seem to be juster than Swinnerton's.

It is, of course, from the standpoint of a Christian, and specifically a Roman Catholic convert, that Chesterton propounds his view of life. He never tires (though a reader sometimes does) of belabouring Darwin and the pretensions of 'science'. Among contemporary liberal intellectuals he was something of an odd man out with his passion for ritual and ceremonial, civic as well as liturgical, and with his loathing of collectivist hopes and Quaker virtues, and refusal to endorse the fatuous expectations invested in the idea of progress. He felt a deep, instinctive distaste for the way the twentieth century was going which enabled him, in his early years of pessimism, to be an impressive prophet. 'The earnest Freethinkers,' he wrote in 1905, 'need not worry themselves so much about the persecutions of the past. Before the Liberal idea is dead or triumphant, we shall see wars and persecutions the like of which the world has never seen.' Stalin, then a young man of 26, and Hitler, ten years younger, were, along with others, to make good his words to a fabulous degree.

It is surprising, in a way, that, when Chesterton has so often been proved right in his judgements, he should still be less seriously regarded than contemporaries like Wells and the Webbs who were almost invariably wrong. This is due partly, I think, to a certain ingrained flippancy in his whole attitude of mind, which, though acceptable to English readers, precludes them from taking him seriously. The Reformation, and a long diet of sermons, has conditioned us to associating wisdom with solemnity. I have to admit that, rereading Chesterton's autobiography, I could not myself get over a feeling that, amiable and gifted as he undoubtedly was, he was also somehow a fraud.

Perhaps it is just the enormous size of him. Those illustrations conveying his colossal bulk weigh on the text. One is oddly conscious of his physical appearance, as with Walt Whitman: a not dissimilar figure in some ways, and greatly admired by Chesterton. Among Whitman's effects when he died was found an artificial butterfly fastened to a ring. It had been worn by Walt for a famous photograph which showed a butterfly delicately and lovingly perched on his finger. One would not be utterly amazed to find that Chesterton had used such props. His get-up, with flowing cloak and wide-brimmed hat, was a kind of fancy dress from whose fastness he proclaimed the glory of the common man. Of Henley and Whistler, 'two very great men', Chesterton wrote that 'nothing strikes one so much about the attitude of both as the fact that a superb melancholy made it necessary for both to take refuge in something outside current life.' However he saw himself as a writer, he was in very much the same case, though he would not have admitted to the superb melancholy, with his macabre 'what larks!' Yet underneath the happy Christian and happy husband, the lover of peasants, Fleet Street roistering and country inns, one senses in Chesterton a brooding, anguished, frightened spirit; a frustrated romantic, a displaced person, a letter delivered at the wrong address.

The only time I ever saw him in the flesh he was seated outside 'The Ship Hotel' at Brighton shortly before he died. His canvas chair looked preposterously small, as did a yellow-covered thriller he was reading. It was a windy day, and I half-expected him to be carried away. Though so huge, he seemed to have no substance: more a balloon than an elephant.

It was, one feels, as a fugitive from an estate agency in Kensington rather than from Victorian agnosticism that Chesterton found his way to Rome; an awareness of the terrors of a larger stage which led him to peer down so assiduously at a toy theatre. The answer to the inadequacies and tedium of Notting Hill is surely not, as Chesterton supposed, to imagine a Napoleon there, but rather to see in that horrible, vainglorious little Corsican all the vulgarity and littleness which so detracts from human life everywhere. Likewise with the notion, to which Chesterton constantly recurred, that the monotony of familiar things and faces can be counteracted by making them seem unfamiliar. This is worked out in, for instance, *Manalive*, in which the hero, Innocent Smith, goes on an enormous, adventurous journey in order to arrive at Clapham Common and home; breaks into his own house in order to taste the excitement of burglary, and ravishes his own wife to experience the delights of seduction—this last, incidentally, a theme which might have made the Marquis de Sade draw back appalled.

The Christian view of the matter I have always assumed to be almost the exact opposite—that going, like Satan, to and fro in the world and up and down in it is a futile proceeding because every part, Clapham Common or Tashkent, is equally illusory; that a Napoleon only projects onto a larger screen the squalid egotism of each human heart; that flesh and spirit pull in contrary directions irrespective of whose flesh. An undercurrent of uneasiness, if not anguish, in Chesterton's writing perhaps derives from an awareness of this contradiction between Christian pessimism and the obligation he felt himself to be under not to shoot the contemporary pianist.

In his play, *The Surprise*, the action is first performed by puppets in accordance with the author's wishes; then it is performed by living men and women with wills of their own, who, of course, create havoc and confusion. 'In the devil's name,' the author cries out from the wings, 'what do you think you are doing to my play? Drop it! Stop! I am coming down.' This, a superb image of the Incarnation, is Chesterton at his best. Such is the occasional yield of precious metal from the vast, disorderly open-cast operations in which he engaged.

CHRISTOPHER HOLLIS
G. K. Chesterton

A LOSS of popularity is the lot of most writers in the years immediately after their death. It is certainly true that Gilbert Keith Chesterton (1874–1936) is not read by the young as he was in the first decade of the century, when he was the popular favourite of journalism.

It was the Boer War which first inspired Chesterton with his romantic love of the small nation fighting for its freedom against the imperialistic octopus. While other pro-Boers championed the cause of the Boers on the pacifist ground that the British were wrong, Chesterton—and his friend Hilaire Belloc—championed it on the ground that the Boers were right. It is true that subsequent experience has shown that the Boers, having gained their own freedom, have been somewhat less deeply concerned with the freedom of others. But that is another story.

When the Boer War was over Chesterton wished to express his idea in a gentler and more whimsical medium. He amused himself by imagining in the future a great revival of local patriotism among the London suburbs, which led to a war between the people of Kensington and the people of Notting Hill. He dedicated his extravaganza with characteristic high spirits to 'the human race to which so many of my readers belong'.

The most interesting point about the book was that, written in 1904, it explains that it is giving the history of what is supposed to be happening eighty years later—that is to say, in 1984. George Orwell had a kind of love-hate relationship with Chesterton. He admired the vigour and humour of his writing, but disagreed very violently with his opinions, and there is no doubt that one of the reasons why Orwell chose 1984 as the date for his vision of the future was in order to make an oblique protest against Chesterton.

Certainly Chesterton's 1984 and Orwell's 1984 differ remarkably from one another. Orwell imagined a future of dreary tyranny where all decency had been stamped out by power-mad lunatics. In contrast, civil war to Chesterton was a light-hearted affair. All his people are decent enough, not very unlike those with whom Chesterton consorted every day. Civil war can rage without any upset to the normal habits of life. People can eat a six-course luncheon in a restaurant. The buses run regularly, and above all men's characters are not at all changed or degraded by their trafficking in violence. It is much the same with Kipling in *Kim*, where the author imagines that people can belong to secret societies and indulge in espionage and killing without any effect on their characters.

In *Nineteen Eighty-Four* Orwell was of course protesting against what seemed to him Chesterton's light-hearted irresponsibility. To Orwell the crust of civilization was a very thin one, and in a world of violence men would easily sink into a sub-human condition. Orwell perhaps overblackened the picture. There is, it seems, an enormous resilience in human nature which makes possible recovery from the most desperate situations. But he was perhaps justified in his protest against the excessive light-heartedness of Chesterton and Kipling.

It is important to remember the different dates at which they wrote. Orwell published his *Nineteen Eighty-Four* in 1950. He had lived through two World Wars, a Bolshevik revolution in Russia, a Nazi revolution in Germany, and the Spanish Civil War. Whether he will prove to be right or not, he certainly had reason for his pessimism. It is hard for us now to recapture the wholly different mood of the last years of the nineteenth century, or the first decade of the twentieth century. The world then, to a comfortable middle-class Englishman, seemed to have achieved stability. It was, if anything, too boringly stable. 'Are we never to shed blood again?' petulantly asked Robert Louis Stevenson. Such wars as could possibly happen would be distant, minor, overseas affairs, sufficient to titillate, but with no danger of upsetting the rhythm of civilized life. Stability was achieved and the future before mankind was one of inevitable progress.

Chesterton, writing in the first decade of this century, was not unique in thinking that nothing catastrophic could ever again happen. He would rather have been unique had he thought otherwise. Even when war broke out in 1914, no one thought of it as the end of civilization. They thought of it rather as an interruption to the inevitable march of progress which would certainly resume its advance as soon as the war was over.

In this same period Chesterton was giving his mind to deeper things. He wrote at that time a very popular column in the *Daily News* in which he tilted at the well-known pundits of the day, amusingly exposing their various contradictions. He published these essays in a book called *Heretics* (1905), and it was G. S. Street, another popular journalist of the day, who, in a criticism of this book, said, 'Mr Chesterton is for ever telling us that everybody else is a heretic, but what is the orthodoxy from which he judged them? What is it that Mr Chesterton himself believes?' This was the kind of challenge to which Chesterton was always very ready to reply, and his reply was the most important book that he had up till then written, *Orthodoxy* (1908).

Chesterton was born of a Unitarian family and had grown up without any definite religious beliefs—certainly without any whiff of Catholic leanings—but it was in stating his own opinions, in writing *Orthodoxy*, that he discovered that his ideas were the same as the ideas of Christianity, that his orthodoxy was the Christian orthodoxy. He was by no means ready at that time to become a Roman Catholic. Sixteen years were to elapse before he took that step, and although when he came to take it he would certainly have said that it was by far the most important step of his life, yet from the literary point of view—with which we are here alone concerned—his argument for the essential truth of Christianity did not differ much as between his pre-Roman Catholic and his Roman Catholic days. His essential argument on that point was perhaps best set out in one of the last, but also one of the best, of his books, *The Everlasting Man* (1925).

The Everlasting Man, which is a long book, falls into two parts. The first part is devoted to showing that, far from being merely a specific sort of animal, differing only in degree from the other animals, man's difference from them is a difference in kind. All men did certain things which no other animal did at all. It was not as if Leonardo da Vinci painted well, and the cave man painted less well, and the rhinoceros painted less well again. Leonardo painted well, and the cave man painted badly, but the rhinoceros did not paint at all.

The second half of the book is similarly devoted to showing that it is equally false to argue that Christ was merely a special sort of man. Again his difference from other men was a difference in kind, not a difference in degree. It was not possible to believe that he was merely a good man. If he was merely man, then he was by no means the best of men. His ethical teaching was muddled up—as we must say on such a hypothesis—with fantastic metaphysical claims that, if they were not true, were hardly sane and which made his teaching less valuable than that of, say, a Socrates or a Confucius, who were able to steer clear of such irrelevancies. Whatever he was, the man who—to take an example from Chesterton—was able to say 'calmly and almost carelessly like one looking over his shoulder, "Before Abraham was I am"' was clearly not merely a valuable ethical teacher. Either he was a lunatic or he was what he claimed to be—the Son of God.

Chesterton was a very prolific writer—probably too prolific. There poured from his pen novels, biographies, works of political criticism, and of apologetics. The best-seller of all his works was the least considered—the detective stories of Father Brown.

He never took his own talent very seriously—probably not as seriously as he should. But it was primarily as a poet that he was at home. His first published works were volumes of poetry—*The Wild Knight* and *Greybeards at Play* (1900). They were devoted to the praise of a perhaps rather facile optimism. An unborn child is imagined dreaming what a wonderful adventure it would be to find his way into a world covered with green hair and warmed by a gigantic ball of fire, and his splendour of delight when, stepping through the door of birth, he finds himself indeed in such a magic world. But later Chesterton came with deeper faith to deeper thinking. Things after all are not quite as simple as all that. Life has its mysteries and its tragedies, and a full philosophy must allow for them. Virtue can conquer but it has to fight for its victory.

ROBERT HAMILTON

The Rationalist from Fairyland

G. K. CHESTERTON is a writer much quoted today, but less read than he deserves. One continually comes across memorable sayings of GKC on many subjects in newspapers and journals of every kind, in books, and reports of speeches. He made few pronouncements on science, but some years ago an eminent scientist quoted him at length to a learned society. Harold Wilson made the famous line, 'We are the people of England, we have not spoken yet', part of his election campaign—though not all of us would agree that Mr Wilson's party stands for the people of England. It is not surprising that Chesterton is so much quoted. His powerful and original mind and range of interests, his arresting style, his warmth and humour and rich humanity make him eminently quotable. But it is time we started to read him again with the serious attention he merits.

Chesterton's supreme gift was a unique power of imaginative reason. (This mark of high intelligence is far more important and rare than the slick, computer-like abilities revealed by IQ tests.) By imaginative reason I mean the power of associating ideas in such a way that new ideas emerge. One might call it the art of thinking. Hegel reflecting on the apparent contradiction of *being* and *nothing* and leaping forward to the idea of *becoming*; Darwin reflecting on variety in species and on Malthus' theory of population, and seeing in a flash how evolution could have come about—these are examples of imaginative reason. Obviously abstract reason must be the basis of all our thinking; but it should be the basis on which imaginative reason builds or the platform from which it takes wing.

Chesterton also possessed great sensibility. He was an artist and poet and saw the world with aesthetic vision; and with all this, as C. S. Lewis observed, he was one of the sanest thinkers of our age. Lewis also greatly admired his humour, which, he argued, was not something added to his work, as with so many writers, but an integral part of it. Humour, the perception and appreciation of the incongruous in life, was bound up with Chesterton's imagination and sense of wonder. He preserved throughout his life the fresh childlike vision (childlike in Christ's sense, 'unless ye become as little children') that sees the world in the morning light of creation. Chesterton was fundamentally a thinking man: even his lightest work is continually informed by thought. But he was a rationalist with a difference—a 'rationalist from fairyland' as he once called himself. He saw that the imaginative world of the fairy tale contained universal truths, and in his most original and challenging books he brought out their meaning with unique insight.

His literary style was, as one critic put it, 'musical and masculine'. It was clear, strong, supple, and rhythmical (rhythm is perhaps the most important aspect of good style), and to all this was added an inexhaustible linguistic range and an ear for subtle nuances and cadences. Himself a poet, he had the poet's gift for the apt and sensitive word. His style was wholly original without being in any way peculiar. It proceeded from an inborn originality of mind and personality. There was no need for him to attract attention to himself by inventing peculiarities of style like some of his less talented but more pretentious contemporaries.

In Chesterton's work there is a perfect fusion of thought and language. But so great is the concentration of language that we often have to read between the lines—unnecessarily in some cases, since a little explanation would have spared us. This concentration of thought and style, and the brilliant paradoxes succeeding each other with cumulative force, can be exhausting; and exhaustion sometimes leads to irritation. He could with profit have spread his thought and style; he could have given more reasons for his assertions and expressed them in more conventional academic language. Sometimes we suspect that with such intellectual and verbal brilliance Chesterton could have made out an equal case for the opposite position to the one he is defending—as indeed he could, witness his astonishing virtuosity in the Father Brown story, *Israel Gow*, where he gives three equally convincing explanations for a random set of unrelated objects. But except in his detective stories he was not clever for cleverness sake. Unlike Wilde and Shaw, he never made idle or merely witty paradoxes. Chesterton's paradoxes were the sparks thrown off by the hammer blows of his convictions.

No book exhibits Chesterton's gifts with greater force than *Orthodoxy*, written at Battersea in the happiest and most creative period of his life, a flowering spring that came at the end of a dark winter of the mind. This period of suffering and trial was hidden from the world. There are hints of it in the autobiography and elsewhere, but it was fully known only to Chesterton's great friend Fr O'Connor ('Father Brown'). From the little we do know it seems that after he left St Paul's (when he was eighteen), probably during his period at the Slade, Chesterton went through a horrific period of doubt, near-despair, and a dangerous tendency to moral anarchy and disintegration. After a fight the magnitude of which we can hardly guess, he emerged triumphant, and there began to build up in his mind that creative philosophy of life which he later came to see was identical with the Christianity of the Apostles' Creed.

Chesterton's happiness was completed by the love and understanding of an exceptional woman; and during the period of their courtship his first work began to appear in the form of highly individual articles and reviews, which attracted the attention of a perceptive few (including Shaw, who so far forgot himself as to ask 'Who is Chesterton?'), and a couple of volumes of verse. Then came marriage and a home of his own in Battersea, and the publication of his first prose work—a book of essays entitled *The Defendant*. It was aptly named, for all his life Chesterton was to defend the eternal things. The essays appeared in October 1901 when he was 27, just after the move to Battersea. In the eight years that followed (until the move to Beaconsfield in 1909) there poured from his fertile mind a series of brilliant and original books on all the subjects he was to make his own—religion, politics, criticism, essays, and fiction. Together with his widely read newspaper articles and criticism, they made him one of the most famous and discussed writers in the land. The twelve books he wrote at Overstrand Mansions, Battersea, in what J. B. Priestley called 'the long golden afternoons of the Edwardian era', are the flowering of his creative life.

Overstrand Mansions, quite unchanged today, is a long four-storey row of late Victorian flats facing Battersea Park.[1] The outlook over the park and across the river to Chelsea is enchanting. On autumn evenings Chesterton must have watched the smoke going up from bonfires in the park—but from the back windows he would have seen the smoky roofs of the town, and this I think would have pleased him more. It was in the streets rather than in the park that he would have found his inspiration, in the sprawling, narrow, crowded, dirty streets with their closely packed little shops and brash noisy pubs. Battersea was a fitting environment for a man who was a born cockney and a convinced democrat. Chesterton loved the streets and the common people, especially the small streets and shops that stood for the freedom and independence he championed so ardently all his life. He saw the odd and unplanned fantasy of London with the eye of an artist: he saw the people with the eye of a democrat, for whom all men are equal, and of a Christian, for whom every man is unique. 'I find the suburbs intoxicating,' he said. Long after he had left Battersea he remarked to his sister-in-law, who was admiring the grounds of his country house, 'I don't really care for gardens,' and she wondered 'if he were thinking of those grey streets of London that blossom like a rose'. (She recalls that at the reception after his funeral she was haunted by a persistent image of the Battersea flat.) In *Tremendous Trifles* Chesterton wrote, 'I am going to wander over the whole world until once more I find Battersea.'

The Battersea period was a time of great happiness for Chesterton, a happiness reflected in the work he produced in those halcyon years. His life was divided between Battersea and Fleet Street, and many of his greatest friendships were made at this time—friendships with Belloc, Shaw, Wells, Barrie, Max Beerbohm, and others. His capacity for making and keeping friends was remarkable, even with men with whom he had nothing in common and who actively disliked his views. Hesketh Pearson was not alone in saying that he was probably the most beloved writer of his day. After his death, in a wireless talk, Wells said with emotion, 'I *loved* Chesterton'; and Shaw, not given to compliments, told T. E. Lawrence that Chesterton was 'a man of colossal genius'. Shaw remained one of his greatest friends and admirers to the end. Belloc felt for him a constant affection and esteem—though I have the impression that the famous Chesterbelloc friendship was greater on Chesterton's side. But for all the social round, the visits to Fleet Street, the lectures and tours, home life was everything to Chesterton, and the Battersea flat was the centre of his most intimate and creative life.

[1] The Chestertons lived at No. 60 from 1901 to 1907, and at No. 48 from 1907 to 1909.

To his neighbours the young man who lumbered up the stairs of Overstrand Mansions, with his great height and girth, his mass of untidy brown hair crowned with a shapeless hat, his ragged drooping moustache and pince-nez usually askew, his benevolent abstracted expression, must have seemed an arresting and picturesque figure. The photographs and portraits of the time reveal two aspects of Chesterton—a happy smiling young man and a rather sombre and much older-looking man. The first has been marvellously captured in the portrait by Rivière with its strange beauty and almost lightsome quality: this is the Chesterton, said Rivière, who wrote *The Napoleon of Notting Hill*. But the sombre and more thoughtful Chesterton has never been so intimately revealed as in the drawing by Rothenstein in which the deepest levels of his mind are brought out in a few masterly lines. One might be tempted to say that this is the man who wrote *Orthodoxy*. But the whole of Chesterton is in *Orthodoxy*. It has the high spirits and almost irresponsible gaiety of youth and the profound insight derived from a seeming infinity of experience.

Orthodoxy is one of the most inspired and original expositions of Christianity ever made, a true 'yea-saying' of life such as Nietzsche sought but never found. But in spite of the force and clarity of every sentence in the book and its powerful cumulative effect, the argument is not easy to sort out. It is a philosophical autobiography rather than a structural discourse ('I am only giving an account of my own growth in spiritual certainty') and this is both a limitation and a liberation. It does not proceed by an obviously built-up, step-by-step thesis. The form of the book is biological rather than chronological—like the rich and varied growth of a living organism. To keep this in mind helps us to understand the book as a whole and enables us to enter sympathetically into Chesterton's mind even where we disagree with some of the details of his argument or on occasion are unconvinced by his analogies and illustrations.

The book was written in answer to a critic who had complained that although Chesterton had told us what he thought was wrong with a lot of things he had not clearly defined his own position. He starts with a question: 'How can we contrive to be at once astonished at the world and yet at home in it?' In a broad sense the basis of the book is an attempt to deal with this question. 'I wish to set forth my faith as particularly answering this double spiritual need.' This sense of wonder and homeliness—the wonder of homely things—is the essence of sanity. Nearly half a century later a psychologist defined sanity as a feeling of being at home in the universe.

I am not attempting to criticize *Orthodoxy* as a whole in this short article. I am only considering, very briefly and inadequately, some of the remarkable passages on the symbolic truths hidden in the fairy tale, and Chesterton's imaginative relating of these truths to Christianity and the world. The long fairyland chapter and the many other allusions to the subject are the most original and arresting things in the book. But *Orthodoxy* is far more than this. Its powerful arguments on logic and imagination, on optimism, pessimism, and loyalty, on the strange contradictions in the criticisms of Christianity that led Chesterton to suspect that it was 'the right shape', the analysis of dogma and heresy, and a hundred things more, transcend and also complement the fairyland motive. The most I can hope for is that these few words of mine will send readers who have forgotten the book on a journey of rediscovery, and induce a few people to read it for the first time and discover its riches for themselves.

Chesterton tells us that his respect for the fairy tales was bound up with his respect for the beliefs of the common people and for that common heritage of all mankind, the tradition of the nursery. This respect for common beliefs was part of his championship of democracy, which he saw as the guardian of freedom against the totalitarianism implicit in materialist thought. How right he was has been tragically proved in our time. The Communist sees godless nature as proceeding dialectically to the classless society via the firing squad and the wastes of Siberia; the Nazi saw godless nature proceeding by survival of the fittest as to the master race via the concentration camp and the gas chamber. These two systems (absurdly called left and right) are united in their ruthless materialism and only differ in their interpretation of nature, e.g. dialectical in Communism; survival of the fittest in Nazism. In spite of their call to revolution they believed that the classless society and the master race were determined from the beginning of time. Fifty years ago Chesterton saw this materialist determinism as malignant nonsense and contrasted it with the sane teaching of Christianity and the fairy tales.

There are certain sequences or developments (cases of one thing following another), which are, in the true sense of the word, reasonable. They are, in the true sense of the word, necessary. Such are mathematical and merely logical sequences. You cannot *imagine* two and one not making three.[2] But you can easily imagine trees not growing fruit ... All the terms used in the science books, 'law,' 'necessity,' 'order,' 'tendency,' and so on, are really unintellectual since they assume an inner synthesis which we do not possess... A

---

[2] In the infinite series of positive integers, of course.

sentimentalist might shed tears at the smell of apple-blossom, because, by a dark association of his own, it reminded him of his boyhood. So the materialist professor (though he conceals his tears) is yet a sentimentalist, because, by a dark association of his own, apple-blossoms remind him of apples. But the cool rationalist from fairyland does not see why, in the abstract, the apple tree should not grow crimson tulips; it sometimes does in his country.

This is reminiscent of Hume's famous criticism of causation. But unlike Hume, it is put in an imaginative and humorous way, all the more arresting for that. The distinction between the necessary and contingent is important, and one should note the qualifying 'in the abstract' in the final sentence.

Chesterton insists that the freedom of Christianity and fairyland is not anarchy. *It depends upon condition rather than causation.* Thus in the Christian story and in fairyland human happiness is not causally determined but is the result of a condition that at first may seem trivial but is of the greatest ultimate significance.

In the fairy tale an incomprehensible happiness rests upon an incomprehensible condition. A box is opened, and all evils fly out. A word is forgotten, and cities perish. A lamp is lit, and love flies away. A flower is plucked, and human lives are forfeited. An apple is eaten, and the hope of God is gone.

This is a striking example of Chesterton's astonishing fusion of imaginative reason and sensibility. In the above quotation (and in thousands of others scattered throughout his work) a powerful and original idea is expressed in language of almost monosyllabic simplicity, and a rhythmical progression of ideas increases cumulatively until the climax of the final sentence on the Fall of Man.

For Christianity the Fall is a basic doctrine, and Chesterton sees an analogy to this great doctrine in Robinson Crusoe. He is a man wrecked on an island, 'but according to Christianity we were indeed the survivors of a wreck, the crew of a golden ship that had gone down before the beginning of the world'. He speaks of the endearingness of little things in the world of a child, and contrasts this with the constant assertion of the materialist that the universe is overwhelmingly large. But, he argues, since size has no meaning without a standard of comparison, we might just as well think of it as small—a much better way if we want to be at home in it. And he returns to Crusoe: 'Crusoe is a man on a small rock with a few comforts just snatched from the sea ... But I really felt as if all the order and number of things were the romantic remnant of Crusoe's ship.'

Chesterton saw that the world of the fairy-tale was in an obscure and yet subtle way an adumbration of paradoxical truths that relate to the great Christian doctrines and to the Christian way of life—adventurous and conventional, challenging and comforting, subtle and elementary, complex and simple, awe-inspiring and homely. But always the world of Christianity and of the fairy-tale astonishes and fills us with wonder. The mind of a child may have its dim and even dark side, but it is a mind that can wonder; and as Einstein was never tired of saying, without the sense of wonder we are nothing. In the power to wonder, to see the world with astonishment and delight, Christianity is at one with the child (and indeed with the great mathematical physicist who looked into the depths of the world and wondered, trying, as he said, to 'think God's thoughts'), and it is in complete opposition to the mechanical materialism that sounds through the marching feet of Moscow and Peking. Chesterton argues that the endless repetitions of nature that seem to the materialist like clockwork are to the child a series of creative acts that never tire. Mere repetition may be dull, but creative repetition, like sunrise and sunset, the tides and the seasons, is an adventure. The view that equates constant repetition with monotony is fallacious 'even in relation to known fact'.

For the variation in human affairs is generally brought into them, not by life, but by death; by the dying down or breaking off of their strength or desire . . . Because children have abounding vitality, because they are in spirit fierce and free, therefore they want things repeated and unchanged. They always say, 'Do it again'; and the grown-up person does it until he is nearly dead. For grown-up people are not strong enough to exult in monotony. But perhaps God is strong enough to exult in monotony. It is possible that God says every morning, 'Do it again' to the sun; and every evening, 'Do it again' to the moon.

Chesterton sees Christianity as an adventure—and yet it is undeniably authoritative. But the two do not cancel out, and indeed he believes that freedom depends on the essential limitation imposed by authority, and that only when there are boundaries can the mind roam.

We might fancy some children playing on the flat grassy top of some tall island in the sea. So long as there was a wall round the cliff's edge they could fling themselves into every frantic game and make the place the noisiest of nurseries. But the walls were knocked down, leaving the naked peril of the precipice. They did not fall over; but when their friends returned to them they were all huddled in terror in the centre of the island; and their song had ceased.

For sheer originality, for intellectual and imaginative virtuosity, the writing on fairyland and the creed in *Orthodoxy* is astonishing. But I must emphasize again that it is only part of the book, and I should be doing Chesterton an injustice if I suggested that it is all. None the less it is the *key* to his outlook, the 'golden key' he spoke of in the *Autobiography*. There he told us that 'the man with the golden key' in the toy theatre of his childhood became for him 'the God with the golden key' of Christian faith. And the toy man and the living God were aspects of the same reality.

To those who read it rightly, with imagination and sympathy, *Orthodoxy* is a liberating experience, a liberation into a world of sanity and essential goodness—not the goodness of stern rectitude that often repels even when we admire, but the deeper goodness that comes from a genuine humility. It is this that endears Chesterton to so many readers and endeared him to his friends. His philosophy of life was won through pain and struggle, and *Orthodoxy* is its fruit. It is an enormously confident book; but this confidence, which in some men might seem arrogance, is tempered by humility and humour. Chesterton was assured but gentle, convinced but never condescending. There is a quality in all his work that is spiritually and psychologically therapeutic—indeed a practising psychiatrist once said that one of the best cures he could offer his discouraged and unhappy patients was to get them to read Chesterton. His writings are a continual affirmation of faith in God and man, and an adventure of the mind.

One of his friends described Chesterton shortly before his death standing at a gate, his figure outlined against the sunset, his hand upraised in farewell. Passing Overstrand Mansions the other day I thought of the man who had lived there in the golden years. I remembered the bright morning of his life, and saw him with his hand upraised, not in farewell but in greeting.

MICHAEL MASON
Chesterbelloc

THE triple knockabout act of Shaw, Chesterton, and Belloc was a certain draw in the many halls of the pre-television era which catered for an educated élite delighting in good talk and debate. 'Television personalities' of today would stand amazed, and it is to be hoped crestfallen, to overhear the brilliant words that were thrown away free of charge to the customers, largely for the fun of it, even if Shaw and Belloc sometimes contrived a contempt for the audience which Chesterton neither affected nor felt.

It was as a result of Shaw's observation of the two friends that he invented a fabulous beast called 'The Chesterbelloc' which in reality largely existed in Shaw's mind. That they were friends and that Belloc wrote several satirical novels which Chesterton illustrated is true. But as a literary or political pantomime horse intent on some goal or pursuing some logical line, the Chesterbelloc was a non-runner. The front half of Belloc, intent on action of some direct kind, would have had an unwilling partner in Chesterton, content to live in the imagination, surveying and creating the world from his own back garden. A glimpse of the letters which passed between the two friends reveals a host of notes from Belloc, largely unreadable, outlining some precise course of action, and the odd letter from Chesterton, calligraphy alpha plus, showing a fine disdain for time and place. The differences between the two were greater than the commonly held virtues of Catholicism, food, drink, conversation, and laughter. It was again Shaw who tried to put his finger on them.

Belloc combines the intense individualism and land hunger of a French farmer with the selfless Catholicism and scholarship of an Aristotelian cardinal. He keeps his property in his own hand, and his soul in a safe bank. He passed through the Oxford rowdyism of Balliol and the military rowdyism of the gunner; and this gave him the super-rowdyism of the literary genius who has lived adventurously in the world and not in the Savile Club. A proletariat of Bellocs would fight, possibly on the wrong side, like the peasants of La Vendée; but the Government they set up would have to respect them, though it would also have to govern them by martial law.

But for Chesterton:

Neither society nor authority nor status are necessary to his happiness; he might be trusted anywhere without a policeman. He might knock at a door and run away—perhaps even lie down across the threshold to trip up the emergent householder; but his crimes would be hyperbolic crimes of imagination and humour, not of malice. He is friendly, easy-going, unaffected, gentle, magnanimous, and genuinely democratic. He can sacrifice a debating point easily; Belloc cannot.

Belloc, the man of action, was politically involved. Chesterton, the artist, was dragged into politics at Belloc's coat-tails. It was Belloc who opened Chesterton's eyes to the political facts of life—or what Belloc thought they were. This involved belief in a number of conspiracies which one feels Chesterton accepted through a delight in the idea of conspiracies and because they satisfied a melancholy pessimism which lurked behind the expansive optimistic front he presented to the world. But it is hard to suppress the feeling that the famous meeting in Soho when Belloc was 30 and Chesterton 26 led Chesterton into wasting his genius in the deserts of politics and the arid soil of Distributism. The 'political facts of life' had better been kept from him so that he could have devoted more time to the better pursuit of literature.

Belloc never ceased to decry the profession of letters, going so far as to assert that the only book he ever wrote for anything but money was *The Path to Rome*. 'The whole art is to write and write and write and write and then offer it for sale, just like butter.' Chesterton had more respect for words, although a life-long shortage of cash led him to have to throw off thousands of squibs for it, torn from him by waiting copy boys at the last minute, with the result that the quality sometimes suffered. The paradoxical tricks would appear to have annoyed Belloc at times, being written off as mere literary conjuring in a trade which demanded no respect. But it is essential to pursue the Chesterbelloc down the political road in order to understand how the Belloc tail wagged the Chesterton dog. In his introduction to *The Servile State*, published two years after his meeting with Chesterton, Belloc set forth both his theme and the manner of his writing when he began:

> This book is written to maintain and prove the following truth: That our free modern society in which the means of production are owned by a few being necessarily in unstable equilibrium, it is tending to reach a condition of stable equilibrium by THE ESTABLISHMENT OF COMPULSORY LABOUR LEGALLY ENFORCEABLE UPON THOSE WHO DO NOT OWN THE MEANS OF PRODUCTION FOR THE ADVANTAGE OF THOSE WHO DO.

To Belloc (and Chesterton) it did not matter whether the masters were labelled 'Socialist' or 'Capitalist', the result for the unfortunate 'people' was the same. The Chesterbelloc answer was that of

Distributism, the redistribution of property and shares, to restore to man the dignity and responsibility of ownership. A similar theme had been propounded in the Papal encyclical *De Rerum Novarum* of 1880 but as Renée Haynes commented in her pamphlet on Belloc in the series *Writers and Their Works*: 'In the England of 1912 the suggestion was considered fantastic, and the remedy as "reactionary" as Cobbett.'

Here was the Distributist dilemma. In the Johnsonian sense both Belloc and Chesterton were Tories but their remedies for the political ills of society were revolutionary. Chesterton regarded himself on the left of communism and Belloc quit parliamentary life disgusted by the belief that the opposing parties in the House of Commons were interrelated and that the political fight was a sham one. But Chesterbelloc continued the fight in the pages of the *New Witness*, which later became *G.K.'s Weekly*, and in a flood of other articles and talks. But despite Chesterton's left-wing paradox and Belloc's radicalism, the Chesterbelloc became regarded by the left as right and by the right as politically powerless and therefore of no use to it. The Distributist Society began to collect cranks of various hues, many devoted to the idea of 'two acres and a cow'. To anyone attending these meetings it was quickly noticeable that few present would have known how to milk a cow or trim the acres—least of all Chesterton.

The effect of the Chesterbelloc on the political thought of the time is difficult to assess, though it was many years later that the Conservative Party hoisted the slogan 'A Property-Owning Democracy' and through the workings of various pension funds and insurances the 'people' have become shareholders in industry though without the power to influence its destiny. In retrospect it must be generally deplored that so much energy was devoted to political matters with so small a result when the creative literary talents of both would have been better employed elsewhere. The continuous strain on their pockets and brains certainly sapped the energies of Chesterton and eventually killed him.

The essential Catholicism of the Chesterbelloc was a thread discernible in everything they wrote. Belloc showed typical impatience with the leisureliness with which his friend abandoned the Anglo-Catholic position in the Church of England and was received into the Church of Rome, but Chesterton had long since committed himself in *Orthodoxy*, a book of such brilliance and imaginative reasoning that its influence still remains strong today. . . . To Chesterton the path to Rome was a homecoming which he celebrated with the ease of a homecomer, whereas Belloc's more strident manner may have raised doubts, in non-Catholics at least, as to whether he ever was really at home.

The Chesterbelloc aroused antipathy during its lifetime by what its enemies called its anti-Semitism. Again it was Belloc's influence which was the greater. A conviction that Jewish financiers played too important a part in the affairs of the State and the Throne led Belloc to be more outspoken but when tackled in 1947 by Hugh Kingsmill and Hesketh Pearson on the subject, Belloc had this to say: 'It was the Dreyfus case which opened my eyes to the Jew question. I'm not an anti-Semite. I love 'em, poor dears. Get on very well with them. My best secretary was a Jewess. Poor darlings! it must be terrible to be born with the knowledge that you belong to the enemies of the human race.' When asked by Kingsmill why he regarded the Jews as the enemies of the human race, Belloc replied: 'The Crucifixion.' Let us remember that all this happened before the setting up of the Race Relations Board, that Chesterton died before the enormities of Hitler were fully exposed to the public, and that Belloc was an old man at the time of the Pearson–Kingsmill interview. But their alleged 'anti-Semitism' was never on the unreasoning emotional level at which such subjects have been conducted in modern times but on the assumption that the Jewish and Christian cultures were radically dissimilar and that this cultural gulf should be recognized. There are those who argue the same about present racial difficulties and have the support of Pakistanis and Africans in this.

The Chesterbelloc was so essentially a journalistic animal that any new assessment of its influence must throw away a great mass of ephemeral material which gave great stimulation at the time of writing. The combined output was colossal and any reference to any of it as 'mere journalism' makes one realize how we could do with a few more mere journalists today. 'The age with nothing to say', said Chesterton, 'invented the loudspeaker.' But they left behind them a by no means derisory amount of genuine literary work which has been too long neglected. In comic verse and poetry, both contributed more than their share at a time when versifiers and poets were fairly thick on the ground. In the field of the essay and belle lettres they excelled. As novelists they were both readable and read. As historians they helped to balance the scale even if Belloc may be said to have overweighted them. To those brought up on the Protestant Whig version of European events it was a refreshing shock to step on the banana skin which Belloc so successfully threw beneath their feet. With a jolt they were made to realize that there was another side to the case. Chesterton's *Short History* showed a greater knowledge of the English people than many a more date-filled volume. Their love of their country and of Belloc's Sussex was well expressed in his sonnet:

> Lift up your hearts in Gumber, laugh the Weald
> And you my mother the Valley of Arun sing.
> Here am I homeward from my wandering,
> Here am I homeward and my heart is healed.
> You my companions whom the World has tired
> Come out to greet me. I have found a face
> More beautiful than Gardens: more desired
> Than boys in exile love their native place.
>
> Lift up your hearts in Gumber, laugh the Weald
> And you most ancient Valley of Arun sing.
> Here am I homeward from my wandering,
> Here am I homeward and my heart is healed.
> If I was thirsty, I have heard a spring,
> If I was dusty, I have found a field.

Belloc would not, and could not, have assessed this in terms of cash. In the event it probably realized very little but it lives and will live. Among the clatter of modern journalism and democratic cant the voice of Chesterton still remains with us:

> Smile at us, pay us, pass us: but do not quite forget;
> For we are the people of England, that never have spoken yet,
> There is many a fat farmer that drinks less cheerfully,
> There is many a free French peasant who is richer and sadder than we.
> There are no folk in the whole world so helpless or so wise.
> There is hunger in our bellies, there is laughter in our eyes,
> You laugh at us and love us, both mugs and eyes are wet:
>   Only you do not know us. For we have not spoken yet.

Chesterton's optimism was philosophically based if not always spiritually felt, and this passage from *Orthodoxy* stresses the point:

> To have fallen into any one of the fads from Gnosticism to Christian Science would indeed have been obvious and tame. But to have avoided them all has been one whirling adventure; and in my vision the heavenly chariot flies thundering through the ages, the dull heresies sprawling and prostrate, the wild truth reeling but erect.

Oecumenical New Year would have met with some derision from Chesterton. One hesitates to think of Belloc's comment.

In a solemn age in which solemnity is mistaken for seriousness much of the output of both Belloc and Chesterton is taken as mere flippancy as in Belloc's comment:

> William, you vary greatly in your verse;
> Some's none too good, but all the rest is worse.

But their wit hit home and is remembered: what better than Belloc's comment on the pacifist position?:

> Pale Ebeneezer thought it wrong to fight,
> But Battling Bill who killed him thought it right.

Neither had any doubts about the lasting qualities of their daily journalism. They were proud to belong to the craft. 'If a thing is worth doing at all it was worth doing badly', said Chesterton when taken to task for some factual inaccuracy which he had lightheartedly ignored.

> When I am dead, I hope it may be said:
> His sins were scarlet, but his books were read.

runs a well-known epigram of Belloc's.

Are they? How do they stand the passage of time? Chesterton died in 1936, Belloc in 1953. The causes of personal liberty, rigid Catholic dogma, and literary light-heartedness, for which they stood, are not popular at this moment. They knew nothing, and cared less, for so-called science and technology. They would not have been impressed by the astronauts nor have preached the cause of East–West trade, but Father Brown is still alive and kicking and Belloc's history more seriously regarded than it used to be. Most of the topical journalism has gone with the newspapers which printed it, but there is still enough gold in the vast combined output to deserve a revival which has happened in the case of many lesser literary figures. Unless writing is not the medium of the message and a new age of illiteracy is on us, they will both survive Marshall Aid.

The Chesterbelloc a fabulous beast? Perhaps Shaw was right after all and something will rise from the ashes to confuse the doubters.

KINGSLEY AMIS
An Unreal Policeman

FATHER BROWN is not an eccentric in the superficial, violin-playing or orchid-rearing sense. But he is extraordinary enough, more so today than when Chesterton was writing, for Brown's extraordinariness is founded in his religion. Whether we like it or not, the little man's devotion, total courage, human insight, and unshakeable belief in reason are at any rate statistically uncommon. Some readers have found too much Roman Catholic propaganda in the stories. I feel that the Chesterton element, which is sometimes built into the plot, is never narrowly sectarian, and that part of it which overlaps with the advocacy of sheer common sense ought to be acceptable to everybody. It would be truer to say that what propaganda there is gets directed against atheism, complacent rationalism, occultism, and superstition, all those shabby growths which the decline of Christian belief has fostered, and many, though perhaps not most, readers will sympathize here, too. My only real complaint is that this bias sometimes reveals the villain too early. We know at once that the prophet of a new sun cult is up to no good, and are not surprised that it is he who allows a blind girl to step to her death in an empty lift shaft.

Brown is made for the situations he encounters and they for him. They embody his love of paradox and of turning things back to front, his gift for seeing what is too obvious for everyone else to notice, his eye for the mentally invisible man. Only Brown could have wandered into a house where the recently dead owner's diamonds lay in full view minus their settings, heaps of snuff were piled on shelves, candles littered the tables, and the name of God had been carefully cut out of the family Bible every single time it occurred. And when the owner is dug up and found to be minus his head, only Brown would have taken this as natural and inevitable, the final clue showing that no crime had been committed after all. (You will have to read 'The Honour of Israel Gow' in *The Innocence of Father Brown* to find the answer.)

Rather too often, Brown runs into impersonations, twin brothers, secret messages, unlikely methods of murder: it would take a lot of good luck to succeed first try in dropping a noose round someone's neck from a dozen feet above him, and again to hit your man on the head with a hammer thrown from a church tower. But good ideas are many and marvellous. The howling dog that gave away a murderer, the trail to an arch crook that began with somebody (guess who) swapping the contents of the sugar and salt containers in a restaurant, the man who seemed to have got into a garden without using the only entrance and the other man who got out of the same garden partly, but not completely—these are pretty standard occurrences in Father Brown's world. That world is vividly atmospheric, thanks to Chesterton's wonderful gift for depicting the effects of light on landscape, so that the stories glow as well as tease and mystify. They are works of art.

DENIS BROGAN
The Chester-Belloc's Better Half

The publication of a new edition of Chesterton's *Autobiography* is beginning to produce a reassessment of his place in English literature. Of course, immediately after his death his reputation began to slump. This book was first published in 1936, the year he died. I remember seeing him a year or two before, coming out of Westminster Cathedral and looking like a parody of the figure I had been familiar with since I was about 10. Chesterton was not merely a fat man, he was a very big man, and by this time he had lost most of his fat and looked as if his clothes and his skin hung loosely over the framework of his bones. When one contrasted him with his dwarfish brother Cecil, it was easy to assert that the material for the Chesterton brothers had been divided in the proportion of one and a half to Gilbert and half to Cecil. No brothers could have looked less alike except in certain details of their features.

GKC's reputation as a writer suffered not merely from the normal deflation that follows the death of a celebrated author (which is now affecting, for example, the reputation of T. S. Eliot). It suffered from the nearly total irrelevance of a great deal of the writing of the Chester-Belloc to Hitler's world. So many things had gone wrong with their picture of the world; so many more things were to go wrong, like the Spanish Civil War; and so many things were to go monstrously wrong, like the Final Solution applied by Hitler to the 'Jewish problem', that there was indifference or hostility to a great deal of what Chesterton and, still more, of what Belloc had written.

But some of Chesterton deserves to live and will, I think, live, although what is his continuous best seller, the *Father Brown Stories*, for all their great merits, are not what I think he would most have liked to be remembered by. But there is not only the famous story of the invisible man (who was the postman), but there is the extremely interesting story of the philosopher Boulnois, who infuriated the great landed magnate on whose estate he had a cottage, by refusing to be jealous about the wicked baronet's ostentatious attentions to the philosopher's wife. There is a great deal of very acute psychology buried in this, and in some of the other stories. The literalness of the Scottish mind is brought out beautifully in the story of the servant who cut out all the gold from the missals because he had been left all the gold in the house and collecting gold was one of the hobbies of the great family he served. After all, they were commemorated in two lines from a mythical ballad:

> As green sap tae the simmer trees
> Is red gold tae the Ogilvies.

Another book that survives, and I think will continue to survive, is his *Short History of England*. *The Times Literary Supplement* recently reprinted the original review of it by A. F. Pollard. Pollard was a great if limited scholar, but he completely missed the point of the book, written by one who was not a scholar at all but had talents very different from and very superior to Pollard's. *A Short History of England* is not only full of brilliant phrases like the famous account of the death of Nelson ('he died with his stars on his breast and his heart on his sleeve'), but there is a great deal of shrewd comment on aspects of English history which the official English tradition did not emphasize. There are many worse ways of introducing intelligent boys and girls to English history.

What has badly affected GKC's reputation was his excessive use of sometimes not very ingenious paradox. Some of his jokes in this *genre* were no better than Philip Guedalla's Oxford Union jokes. Some were as good as Ronald Knox's much superior Oxford Union jokes. But at times they were better than anything that either Guedalla or Knox could possibly have produced.

It is possible that GKC will be largely remembered for his often admirable verses. I say 'verses' deliberately, not poems. I think it could be reasonably said of Belloc that he was a genuine minor poet, and we should remember that Augustine Birrell said when one described a man as a minor poet, the word to emphasize was 'poet'. Belloc was sometimes a poet; I don't think GKC ever was. But is there a better satiric poem in the English language than his attack on F. E. Smith with its famous refrain. 'Chuck it, Smith!' (I am told that the second Earl of Birkenhead does not refer to this poem in his life of his father.)

It is far more effective and basically far more savage than Belloc's celebrated anti-semitic poem about the Rand gold lords. And there is 'Lepanto'. This is very much over-written in places, but it is full of admirable verses to chant aloud.

I never took much interest in *The Ballad of the White Horse*, but some of the shorter poems are excellent as rhetorical statements of a position, and possibly 'The Donkey' is more than that. But Chesterton was a polemical writer with something serious to say. In his *Autobiography*, which shows so many signs of the falling off of his physical and literary powers, he is still very well worth reading for its capture of the atmosphere of the very prosperous Liberal middle class in the London of his youth. I think Chesterton is wrong in writing off Hardy's poetry, but his account of Hardy is curiously moving, and he denied the hostile implications of his verdict on Hardy in *The Victorian Age in Literature*. This book was written hastily for the Home University Library, but it is a very good book still, not only highly readable but with the great additional merit of being written by a man with his own independent tastes, a man of great intelligence and a man who had something more to say about literature than mere points of style or formal organization. There are other books still worth re-reading.

But unfortunately Chesterton fell more and more under the influence of Belloc, and I think that influence was almost totally bad. For one thing, Chesterton took Belloc's claims to scholarship far too seriously. GKC *did* recognize that there was something special about the knowledge of his great friend and my old teacher, John Swinnerton Phillimore, but he had no real idea of how remote from Phillimore's subtle academic mind were Belloc's hasty dashes across history, literary criticism, religion, and the like. GKC, however, makes a good point when he tells us that although his family were the agents or, we should say in Scotland, the factors for the Phillimore estate, he had never met the son of Admiral Phillimore until he met him through Belloc. The upper middle class to which Chesterton belonged did not expect to meet socially, apart from professionally, families like the Phillimores, and GKC, who knew that the most important thing about the Duke of Argyll was that he was McCallum More, would no more have thought of speaking to the Duke or being introduced to him than he would have thought of being introduced to Queen Victoria. (Later on, he once met George V of whom he gives a very lively, amusing and convincing account.)

The damage done by Belloc was not merely the damage done by inserting a great deal of historical nonsense into Chesterton's head. It involved him in the great Marconi controversy, to his loss in many ways. Maisie Ward has told us how shocked she was when she discovered how irresponsible Belloc and Cecil Chesterton were in the Marconi affair. Yet though they were not irresponsible in their attack on Lloyd George, they were very irresponsible in attacking other people: for example, the future Lord Samuel. One of the largest bricks I have ever dropped (and I drop a great many) was at a great official party in the Guildhall where I saw Lord Samuel who by this time was Visitor of Balliol. I knew—and know—his son Godfrey extremely well, and I went up to Lord Samuel and said I had been discussing with his son Belloc's ambition to be made an honorary Fellow of Balliol. I had quite forgotten what very good reasons Samuel had for disliking Belloc. I can remember the cool severity with which he said: 'As long as I am Visitor, Belloc will not be an honorary Fellow of Balliol', an answer which, overheard by Norman Robertson, then Canadian High Commissioner in London, sent him into roars of laughter.

The *New Witness* not only drew money from Chesterton, it drew energy. Most of the writers were rather incompetent parodies of GKC at best. As in Mr Buckley's *National Review* in America today, there was only one writer in the *New Witness* always worth reading, and that was GKC himself. It was like him to undertake this 'chore' if only as a memorial to his dead brother Cecil. But nearly all the best of GKC was written before 1914, and some of it still seems to me very good indeed.

ANTHONY BURGESS
The Level of Eternity

THE works of Gilbert Keith Chesterton have never lacked praise, but the praise has never lacked qualification. The qualification has usually been a kind that a serious author might be expected to resent, best expressed as a humorously affectionate headshake at *idées fixes* that, though amusing, are becoming a bit of a bore, at exhibitions of clowning that, though highly professional, go on a bit too long, at the tricks of an *enfant terrible* who, though undeniably brilliant, ought to consider growing up. But Chesterton, serious as he was, never took himself seriously: he anticipated the parodist and the satirical cartoon. The take-off by Jack Squire of the Chesterton poetic style is hardly distinguishable from the real thing. In H. G. Wells's *The Bulpington of*

Blup, two spinsters open a bottle of Liebfraumilch. 'Milk of the Blessed Virgin,' one of them says. '*So* reminiscent of dear GKC.' Indeed. If Chesterton never wrote a mariolatrous ballade on that hock, it was only because he didn't get round to it. Philip Guedalla, seeing Chesterton's corpulence and optimism as cognate, alleged that he suffered from 'hearty degeneration of the fat'. I can't help feeling that Chesterton thought of that first, and that Guedalla stole it.

Chesterton belonged to an age when literary men could be public figures, just like politicians. His girth and pince-nez were a gift for the cartoonist, as were Shaw's beard, eyebrows and Jaeger. But no writer becomes a public figure merely because he writes well or much (Chesterton did both): he has to be a sort of eccentric schoolmaster, dinning mad ideas into pupils who grin more than they yawn. Shaw preached rationalism and the Life Force; Chesterton merely preached the joys of medieval Christianity: both must have seemed equally iconoclastic and revolutionary, though both ended up—without modifying their ideas one whit—as part of the safe and comfortable British Way of Life.

One of the first of the notable modern British converts to Catholicism, Chesterton found in the faith neither the agonized Jansenism of Graham Greene nor the hopeless chivalry of Evelyn Waugh (or, for that matter, Ford Madox Ford). He found a certitude which justified a powerful and, to the generations that have lived through the time of Auschwitz and Hiroshima, embarrassing optimism. The natural order was the work of God; the very colours of the earth were the pennants of divinity; to live in the flesh, however much you carried, was glorious. Greeneans, or Jansenists, must shudder at this, but they are heretics, while Chesterton is orthodox. His orthodoxy enabled him to concentrate on the liturgy rather than worry about theology: he is rhetorical, full of primary colours, and much given to delight in seeming paradox and true mystery.

Chesterton's joy in life seems nowadays a little too extrovert to be acceptable. It owed something to the English Frenchman Hilaire Belloc, his close friend, and it was promoted by such properties as ale, the Sussex downs, the companionship of Fleet Street and the conviviality of stag-parties. When, in the *Autobiography*, we read about Belloc shouting for bacon and beer in the gardens of Lamb House, Rye, where Henry James was keeping solemn state, we are perhaps less amused than Chesterton is. The heartiness seems somewhat contrived; one fancies that both Chesterton and Belloc would have been appalled by the real medieval England or Rabelaisian France that the loudness and the swilling attempted to evoke. There's a dream of a

Merry England that perhaps never existed and certainly could never be realised even in Chesterton's twentieth century. Graham Greene's Catholicism, heterodox though it is, knows more about the facts of life than does Chesterton's. Chesterton isn't sufficiently interested in the Fall. Thus, though he belongs to the same generation as Yeats and Ford, and is only a decade older than Joyce, Pound and Eliot, he seems to have less to say to our world than they have. There is rather more eupepsia than we can stomach.

This is not to suggest that Chesterton is a mere historical figure, more read about than read. He wrote too well, too sincerely and vigorously, to earn a mere niche in a museum. His best novels—*The Napoleon of Notting Hill*, *The Man Who Was Thursday* and *The Flying Inn*—are as entertaining as when they were first written, and the substructure of the farce and fantasy—a concern with free will, Western civilization, and the ultimate mysteries of religion—is not less valid in the age of superstates and nuclear deterrents and brainwashing than it was in Chesterton's more innocent heyday. To teach and please at the same time is given to few. In the twentieth century, perhaps only Shaw, Wells, Aldous Huxley, and Chesterton have had the faculty.

It is a faculty that has to be nurtured, and it grows best when the writer has to write for a living. Chesterton worked as a journalist in the flourishing romantic days of Fleet Street, when it still possessed some of the dramatic colour of Dr Johnson's time. A man could be a success, but he could also starve. There was no room for bad craft or the superciliousness of the aesthete: one had to do everything well and, if need be, do it quickly. Chesterton never became deadened to the glamour of seeing his name in print, and he was proud of the name of journalist. His immense range was essentially that of the professional writer who would be ashamed to reject any literary challenge: in this respect he is closer to Dr Johnson than any of his contemporaries except Belloc. He could write biography, *belles lettres*, literary criticism, history, philosophy, drama, as well as novels, detective fiction, and verse. He could also draw: his illustrations to his friend Bentley's clerihews are a minor delight. He could also speak. I remember him on the radio in 1936, the year he died, bronchially eloquent in a debate. 'They talk about things being as dull as ditchwater,' he said. 'For my part, I always think of ditchwater as teeming with quiet fun.' He was spontaneously witty, but he could also be carefully epigrammatic. He thought of words not as neutral rational counters, but as confetti, bonbons, artillery.

His novels are read, his Father Brown detective stories devoured, his poems recited. He was not a great poet, but he was incapable of turning a mean rhyme, and he was too fond of language ever to admit the bathetic or pedestrian. If his poems are unsubtle, this is in keeping with the 'public' quality of all his writing—a concern with broad strokes and bold colour, a love of the oratorical. Poems like 'The Rolling English Road' and 'The Donkey' remain superb recitation pieces, and Chesterton would be pleased to know that public bar audiences are better prepared to hear them than are Hampstead drawing rooms. He had an easy mastery of traditional verse-forms, including difficult ones like the ballade, and he learned from such old practitioners as Villon (perhaps through Belloc) how to make his rhymes and images bite. Chesterton never had a moonlight or twilit phase in his poetic development: there was always boldness, wit, even anger. And he could write light verse which remains funny without being facetious.

Triumphant in so many branches of his craft, Chesterton was essentially modest, totally opposed to exhibitionism of the Frank Harris kind, reticent about intimate matters and happier to write about his friends than himself. Nevertheless, there is much pleasure to be gained from this engaging chronicle of a life dedicated to worthy and harmless causes, such as learning to write well and to enjoy the varied solace of an imperfect world. There is also quixotry and a humorous stoicism. Above all, there is faith. The newcomer to his work may wonder how that faith can hold so well in the face of the Spanish Civil War and the persecution of the Jews. He has to be told that Chesterton's faith was of the kind that moves mountains, and that his optimism transcended history. Solid and earthbound as he seemed, Chesterton knew what it was like to live on the level of eternity.

JOHN GROSS
A Man of Letters

'THE world was very old indeed, when you and I were young.' Of all the Edwardians, G. K. Chesterton, who was still only a schoolboy at the beginning of the 1890s, was the one who felt the Decadence most directly as a personal burden. By contrast, Fleet Street seemed to offer hope and freedom. He began his literary career as a journalist, and a journalist he remained. This, for some people, has been enough in itself to condemn him outright; but even the most zealous Chestertonians must often have wished that he had not written so much, so fast, under such pressure. He was a man of remarkable gifts, far more remarkable than his present overclouded reputation would suggest. At his finest, for instance, he was a wittier writer than Oscar Wilde or Max Beerbohm (as Beerbohm himself readily acknowledged): wittier because he had a deeper understanding of life. But as he turns the barrel-organ, and the paradoxes come thumping out, the wit can easily pall and the profundity can get completely overlooked. The best is so very good, the worst is so flashy and cheap. And in between there is that great sprawl of ready-made Chesterton humour, amusing, readable, but never saying quite as much as it seems to, and easy to forget.

Confronted with an output of such variable quality, admirers have naturally sometimes tried to boil Chesterton's work down in an attempt to isolate its essential virtue. The most determined effort along these lines of which I know is Hugh Kenner's *Paradox in Chesterton* (1948). Kenner has very little use for what he calls the toby-jug side of his hero; the Chesterton who counts in his eyes is the writer who was a Neo-Thomist by instinct long before he had actually read Aquinas, whose paradoxes and analogies, far from being mere verbal conjuring-tricks, expressed a consistent vision of the paradoxical nature of reality itself. A traditional Christian vision, which would have been wholly intelligible to such masters of paradox as the Church Fathers, the Schoolmen, the Metaphysical poets. If he had been born in a better age, the author of *Orthodoxy* and *The Everlasting Man*, instead of dissipating his energies in ephemeral newspaper work, might have been 'a principal ornament of the medieval Sorbonne'. Kenner's position is stated even more sternly in the essay by Marshall McLuhan which serves as an introduction to his book. According to McLuhan, we must abandon the literary and journalistic Chesterton to his fate, and forget about his idiosyncrasies, his day-to-day interests, the accident of his having happened to grow up in Late Victorian England. The only claim he finally has on our attention is his achievement as 'a metaphysical moralist'. Few things could be more misleading, it would seem, than the genial popular legend of GKC. Inside him all the time there was a very thin man indeed struggling to get out.

Whether or not one accepts the Kenner–McLuhan thesis is ultimately a question of religious belief, but there are aspects of it with which one does not have to be a Catholic to agree. Chesterton's paradoxes usually make a serious point, and the best of them often have the quality of theological wit. When he protested that he disliked nothing more than light idle sophistry, he was not being sophistical. Nevertheless, Kenner's portrait seems to me to misrepresent the essential Chesterton. It turns him into the kind of solemn theoretician whom he would have been the first to find irksome. It gives no real idea of his recklessness, his lightheartedness, the extent to which he was immersed in the present. Above all it leaves out of account his democratic temper. Many other English writers have preached Democracy in the abstract; Chesterton is one of the very few who genuinely liked the common man (because he was sure there was no such thing). And then, like any writer, he has his own rhythm. When one actually reads him, his strengths and weaknesses are mixed up together—it is impossible to prise the pearl from the oyster. Moreover, as Kenner makes clear, Chesterton's philosophy is based on celebrating the uniqueness of things, and it is a contradiction to celebrate *him* by draining his work of its precise flavour and individuality.

Had he been meant by nature for a theologian, he could have become one, even without the medieval Sorbonne at his disposal. He dissipated himself in journalism because he wanted to. Originally, it is true, he had vague hopes of becoming an artist, and he studied at the Slade for several years, without making much headway. While there, he also attended lectures at University College, where he met Ernest Hodder Williams, whose family published the *Bookman*, and he began by reviewing art books for that magazine:

> I need not say that, having entirely failed to learn how to draw or paint, I tossed off easily enough some criticisms of the weaker points of Rubens or the misdirected talents of Tintoretto. I had discovered the easiest of all professions; which I have pursued ever since.

It was his work for the *Speaker* which first made his reputation, however. This was a weekly which, after J. L. Hammond took over as editor in 1899, became for a time the leading paper of advanced young

Liberal intellectuals. And it was on another, much more popular Liberal paper, the *Daily News*, that he reached his largest audience, writing a regular Saturday article from 1903 to 1912. This was his best-known platform, but he wrote for dozens of other papers as well. In the years before 1914 he was one of the sights of Fleet Street. He was also one of the great exponents of the Fleet Street myth. He habitually portrayed it as a larger-than-life bohemia, a haunt of self-destructive minor genius, a warren of grotesques; and he could be as romantic as Pendennis about the grandeur of the Press:

> A poet writing his name upon a score of little pages in the silence of his study may or may not have an intellectual right to despise the journalist: but I greatly doubt whether he would not be morally the better if he saw the great lights burning on through darkness into dawn, and heard the roar of the printing wheels weaving the destinies of another day. Here at least is a school of labour and of some rough humility, the largest work ever published anonymously since the great Christian cathedrals.

By the time of the First World War, however, the intoxication had worn off. From 1911 onwards, although he still contributed to many other periodicals, he became increasingly associated with his brother's paper the *New Witness*—originally the *Eye Witness*, and continued by him after Cecil Chesterton's death in the War as *G.K.'s Weekly*. This withdrawal was a result, first of his bitter falling-out with the Liberals, brought to a head by the Marconi affair, and then of his disillusionment with the party system itself. Chesterton's hatred of capitalism and his dread of the monolithic state were the generous responses of a man who saw the sickness of his society far more clearly than the ordinary Liberal and felt it far more deeply than the self-confident Fabian social engineers. Unfortunately, though, a sense of outrage often proved as bad a counsellor in his case as it had in Carlyle's. His diatribes against usury and corruption were those of a man on the edge of hysteria; his anti-semitism was an illness. Despite this, his fundamental decency is never obscured for long. He hated oppression; he belonged to the world before totalitarianism. But the positive side of his politics— Distributism, peasant smallholdings, Merrie Englandism—led him into a hopeless cul-de-sac.

All this is now a matter of history. His critical writings, on the other hand, are still widely known, although they have long been excluded from the official canon of modern literary criticism. The reasons are plain enough. His methods are everything that our schoolmasters have brought us up to abjure. He wanders from the text and generalizes lavishly. He is too excited by large conceptions to pay very much attention to accuracy in small ones. He is often content to make his point through a mere phrase, or a joke or an unexpected adjective. He would hardly have known how to begin 'erecting his impressions into laws'. He is extravagant, and he relished extravagance in others. Much of what he wrote was unashamed popularization. He is casual, unguarded, unsystematic. He plays with words, and he would rather parody an author than tabulate his faults. He contradicts himself. While he is working out his own ideas he is never afraid to get in the way of his author. In a word, he is a stimulating and at times an inspired critic.

What exactly do we learn from him? A Chesterton essay is a performance, a furious display which is meant to leave us with a warm diffuse feeling that life has more possibilities than we had realized, or that a familiar author is more interesting, or than an unfamiliar author is more enticing. He does not so much make out a case as work himself up, and his enthusiasm is usually intelligently enough expressed to carry the reader along with him. For the time being, at least. The overall effect is to animate rather than inform, and it is an effect which can easily wear off. But the sum of the parts if often greater than the whole: Chesterton's incidental remarks are more important than his grand conclusions. They are also more startling, not because they are unexpected—anyone can stand a platitude on its head—but because they are unexpectedly true. And a good Chesterton paradox is the reverse of a neat self-contained epigram: it suggests ideas rather than clinches them. When he says, for instance, that Herbert Spencer's closed intellectual system made him more truly medieval than Ruskin, he is starting a train of thought which may not necessarily take in Spencer *and* Ruskin *and* the Middle Ages, but which does cut provocatively—'bisociatively'—across conventional assumptions. He once described Shaw's plays as expanded paradoxes, and his own paradoxes have the power to gather force and expand in the mind.

Few charges have been levelled at him more frequently than that of verbalism. He himself would hardly have come to the defence of every last pun and quibble in his work, but while pleading guilty on minor counts he would probably have added something to the effect that taxing a writer with verbalism was like reproaching a painter for pigmentism: his whole business was with words. And it could fairly be claimed that his excess ingenuity represents the overflow of a remarkable insight into language and its workings. William Empson has commented on his great powers as a verbal critic, and as with Empson the ambiguities which he unravels are functional rather than decorative. He focuses on a cliché or a battered simile until it begins to recover its original brightness. He brings out the wealth of implication in everyday speech, and also its inadequacy: we habitually say more than we realize, but less than we intend. And he scrapes away some of the prejudices that have accumulated around the big abstract words. True imagination, he reminds us, is intensely materialistic; a sentimentalist is in a sense more realistic than a scientific enquirer, because, though he may distort situations, he is incapable of reducing them to statistics. A master of language himself, Chesterton is exceptionally alive to verbal felicity in others. He also has an outstanding talent for clear exposition. Many of the mots which he flings out—'Tennyson could not think up to the height of his own towering style', 'the fighting of Cobbett was happier than the feasting of Walter Pater'—are perhaps only neat journalistic devices, though they are very effective ones of their kind; but where he comes into his own is in describing the effect of an author evocatively—and his comments are no less critical for being picturesque. On the contrary, a serious judgement on a serious author was too subtle a thing to be definitely summed up in a formula. Judgements, like scenes, had to be *rendered*. George Eliot, for instance:

> In her best novels there is real humour, of a cool sparkling sort; there is a strong sense of substantial character that has not yet degenerated into psychology; there is a great deal of wisdom, chiefly about women; indeed there is almost every element of literature except a certain indescribable thing called *glamour*; which was the whole stock-in-trade of the Brontes, which we feel in Dickens when Quilp clambers amid rotten wood by the desolate river; and even in Thackeray when Esmond with his melancholy eyes wanders like some swarthy crow about the dismal avenues of Castlewood. Of this quality (which some have called, but hastily, the essential of literature) George Eliot had not little but nothing. Her air is bright and intellectually even exciting; but it is like the air of a cloudless day on the parade of Brighton. She sees people clearly, but not through an atmosphere. And she can conjure up storms in the conscious, but not in the subconscious mind.

He makes elaborate judgements, and at the same time he responds to simple qualities. His writings on Dickens are the most notable example here. What he is best at is rejuvenating the original popular idea of Dickens, the Dickens whose characters have escaped from literature into folklore and whose jokes are self-explanatory. Everyone now recognizes that he makes far too much of the Christmassy Dickens, and modern critics have revealed levels of symbolism and psychological depths of which he was very largely unaware. But arguably he was doing something more difficult than his successors. Most of them in fact agree that within his limits he was concerned with essential aspects of Dickens—aspects which they seldom try very hard to convey in their own words. One suspects that they are secretly rather grateful to him for having saved them the task. His strength was that he was unembarrassed by the childish element in art, and would no more have condescended to it than he would have condescended to an actual child. He acknowledged the primary claims of myth: 'art is a luxury, fiction is a necessity'. And he was able to deal uninhibitedly (*not* indiscriminately) with popular literature because he believed that everything should be judged first as an example of its own kind. Tom Hood was no more to be blamed for not being Wordsworth than Gilbert and Sullivan were to be criticized for having failed to grapple with the ethical problems of Norwegian idealists. A writer's chief duty was to obey that most difficult of injunctions and to be himself. Possibly this large tolerance of Chesterton's is a fault in him as a critic. But it is one of the things that helps to make him a popularizer of genius. How pale and constricted most professional advocates of literature seem by comparison.

Like all platform performers, he runs a constant risk of being trapped by his own style. Opinions get fed into the machine, and what emerges is not so much inaccurate as inappropriate. Everything takes on the same slightly hectic tone. But this is as much a basic characteristic of Chesterton's imagination as the result of journalistic habit, an aesthetic rather than an intellectual defect. He sees the world in terms of loud contrasts and garish colours; the picture has the boldness of a cartoon, but it lacks light and shade.

And yet though these are faults, they are the defects of his qualities, of an approach which is admirably personal and direct. Looking back in his autobiography on his first critical study, *Browning* (1903), Chesterton ruefully recalled all the errors of fact which it contained—'but there is something buried somewhere in the book; though I think it is rather my boyhood than Browning's biography'. What this means in practice is that though his comments are often technically irrelevant, unlike more single-minded 'responsible' critics he moves easily on the same emotional level as his poet. And when he digresses—talking about a stanza from 'Childe Roland', for instance—he does not simply wander off, but looks to a larger context:

> This is a perfect realization of that eerie sentiment which comes upon us, not so often among mountains and water-falls, as it does on some half-starved common at twilight, or in walking down some grey mean street. It is the song of the beauty of refuse; and Browning was the first to sing it. Oddly enough it has been one of the poems about which most of those pedantic and trivial questions have been asked, which are asked invariably by those who treat

Browning as a science instead of a poet, 'What does the poem of "Childe Roland" mean?' The only genuine answer to this is, 'What does anything mean?' Does the earth mean nothing? Do grey skies and wastes covered with thistles mean nothing? Does an old horse turned out to graze mean nothing? If it does, there is but one further truth to be added—that everything means nothing.

These are the words of the potential convert to Catholicism, but one hardly has to share Chesterton's doctrines to be stirred by his questions.

Living on into a vastly altered post-war world, he declined the consolations of old fogeydom. But while his comments on Eliot, Aldous Huxley, and other writers whom he saw emerging in the 1920s are respectful, and even friendly, he couldn't pretend that he felt at home with them. In a broadcast talk which he gave shortly before his death in 1936 he summed up the whole trend of the times as 'intellectually irritated'. And he could equally have characterized the Edwardian age, or at least the Edwardian literary scene, by its comparative lack of irritability. For better or worse, the writers who held the stage before 1914 were thicker-skinned than their successors. They were expansive; they believed (not too fanatically) in their schemes for saving the world; they didn't feel compelled to write as though they were always on oath. If there was such a thing as a dominant Edwardian note, it was one of confident give-and-take. It is a note which has largely disappeared; but the fact that it would ring false if anyone tried to revive it today shouldn't mislead us into supposing that it was not once natural and spontaneous.

W. H. AUDEN
Chesterton's Non-Fictional Prose

I HAVE always enjoyed Chesterton's poetry and fiction, but I must admit that, until I started work on [a] selection, it was many years since I had read any of his non-fictional prose.

The reasons for my neglect were, I think, two. First, his reputation as an anti-Semite. Though he denied the charge and did, certainly, denounce Hitler's persecution, he cannot, I fear, be completely exonerated.

I said that a particular kind of Jew tended to be a tyrant and another particular kind of Jew tended to be a traitor. I say it again. Patent facts of this kind are permitted in the criticism of any other nation on the planet: it is not counted illiberal to say that a certain kind of Frenchman tends to be sensual ... I cannot see why the tyrants should not be called tyrants and the traitors traitors merely because they happen to be members of a race persecuted for other reasons and on other occasions.

The disingenuousness of this argument is revealed by the quiet shift from the term *nation* to the term *race*. It is always permissible to criticize a nation (including Israel), a religion (including Orthodox Judaism), or a culture, because these are the creations of human thought and will; a nation, a religion, a culture can always reform themselves, if they so choose. A man's ethnic heritage, on the other hand, is not in his power to alter. If it were true, and there is no evidence whatsoever to suppose that it is, that certain moral defects or virtues are racially inherited, they could not become the subject for moral judgement by others. That Chesterton should have spoken of the Jews as a race is particularly odd, since few writers of his generation denounced with greater contempt racial theories about Nordics, Anglo-Saxons, Celts, etc. I myself am inclined to put most of the blame on the influence of his brother and of Hilaire Belloc, and on the pernicious influence, both upon their generation and upon the succeeding generation of Eliot and Pound, exerted by the *Action Française* Movement. Be that as it may, it remains a regrettable blemish upon the writings of a man who was, according to the universal testimony of all who met him, an extraordinary 'decent' human being, astonishingly generous of mind and warm of heart.

My second reason for neglecting Chesterton was that I imagined him to be what he himself claimed, just a 'Jolly Journalist', a writer of weekly essays on 'amusing' themes, such as 'What I found in my Pockets', 'On Lying in Bed', 'The Advantage of having one Leg', 'A Piece of Chalk', 'The Glory of Grey', 'Cheese', and so forth.

In his generation, the Essay as a form of *belles-lettres* was still popular: in addition to Chesterton himself, there were a number of writers, Max Beerbohm, E. V. Lucas, Robert Lynd, for example, whose literary reputations rested largely upon their achievements in this genre. Today tastes have changed. We can appreciate a review or a critical essay devoted to a particular book or author, we can enjoy a discussion of a specific philosophical problem or political event, but we can no longer derive any pleasure from the kind of essay which is a fantasia upon whatever chance thoughts may come into the essayist's head.

My objection to the prose fantasia is the same as my objection to 'free' verse (to which Chesterton also objected), namely, that, while excellent examples of both exist, they are the exception not the rule. All too often the result of the absence of any rules and restrictions, of a metre to which the poet must conform, of a definite subject to which the essayist must stick, is a repetitious and self-indulgent 'show-off' of the writer's personality and stylistic mannerisms.

Chesterton's insistence upon the treadmill of weekly journalism after it ceased to be financially necessary seems to have puzzled his friends as much as it puzzles me. Thus E. C. Bentley writes:

> To live in this way was his deliberate choice. There can be no doubt of that, for it was a hard life, and a much easier one lay nearby to his hand. As a writer of books, as a poet, he had an assured position, and an inexhaustible fund of ideas: the friends who desired him to make the most of his position were many. But G. K. Chesterton preferred the existence of a regular contributor to the Press, bound by iron rules as to space and time. Getting his copy to the office before it was too late was often a struggle. Having to think of a dead-line at all was always an inconvenience.

Whatever Chesterton's reasons and motives for his choice, I am quite certain it was a mistake. 'A journalist', said Karl Kraus, 'is stimulated by a dead-line: he writes worse if he has time.' If this is correct, then Chesterton was not, by nature, a journalist. His best thinking and best writing are to be found, not in his short weekly essays, but in his full-length books where he could take as much time and space as he pleased. (In fact, in this selection, I have taken very little from his volumes of collected essays.) Oddly enough, since he so detested them, Chesterton inherited from the aesthetes of the 1880s and 1890s the conviction that a writer should be continuously 'bright' and epigrammatic. When he is really enthralled by a subject he is brilliant, without any doubt one of the finest aphorists in English literature, but, when his imagination is not fully held he can write an exasperating parody of himself, and this is most likely to happen when he has a dead-line to meet.

It is always difficult for a man as he grows older to 'keep up' with the times, to understand what the younger generation is thinking and writing well enough to criticize it intelligently; for an overworked journalist like Chesterton it is quite impossible, since he simply does not have the time to read any new book carefully enough.

He was, for example, certainly intelligent enough and, judging by his criticisms of contemporary anthropology, equipped enough, to have written a serious critical study of Freud, had he taken the time and trouble to read him properly: his few flip remarks about dreams and psycho-analysis are proof that he did not.

Chesterton's non-fictional prose has three concerns, literature, politics and religion.

Our day has seen the emergence of two kinds of literary critic, the documentor and the cryptologist. The former with meticulous accuracy collects and publishes every unearthable fact about an author's life, from his love-letters to his dinner invitations and laundry bills, on the assumption that any fact, however trivial, about the man may throw light upon his writings. The latter approaches his work as if it were an anonymous and immensely difficult text, written in a private language which the ordinary reader cannot hope to understand until it is deciphered for him by experts. Both such critics will no doubt dismiss Chesterton's literary criticism as out-of-date, inaccurate and superficial, but if one were to ask any living novelist or poet which kind of critic he would personally prefer to write about his work, I have no doubt as to the answer. Every writer knows that certain events in his life, most of them in childhood, have been of decisive importance in forming his personal imaginative world, the kinds of things he likes to think about, the qualities in human beings he particularly admires or detests. He also knows that many things which are of great importance to him as a man, are irrelevant to his imagination. In the case of a love-poem, for example, no light is thrown upon either its content or its style by discovering the identity of the poet's beloved.

This Chesterton understands. He thought, for example, that certain aspects of Dickens's novels are better understood if we remember that, as a child, Dickens was expected to put on public performances to amuse his father, so he informs us of this fact. On the other hand, he thought that we shall not understand the novels any better if we learn all the details about the failure of Dickens's marriage, so he omits them. In both cases, surely, he is right.

Again, while some writers are more 'difficult' than others and cannot therefore hope to reach a very wide audience, no writer thinks he needs decoding in order to be understood. On the other hand, nearly every writer who has achieved some reputation complains of being misunderstood both by the critics and the public, because they come to his work with pre-conceived notions of what they are going to find in it. His admirers praise him and his detractors blame him for what, to him, seem imaginary reasons. The kind of critic an author hopes for is someone who will dispel these pre-conceived notions so that his readers may come to his writings with fresh eyes.

At this task of clearing the air, Chesterton was unusually efficient. It is popularly believed that a man who is in earnest about something speaks earnestly and that a man who keeps making jokes is not in earnest. The belief is not ill-founded since, more often than not, this is true. But there are exceptions and, as Chesterton pointed out, Bernard Shaw was one. The public misunderstood Shaw and thought him just a clown when, in fact, he was above all things a deadly serious preacher. In the case of Browning, Chesterton shows that many of his admirers had misunderstood him by reading into his obscurer passages intellectual profundities when in fact the poet was simply indulging his love of the grotesque. Again, he shows us that Stevenson's defect as a narrator was not, as it had become conventional to say, an over-ornate style but an over-ascetic one, a refusal to tell the reader anything about a character that was not absolutely essential. As a rule, it is journalism and literary gossip that is responsible for such misunderstandings; occasionally, though, it can be the author himself. Kipling would certainly have described himself as a patriotic Englishman who admired above all else the military virtues. In an extremely funny essay, Chesterton convincingly demonstrates that Kipling was really a cosmopolitan with no local roots, and he quotes in proof Kipling's own words.

> If England were what England seems,
> How soon we'd chuck her, but She ain't.

A patriot loves a country because, for better or worse, it is his. Kipling is only prepared to love England so long as England is a Great Power. As for Kipling's militarism, Chesterton says:

Kipling's subject is not that valour which properly belongs to war, but that interdependence and efficiency which belongs quite as much to engineers, or sailors, or mules, or railway engines.... The real poetry, the 'true romance' which Mr Kipling has taught is the romance of the division of labour and the discipline of all the trades. He sings the arts of peace much more accurately than the arts of war.

Chesterton's literary criticism abounds in such observations which, once they have been made, seem so obviously true that one cannot understand why one had not seen them for oneself. It now seems obvious to us all that Shaw, the socialist, was in no sense a democrat but was a great republican; that there are two kinds of democrat, the man who, like Scott, sees the dignity of all men, and the man who, like Dickens sees that all men are equally interesting and varied; that Milton was really an aesthete whose greatness 'does not depend upon moral earnestness or upon anything connected with morality, but upon style alone, a style rather unusually separated from its substance'; that the Elizabethan Age, however brilliant, was not 'spacious', but in literature an age of conceits, in politics an age of conspiracies. But Chesterton was the first critic to see these things. As a literary critic, therefore, I rank him very high.

For various reasons I selected very little from his writings on historical and political subjects. Chesterton was not himself an historian, but he had both the gift and the position to make known to the general public the views of historians, like Belloc, who were challenging the Whig version of English History and the humanists' version of cultural history. It must be difficult for anyone under 40 to realize how taken for granted both of these were, even when I was a boy. Our school textbooks taught us that, once the papist-inclined and would-be tyrants, the Stuarts, had been got rid of, and the Protestant Succession assured, the road to Freedom, Democracy and Progress lay wide open; they also taught us that the civilization which had ended with the fall of the Roman Empire was re-born in the sixteenth century, between which dates lay twelve centuries of barbarism, superstition, and fanaticism. If today every informed person knows both accounts to be untrue, that the political result of the Glorious Revolution of 1688 was to hand over the government of the country to a small group of plutocrats, a state of affairs which certainly persisted until 1914, perhaps even until 1939, and that, whatever the Renaissance and the Reformation might signify, it was not a revolt of reason against fanaticism—on the contrary, it must be more fairly described as a revolt against the over-cultivation of logic by the late Middle Ages—Chesterton is not the least among those persons who are responsible for this change of view. The literary problem about any controversial writing is that, once it has won its battle, its interest to the average reader is apt to decline. Controversy always involved polemical exaggeration and it is this, of which, once we have forgotten the exaggerations of the other side, we shall be most aware and critical. Thus, Chesterton's insistence, necessary at the time, upon all that was good in the twelfth century, his glossing over of all that was bad, seems today a romantic day-dream. Similarly, one is unconvinced by Belloc's thesis in *The Servile State*, that if, when the monasteries were dissolved, the Crown had taken their revenues instead of allowing them to fall into the hands of a few of its subjects, the Crown would have used its power, not only to keep these few in order, but also for the benefit of the common people. The history of countries like France where the Crown remained stronger than the nobility gives no warrant for such optimism. Absolute monarchs who are anxious to win glory are much more likely to waste the substance of their country in wars of conquest than plutocrats who are only interested in making money.

Chesterton's negative criticisms of modern society, his distrust of bigness, big business, big shops, his alarm at the consequences of undirected and uncontrolled technological development, are even more valid today than in his own. His positive political beliefs, that a good society would be a society of small property-owners, most of them living on the land, attractive as they sound, seem to me open to the same objection that he brings against the political ideas of the Americans and the French in the eighteenth century: 'Theirs was a great ideal; but no modern state is small enough to achieve anything so great.' In the twentieth century, the England he wanted would presuppose the strictest control of the birth-rate, a policy which both his temperament and his religion forbade him to recommend.

On the subject of international politics, Chesterton was, to put it mildly, unreliable. He seems to have believed that, in political life, there is a direct relation between Faith and Morals: a Catholic State, holding the truth faith, will behave better politically than a Protestant State. France, Austria, Poland were to be trusted: Prussia was not. It so happened that, in his early manhood, the greatest threat to world peace lay, as he believed, in Prussian militarism. After its defeat in 1918, he continued to cling to his old belief so that, when Hitler came to power in 1933, he misread this as a Prussian phenomenon. In fact, aside from the economic conditions which enabled it to succeed, the National Socialist Movement was essentially the revenge of Catholic Bavaria and Austria for their previous subordination to Protestant Bismarckian Prussia. It was not an accident that Hitler was a lapsed Catholic. The nationalism of the German-speaking minority in the Hapsburg Empire had always been racist, and the hot-bed of anti-Semitism was Vienna not Berlin. Hitler himself hated the Prussian

Junkers and was planning, if he won the war, to liquidate them all.

Chesterton was brought up a Unitarian, became an Anglican and finally, in 1922, was converted to Roman Catholicism. Today, reading such a book as *Heretics*, published in 1905, one is surprised that he was not converted earlier.

If his criticisms of Protestantism are not very interesting, this is not his fault. It was a period when Protestant theology (and, perhaps Catholic too) was at a low ebb, Kierkegaard had not been rediscovered and Karl Barth had not yet been translated. Small fry like Dean Inge and the ineffable Bishop Barnes were too easy game for a mind of his calibre. Where he is at his best is in exposing the hidden dogmas of anthropologists, psychologists and their ilk who claim to be purely objective and 'scientific'. Nobody has written more intelligently and sympathetically about mythology or polytheism.

Critical Judgement and Personal Taste are different kinds of evaluation which always overlap but seldom coincide exactly. On the whole and in the long run, Critical Judgement is a public matter; we agree as to what we consider artistic virtues and artistic defects. Our personal tastes, however, differ. For each of us, there are writers whom we enjoy reading, despite their defects, and others who, for all their virtues, give us little pleasure. In order for us to find a writer 'sympathetic', there must be some kinship between his imaginative preferences and our own. As Chesterton wrote:

> There is at the back of every artist's mind something like a pattern or a type of architecture. The original quality in any man of imagination is imagery. It is a thing like the landscape of his dreams; the sort of world he would wish to make or in which he would wish to wander; the strange flora and fauna of his own secret planet; the sort of thing he likes to think about.

This is equally true of every reader's mind. Our personal patterns, too, unlike our scale of critical values, which we need much time and experience to arrive at, are formed quite early in life, probably before the age of 10. In 'The Ethics of Elfland' Chesterton tells us how his own pattern was derived from fairy-stories. If I can always enjoy reading him, even at his silliest, I am sure the reason is that many elements in my own pattern are derived from the same source. (There is one gulf between us: Chesterton had no feeling for or understanding of music.) There are, I know, because I have met them, persons to whom Grimm and Andersen mean little or nothing: Chesterton will not be for them.

KINGSLEY AMIS
The Poet and the Lunatics

MY excuse for making this a more than usually personal reappraisal is that, as it happens, *The Man Who Was Thursday* was the first grown-up novel I remember reading outside school. (*The War of the Worlds* may have preceded it, but I did not then count that as grown-up.) I had reached *Thursday* by way of the Father Brown stories, and went on from it to Chesterton's first novel, *The Napoleon of Notting Hill*. I found, as I have continued to find, that I could not happily get very much further with the rest of his fiction; though I now reflect, a generation later, that a total score of seven lasting volumes out of seventeen is rather enviable. Anyway, their author had impressed himself upon me to the extent that I can still remember how I felt at the news of his death in 1936, the first total stranger's death that meant anything to me personally.

The foregoing is not intended as just an account of boyhood sensitivity. *Thursday* is a boys' book as well as a grown-ups' book, along with *Hamlet* and *The Mill on the Floss*, and not along with *King Lear* and *Middlemarch*. The degree of overlap between the adolescent and the adult response must vary from case to case, and will remain very hard to chart while we know so little about the former. If we could find out by some more reliable means than the adult rereader's unreliable memory, how adolescents respond to works of literature (beyond categorizing them as terrific or tripe), then we would have found out a lot about literature. However, I had better shut down this line of speculation: I can think of nobody I could trust to pursue it who would also want to.

Anyhow, a boy either gobbles a book up or throws it away. I gobbled up *The Man Who Was Thursday*. The opening caught me by the scruff: two poets discussing art and anarchism at an open-air party in a romantic setting under an extraordinary sunset. Art I knew I was for; anarchism seemed a sinister and alluring fantasy, dreamed up by the author himself from nothing or from some sliver of a remote past—I may have noticed indifferently that the book was first published in 1908. Today, anarchism is something else, but lapse of time and events has done nothing to date Chesterton's insight into it: a little more of this later. What I find unchanged and wonderful is the setting and the sunset. Here, as almost throughout, the novel reveals a characteristic it shares with some—though by no means all—greater works: an irresistible power of suggestion that the extraordinary is, if not the most ordinary thing in the world, as GKC might have put it, then at least almost literally round the corner. I did not know on my first readings, and do not care now, that the suburb of 'Saffron Park' stands for Bedford Park, a vanished name for the network of streets just north of Turnham Green station, probably at no time long on romantic potential; I still step, with the first sentence, into village gardens glowing with Chinese lanterns and overhung with a heaven full of red-hot plumes. No writer has ever excelled Chesterton, that drop-out art student, in descriptions of skies and the effects of light on landscape.

After the party, the story takes an apparently more ordinary turn (more like Bulldog Drummond, I thought originally), whereby the poet who has talked against anarchism, Gabriel Syme, gets himself elected to the Central Anarchist Council, the members of which are named after the days of the week. Syme is not merely an anti-anarchist, but a member of the New Detective Corps formed expressly to fight anarchism, and carries credentials franked 'The Last Crusade'—not quite like Bulldog Drummond. The flashback account of Syme's enlistment, in a totally dark room at (so it says) Scotland Yard, deserves a quotation: Chesterton could elevate melodrama, or prose opera, to the threshold of poetry:

'Are you the new recruit?' asked a heavy voice.

And in some strange way, though there was not the shadow of a shape in the room, Syme knew two things: first, that it came from a man of massive stature; and second, that the man had his back to him.

'Are you the new recruit?' asked the invisible chief, who seemed to have heard all about it. 'All right. You are engaged.'

Syme, quite swept off his feet, made a feeble fight against this irrevocable phrase.

'I really have no experience,' he began.

'No one has any experience,' said the other, 'of the battle of Armageddon.'

'But I am really unfit—'

'You are willing, that is enough,' said the unknown.

'Well, really,' said Syme, 'I don't know any profession of which mere willingness is the final test.'

'I do,' said the other—'martyrs. I am condemning you to death. Good day.'

Hardly like Bulldog Drummond at all.

In due course, Syme meets the other members of the Council, an array of carefully-described and sharply-differentiated grotesques. The chair of Friday is filled by the ancient Professor de Worms, who seems to have moved beyond decrepitude into actual corruption, so that, whenever he moves, Syme is afraid a leg or arm might fall off. Saturday is young Dr Bull, coarse and commonplace but for his black glasses; Syme thinks his eyes might be covered up because they are too frightful to be seen. But, as is fitting, the President, Sunday, outclasses them all with his unnatural bulk, such that his face, seen close to, might be too big to be possible. Melodrama again, if you like, but done with such pictorial, concrete energy and conviction that I believe every detail as firmly as I ever did, even though I now know almost by heart what is to follow.

What follows is the revelation of first Tuesday, then Friday and Saturday as co-members of the anti-anarchist constabulary, a trip to France to frustrate a bomb-throwing by Wednesday, a duel with swords, a pursuit on foot, on horses, by motor-car, a final confrontation on the seashore. The reader of this article can guess, only a little more easily than the reader of the book, what Wednesday and finally Monday, the fanatical Secretary, turn out to be; but that confrontation, in which four detectives, believing themselves to be the last champions of Christendom, turn at bay against what seems a world given over to anarchy, remains undimmed, full of fine, highly artificial rhetoric and gestures both theatrical and dignified, enough on its own to earn the novel its subtitle, A Nightmare.

What Sunday turns out to be—apart, of course, from having doubled as the man in the dark room—is not foreseeable. Unfortunately it is also not quite satisfactory or even clear. Near the end, in flight from the detective posse, he turns practical joker or comic ogre, taking rides on a Zoo elephant and a balloon from the Earl's Court Exhibition. (This part is shot through with facetiousness that will make most people wriggle; as a lad I thought the fault was in me—a sadly dated reaction—or was reading too fast to notice.) Right at the end, Sunday is revealed as something like Pan, or Nature, or God. Just before his death, Chesterton disclaimed any such intention, though without citing any alternative; some might see this as a pious disavowal by one who had joined the Roman Church years after the book was written. At any rate, one is finally left thrilled and baffled, peering at a half-decipherable message about life being a bewildering but good-natured and all-reconciling joke.

Boy Amis and man Amis differ most sharply over some possible defects in presentation: what the latter, in his know-all way, might describe as on-the-face-of-it unlikelihoods or impossibilities (which the former took in his stride). Thus, half the plot hangs on the detectives having given, and continuing to keep, crazy promises not to go to the police. Defence: the promises were necessary in order to get on to the Council, and to break them would have fatally dishonoured the Last Crusade. Well yes, but these are inferences, not in the book. Again, the seasons are muddled up, so that the Council breakfasts on an open balcony in Leicester Square just before a February snowstorm, and two days later the countryside near Calais is hot at seven in the morning. Here a juvenile defence works best: the open-air breakfast is fun, and the snowstorm is exciting, and every schoolboy knows (what grown-ups have forgotten) that France is much hotter than England.

But thirdly, time is hectically telescoped, so much so that, as recounted, with no pauses in conversation, etc., the Channel crossing evidently takes about five minutes. Defence: the whole thing is a nightmare. This will not quite do: it is a novel as well as a nightmare, and so must observe some limitations; worse, Chesterton telescopes time in other, non-nightmarish stories. Does it matter whether he knew what he was doing, whether all these oddities are skilful nightmare adjuncts or sheer carelessness?—he certainly shows carelessness in allowing his own narrative style to infiltrate dialogue. But is the question a real question? The narrative-into-dialogue business apart, everything fits well enough into *this* story, and, more generally, writers, like other people, know what they are doing when they may seem not to know they know.

The Man Who Was Thursday is not a political novel, but it has political implications, and from this point of view the 1971 reader is better served than the 1935 reader, perhaps even the 1908 reader. Rather as Jack London foretold the rise of fascism in *The Iron Heel* (published the previous year), Chesterton foretold the shape of some of our present discontents. He saw destructive forces in our society that would be nothing but destructive, 'not trying to alter things, but to annihilate them', basing themselves in the first place on an inner anarchy that denies all the moral distinctions 'on which mere rebels base themselves'. 'The most dangerous criminal now is the entirely lawless modern philosopher.' The enemy arises not from among the people, but from the educated and well-off, those who unite intellectualism and ignorance and who are helped on their way by 'a weak worship of intellect and force'—near the knuckle, that one. Most specifically, Chesterton, or the man in the dark room, 'is certain that the scientific and artistic worlds are silently bound in a crusade against the Family and the State'. By 'silently' he can hardly have meant more than 'secretly'—if he had meant 'by an unspoken consensus' he would have been not just a remarkable prophet but a terrifying one.

All in all, I rather envy that younger self of mine who thought that the bad men in *Thursday* were anarchists, or bomb-throwers, or political assassins, chiefly because they had to be something.

KINGSLEY AMIS
Four Fluent Fellows

CHESTERTON's first book of fiction, *The Napoleon of Notting Hill* (1904), is in outline, and on the face of it, a romance about the future. The date indicated is 1984, but this coincidence, though doubtless odd, is unilluminating. Those twin concerns, to diagnose the contemporary world and through doing so to sound a warning about what it may turn into, characterize Orwell's novel and much orthodox science fiction besides; they find no more than a very incidental place in the Chesterton work. This indeed opens with an attack on prophecy and shows us, first, a society 'almost exactly like what it is now', then, later, a creation of pure and free fancy. (We could capture the book for science fiction only by taking what C. S. Lewis called 'the German view' that any and every romance about the future must fall within that category.)

The product of Chesterton's fancy is a London in which, by a kind of reversion to a medievalism that never existed, the various boroughs, while owing ultimate allegiance to the Crown, become independent city-states. Each has its Provost with his attendant group of heralds, its flag, its citizen-soldiery—armed with no more than sword and halberd—in their distinctive uniforms, its manufactured traditions and mottoes, its ambition and honour. Notting Hill, having ignominiously defeated an aggressive coalition of Bayswater and the Kensington boroughs, acquires a twenty-years' hegemony over all London, only to perish by a combined final onslaught, a set battle in Kensington Gardens.

This sounds like a straightforward, if idiosyncratic, chronicle of adventure and action. In fact, it is both more and less than that, a verdict that may prove not quite so hideously dull as it sounds if I amplify it by suggesting that in this first novel are to be found all its author's important concerns as a writer of fiction, concerns at times cumulative and mutually helpful to marvellous effect, now and then disastrously at odds, but concerns that always recur in his tales and give them their unique flavour and place in the canon.

The preludial attack on prophecy mentioned above consists of some sensible and provocative remarks couched in, often buried under, a style that wavers from the jocular to the facetious. Two concerns are at work here, or, to put it perhaps more appropriately, two men: Chesterton the Polemicist and Chesterton the Buffoon.

The Polemicist, thickly or thinly disguised, turns up virtually everywhere in Chesterton's fiction, and it must be said of him at once that he is rarely less than entertaining and often lends argument an elegance and an epigrammatic sting worthy of the best of the author's avowed polemical writing. Here, the argument is about nationalism and politics generally, it is conducted from more than one point of view, and most of it has nothing to do with the rest of the story; indeed, after the principal debate one of its chief participants, Juan del Fuego, a Nicaraguan grandee about as authentic as his name, is reported to have dropped dead. Right at the end of the book, the Polemicist puts in a secondary appearance to advance the not very inflammatory point that the humorist and the idealist, or the clown and the fanatic, are the two essential parts of the whole, sane man.

The clownish half of this synthesis corresponds to the figure I have called the Buffoon, to some tastes a mildly dismaying companion. He is actually incarnated in the character of Auberon Quin, an owlish minor civil servant who finds himself elected King—by what must have been a very arbitrary process. Until then he has been able to do little more than mystify and bore his friends with a string of elaborately pointless anecdotes; now he can mystify and bore all the people of London by decreeing that they build walls round their municipalities and parade in grotesque costumes sounding tocsins. At last—rather late, in fact, there appears a man who takes the whole charade seriously.

This is Adam Wayne, the youthful Provost of Notting Hill, tall, blue-eyed and red-haired. Red hair in Chesterton's men is a badge of unworldliness and chivalry (in his women it belongs to the sedate and serious-minded). Here we have the fanatical half of the synthesis, and at this point, too, a third Chesterton takes a hand in the shaping of the story. Aware of the deficiencies of the title, I dub the newcomer Chesterton the Melodramatist, meaning no disparagement, intending only to allude to that fusion of the grand and the histrionic, the magnificent and the magniloquent, which we find in the poems of Housman or the music of Tchaikovsky, and which we can respond to, even be deeply moved by, without necessarily ranging it alongside the work of Tennyson or Beethoven.

But let the Melodramatist speak: he does much of his work in dialogue, though he also operates through the heroic gesture, the stroke of symbolism, through *coup de théâtre* and *peripeteia* and transformation scene. (Wayne has interrupted a royal audience where the provosts of the Kensington boroughs—Buck, Barker, *et al.*—are petitioning the King to sanction a new development scheme which will swallow up Pump Street, an unimportant corner of Notting Hill cluttered with dingy toyshops, etc.)

... Buck said, in his jolly, jarring voice: 'Is the whole world mad?'
The King sprang to his feet, and his eyes blazed.
'Yes,' he cried, in a voice of exultation, 'the whole world is mad, but Adam Wayne and me ... He has answered me back, vaunt for vaunt, rhetoric for rhetoric. He has lifted the only shield I cannot break, the shield of an impenetrable pomposity. Listen to him. You have come, my Lord, about Pump Street?'
'About the city of Notting Hill,' answered Wayne proudly. 'Of which Pump Street is a living and rejoicing part.'
'Not a very large part,' said Barker, contemptuously.
'That which is large enough for the rich to covet,' said Wayne, drawing up his head, 'is large enough for the poor to defend.'
The King slapped both his legs, and waved his feet for a second in the air.
'Every respectable person in Notting Hill,' cut in Buck, with his cold, coarse voice, 'is for us and against you. I have plenty of friends in Notting Hill.'
'Your friends are those who have taken your gold for other men's hearthstones, my Lord Buck,' said Provost Wayne. 'I can well believe they are your friends.'
'They've never sold dirty toys, anyhow,' said Buck, laughing shortly.
'They've sold dirtier things,' said Wayne, calmly; 'they have sold themselves.'
'It's no good, my Buckling,' said the King, rolling about on his chair. 'You can't cope with this chivalrous eloquence. You can't cope with an artist ...'[1]

Even the couple of intrusions by the Buffoon, even the occasional lapses into the idiom of a lower kind of melodramatic writing, cannot hold back the speed and zest of that exchange, the confidence with which the unvarnished adverbs—proudly, contemptuously—are slapped down, and Buck and Barker are made to toss up their slow lobs for instant, unreturnable despatch. At this stage, the nature of the book changes; a little further on, the King tells Wayne, 'I tried to compose a burlesque, and it seems to be turning half-way through into an epic.' After making every allowance for the fact that novelists move by unanalysable instinct at least as much as by conscious thought, it still

---

[1] I suspect that Chesterton meant to give the first speech quoted to Barker and the fourth to Buck. He might mix up names but he would probably not have given the same character two incompatible voices.

remains tempting to hear the author's voice in the King's remark, owning up to a belated change of mind.

The epic admittedly ticks over for a few pages while something is said about Wayne's 'mental condition'; not a great deal, hardly more in quantity than is needed to show what it is and how he attained it. No one doubts that Chesterton was deeply interested in people and their relations with one another, but at least in his fiction—parts of his literary criticism may be a different matter—he does not go in much for 'characterization' as it is normally thought of, for differentiating, developing, showing response to circumstance. Adam Wayne has just got to be extraordinary so that he will speak and act as he does.

He acquired his fanatic's brand of local patriotism as a child, we are told: not in the humdrum sense that it all started early, but in that he was most fully a child when he discovered what Notting Hill meant to him. Chesterton never developed a character who is and remains a child, but his tales are full of insights into childhood, celebrations of it and of the disregarded truth that the adult who has missed or rejected any part of what it is to be a child is a sad, stunted creature. In another novel he speaks of 'that concrete and material poetry which a child feels when he takes a gun upon a journey or a bun with him to bed'. (I can still remember the gentle shock of reading that sentence for the first time, the flash of realization, not that you fully understand the author, but, far rarer and more memorable, that the author knows all about you.) More directly, Father Brown explains why a naval lieutenant in full dress, and in suspicious circumstances, was brandishing his sword.

'. . . He thought he was quite alone on the sands where he had played as a boy. If you don't understand what he did, I can only say, like Stevenson, "you will never be a pirate." Also you will never be a poet; and you have never been a boy.'

Adam Wayne is a poet all right, in fact the poet, as well as the Napoleon, of Notting Hill: a poet, moreover, not only in the rather shifty, late meaning of 'one with strong (or refined) feelings' about this and that. He also, it turns out, once published a book of actual poems conveying his sense of the beauty and mystery of his urban surroundings, in the spirit of more conventional poets whose subject is rural nature. The Buffoon has a finger or two in this pie, but it is another Chesterton who says of Wayne that 'twenty feet from him (for he was very short-sighted) the red and white and yellow suns of the gaslights thronged and melted into each other like an orchard of fiery trees, the beginning of the woods of elf-land'.

The same Chesterton is at work in an earlier passage, long before the advent of Wayne:

> The morning was wintry and dim, not misty, but darkened with that shadow of cloud or snow which steeps everything in a green or copper twilight. The light there is on such a day seems not so much to come from the clear heavens as to be a phosphorescence clinging to the shapes themselves. The load of heaven and the clouds is like a load of waters, and the men move like fishes, feeling that they are on the floor of a sea. Everything in a London street completes the fantasy; the carriages and cabs themselves resemble deep-sea creatures with eyes of flame...

Here, obviously, is our fourth Chesterton, one whom, again, I am chary of labelling. 'Chesterton the Painter' has its attractions. It does the service of reminding us that the author started his career not as an author, but as a painter in the proper sense, or at any rate as an art student at the Slade. I wonder very much whether the pictures he produced at that time, if any, show the fascination with the effects of light which can be seen in what I have quoted and which constantly reappears throughout his novels and stories. Every such reappearance seems fresh, different from all the others; we see not only London (though we see London most and perhaps best), not only the town or the village or the castle or the inn, but countryside from the lush to the stark, salt-flat, seashore, estuary, river (light on water, as might be expected, particularly attracted him), at all times of day but with a preference for dusk and dawn, in fog, in rain, in lightning, under snow. Chesterton the Impressionist? That is about the best I can do; the title has thin but legitimate roots in the writing of the 1880s, as a glance at, say, some of Wilde's shorter poems will show.

In *The Napoleon of Notting Hill*, the status of Chesterton the narrator is sufficiently indicated by denying that role an initial capital. There is a story holding at any rate its second half together, there are some fine moments of military action and expectation, but the main job of the narrative is to bridge the gaps between the returns of the various other roles. The closing scene, at the end of which Wayne and Quin set off like—perhaps a bit too much like—Don Quixote and Sancho Panza to roam the world, is right, is a triumph, but it is the triumph of the Melodramatist with some useful support from the Polemicist. No self-respecting, or mere, storyteller would have permitted himself a finale so blatantly implausible; implausible not just by ordinary commonsense standards, but even by the bizarre ones of what has gone before. (A look at the text compels the unanswerable question: Where had everyone else disappeared to?)

There is the sufficiently obvious view that some narrative power must reside in any work of fiction that keeps the reader reading; I cannot decide how far it is vulnerable to the demurrer that readers are kept reading by plenty of things that are not fiction—much of Chesterton's verse, for instance. However, on this view, the Narrator earns his capital letter, and to spare, in the second novel, *The Man Who Was Thursday*, which is the author's fictional masterpiece. To revert to Father Brown's dictum, and perhaps to take a slightly menacing tone for the moment, if you are not kept reading by it you will never be a poet (in any sense I would accept) and you have never been a boy. At any rate, saying so gives me the chance of saying too that it would not be in order to add 'or a girl' at the end of that last sentence. There are no unsympathetic female characters in Chesterton's tales; there are no developed female characters either. Sexual love, marriage, domesticity stand high and firm in the Chestertonian set of values, but they are implied, invoked, used as unargued motives or goals, not explored.

The solitary girl in *The Man Who Was Thursday* comes in in the first chapter and on the last page. At the start, she is doing what Chesterton's women spend a good half of their time doing: listening to the men talk. The men in this case are two poets called Gabriel Syme and Lucian Gregory, and they are talking about anarchism. As might be expected, the Polemicist gets a good innings here, but by the end of the chapter he has merged into the Melodramatist. The setting is a suburban garden transfigured by Chinese lanterns that glow 'in the dwarfish trees like some fierce and monstrous fruit' and a strange evening sky that seems to be 'full of feathers, and of feathers that almost brushed the face'. The Impressionist goes on as he begins.

What our three collaborators produce, working so closely in harmony that there are no incongruities, no gaps to be bridged by orthodox exposition, is of course a story, one of a kind that eludes categorization, like Chesterton's other novels. *The Man Who Was Thursday* is not quite a political bad dream, nor a metaphysical adventure, nor a cosmic joke in the form of a spy thriller, but it has something of all these. It is *sui generis*.

Syme, besides being a poet who talks against anarchism, is a policeman with the job of fighting it. His credentials are franked 'The Last Crusade'—I would love to see that emblazoned on some Whitehall office door. In pursuance of a sort of bet, he allows Gregory to take him to dinner in a greasy riverside pub that turns out to serve champagne and excellent lobster mayonnaise. More surprisingly, dinner-table and diners presently descend *en bloc* into an underground chamber: shades, or rather anticipations, of Ian Fleming's *Live and Let Die*.

Here a secret anarchist meeting is held; not very plausibly, but after some enjoyable and perverse rhetoric, Syme gets himself elected in place of Gregory to the Central Anarchist Council, whose members are named after the days of the week. So the man who has become Thursday sets out by moonlight, in a heavy cloak and bearing a swordstick, to carry the war to the enemy.

At his introduction to the others on the Council, the novel starts to earn its subtitle, *A Nightmare*. They are not only hateful, they are physically monstrous in a way that seems to body forth the evil in their souls. The eyes of Saturday are covered by black glasses, as if they were too frightful to see. Monday, the Secretary, has a beautiful face with a smile that goes up in the right cheek and down in the left and is wrong. The face of Sunday, the gargantuan President, is so huge that Syme is afraid that at close quarters it will be too big to be possible, like the mask of Memnon he remembers seeing as a child in the British Museum, and he will have to scream. The Melodramatist breaks new ground here.

When the spectacles and the mask of anarchy are removed, however, Saturday is human enough, in fact another paladin of the Last Crusade. One by one the other Days declare themselves, until only the Secretary stands between police and President. But Monday is Sunday's agent, with the power of turning the world itself against the crusaders. The chase sweeps across a France of sturdy peasants and soldierly patriots and cultured men of wealth (more fun to read about, at least, than most real Frenchmen) transformed one after the other, in true nightmare fashion, into fanatical allies of anarchy. At last the only surviving champions of Christendom, and of reason and order, turn at bay.

The end is fantasy on a different level. In a second chase, it is Sunday who is pursued. He is caught up with but not caught; you cannot catch a being that is more than human, that represents what, were it not for Chesterton's well-known abhorrence of Shaw, I would be tempted to call the Life Force. This end sounds less than satisfactory, and it is, though not so unsatisfactory as it sounds. I can swallow it.

What I find indigestible in the closing scenes is the fortunately brief reappearance of the Buffoon in the person of the fleeing Sunday, who at one point makes off mounted on a Zoo elephant and who bombards the pursuit with messages of elephantine facetiousness. I have one further cavil, in that the Impressionist sometimes gets out of hand. He is itching to describe a breakfast-party on an open balcony in Leicester Square and also a snowstorm in the London streets, so the latter is made to follow the former on the same morning. And chronology is telescoped, as it often is in Chesterton's tales, just to save time. He could not be bothered to introduce what we might now call cut-away material to imply an interim, nor even to throw in the odd dull but useful sentence merely mentioning a pause in talk, a couple of hours on the road in which nothing of interest took place. This is the sort of thing critics mean when they accuse Chesterton of not being a proper novelist. In a nightmare, the defect perhaps matters less; at any rate, unlike other such defects, it seems to matter less as soon as its presence has been fully noted.

In the later novels, the Melodramatist bows himself out and, *pari passu*, the Impressionist is reduced to the status of a scene-painter: any old time of day and most places will do as background to conversation. It is the Polemicist who comes increasingly to the fore. Part of him is propagandist, the champion of Roman Catholicism against its various foes, but it is a much smaller part than people with a sketchy knowledge or recollection of the real Chesterton would have us believe. And those foes embrace, not any kind of Protestantism, but unreason, superstition, diabolism, militant rationalism, moral neutralism, Tolstoyism, anarchism, materialism, mystical cults. To enjoy the spirited peppering of most of these targets, nothing more doctrinal than a belief in reason is required. Further, the shots fly in both directions: if the outcome of the battle is never in doubt, the enemy is allowed some not always ineffectual return fire.

This interest in dialogue as against monologue comes out strongly in the third novel, *The Ball and the Cross* (1910), an undervalued and, I find, largely overlooked work. Just what sort of novel it is is once more not easy to tie down. It opens like a kind of science fiction, in the marvellous flying-machine of Professor Lucifer; its middle and main part is a blend or alternation of Polemicist material and a simple but vigorous action story. Atheist and Christian—the one is made as attractive a character as the other—fight it out in words and, when not prevented by officious interveners, notably policemen, with swords as well. The physical duel is inconclusive; the metaphysical comes out predictably but acceptably.

The final section is as fantastic as its counterpart in *The Man Who Was Thursday*, but this is more overtly theological: Lucifer lives up to his name in the role of superintendent of a gigantic lunatic asylum in which, were he to triumph, the mass of mankind would end up incarcerated. Chesterton's interest in madmen, which pervades his fiction, is at more than first sight surprising in so sane a writer and human being. Some of it serves his love of paradox, furnishing the man whose devotion to rationalism deprives him of his reason, the man whose devotion to reason makes him appear insane in the eyes of the illogical or the incurious. This is not the whole story, the rest of which I confess I cannot for the moment discover. But it must be stressed that Chesterton is not anticipating, would on the contrary have found devilish, current trendy notions that sanity is a relativist or quasi-political label, that there are insane societies but no insane individuals, etc.

By the time of *Manalive* (1912), the novels are in sad case. The hero, Innocent Smith—a name, like the title, that bodes no good—spends his inexhaustible leisure on projects like dropping his wife off at boarding-houses or places of business in order to re-encounter and re-woo her. As a long lecture near the close informs us flatly, this is his way of demonstrating (to whom?) his belief in the perpetual freshness of marriage. The author has to keep telling us that Smith is irrepressible and a harlequin, that 'he filled everyone with his own half-lunatic life.' These assertions are never demonstrated; Chesterton could not create comic incident or write funny dialogue.

The same primacy of the Buffoon—for it is he—disfigures collections of stories like *Tales of the Long Bow* (1925) and *The Paradoxes of Mr Pond* (1937, posthumous). In the first, a group of buffoons conspires to produce as literal facts a series of common phrases habitually used to express unreality or impossibility. So Commander Blair builds a castle in the air—a doctored balloon; Owen Hood sets the Thames on fire—or rather ignites a layer of petrol floating on part of its surface; and what of it? The eponymous narrator of the other book introduces his yarns with laborious verve, such that his 'paradoxes', far from being the unlooked-for truths of GKC at his best, are no more than irresponsible and pointless mystifications.

Enough of the Buffoon: significantly, he makes not even the most fleeting appearance in any of the Father Brown stories. That incisive little cleric does occasionally and briefly pose as a simpleton, but always for a dead-serious purpose, and any full-grown specimen of the type he would have booted off the stage in short order.

With no more than three or four exceptions, the best of Chesterton's work in the shorter form is to be found in the Brown saga. This consists of what are much closer to being detective stories than most of his other fiction is close to any accepted category. Before turning to it, I might just remark on the pertinacity, or desperation, with which he clung to the detective form in nearly all his short pieces. For a writer with little interest in, or aptitude for, narration as such, a crime or mystery plus investigation plus dénouement is a convenient clotheshorse over which all manner of diverse material can be draped. So it is with 'The Trees of Pride',[2] the sixty-five pages of which involve a disappearance and a reappearance, but really describe a glorified nature-ramble.

The Impressionist (to resume) is in top form throughout the Brown stories, even those written after, say, 1918, when he had all but disappeared from the novels. In 'The Sins of Prince Saradine' and 'The Perishing of the Pendragons' and elsewhere he achieves some of the finest, and least regarded, descriptive writing of this century. And it is not just description; it is atmosphere, it anticipates and underlines mood and feeling, usually of the more nervous sort, in terms of sky and water and shadow, the eye that sees and the hand that records acting as one. The result is unmistakable: 'That singular smoky sparkle, at once a confusion and a transparency, which is the strange secret of the Thames, was changing more and more from its grey to its glittering extreme . . .' Even apart from the alliteration, who else could that be?

In these stories, the Polemicist and the Melodramatist are often hard to distinguish from the Impressionist and from each other. Between them, the three produce wonderfully organized puzzles that tell an overlooked truth, parables that touch the emotions, syllogisms that thrill the senses. No author ever invented titles that more exactly prefigure what is to follow them—'The Salad of Colonel Gray', 'The Song of the Flying Fish', 'The Point of a Pin', 'The Wrong Shape', 'The Absence of Mr Glass'. There is a glint of alloy there among the pure metal, yet it seems to hold everything together in a way that nothing else could.

Now and again a solution is fudged, the criminal is unbelievably lucky, a witness is unbelievably unobservant, Father Brown sits on a clue. Now and again (a mannerism not uncommon in the other fiction) he pauses pregnantly and a feed-man obligingly throws up a 'By which you mean . . .?' or a 'But then you knew . . .?' so that Brown can drop his revelatory blockbuster in a voice like the roll of a drum. I admit to surges of irritation at times like these; they soon die down.

Of Chesterton's seventeen volumes of fiction, seven or eight, with another half-book or so of odds and ends, are worth keeping and rereading; not a bad score. Very well, he remains a minor master in the genre, but there he remains, in spite of the glaring fact that so much of

[2] From *The Man Who Knew Too Much* (1922). Inexplicably omitted, along with three shorter stories, in the 1961 Darwen Finalyson reprint of the volume.

what interested him was irrelevant or even directly inimical to fiction as we usually think of it. His reader can promise himself, at the lowest count, what Lucian Gregory promised Gabriel Syme near the beginning of *The Man Who Was Thursday*. And that is ...? A very entertaining evening.

IAN BOYD
Chesterton and Distributism

CHESTERTON never gave a systematic account of what he meant by Distributism anywhere in his writing, but the outline of this sociopolitical philosophy is clear to anyone who is familiar with his work and that of the circle of writers to which he belonged. As the name implies, Distributism meant first of all that property should be distributed in the widest possible way. Belloc stated the case for this policy in *The Servile State*, which he published in 1912 and which became the textbook of the movement. He argued that Socialism and State Capitalism were helping to create the same kind of society in which power would be concentrated in the hands of a small ruling-class and security would be given to a permanent proletariat whose economic position would be fixed by law. The only alternative to the 'slave' state was the Distributist state of small peasant ownership and workers' guilds. The nearest approximation to this simple society was found in medieval times. Consequently Distributists must be prepared to repudiate modern industrialism in its present form and work for a return to the past. The way in which this theory was interpreted among Belloc's followers is best illustrated by a quotation from a Distributist manifesto published twenty-five years later:

> Distributists agree with Socialists in their condemnation of the present system of society, but they think the evil is far more deeply rooted than socialists suppose ... Distributists propose to go back to fundamentals, and to rebuild society from its basis in agriculture, instead of accepting the industrial system and changing the ownership, which is all that Socialists propose. Apart from their conviction that industrialism is essentially unstable and cannot last, Distributists refuse to accept it as a foundation upon which to build, because they believe that large scale industry may be as great a tyranny under public as under private ownership. They therefore seek to get the smallholder back into

industry as they seek to get him back on the land; and they accept all the implications which such a revolutionary proposal involves.[1]

Chesterton's own interpretation of this theory is more difficult to determine. Involvement in what might be called Distributist politics dominated the latter part of his life, but it is doubtful whether the gradual change in political emphasis which this involvement represents can be called a conversion to Distributism. As a schoolboy he regarded himself as a Socialist. But his writing career began in association with a pro-Boer group of Liberals and he published some of his first essays and verse in their journal *The Speaker*. A growing interest in Anglo-Catholic theology and a growing concern with social problems led him to join the Christian Social Union and in the early part of the century he was busy speaking at their meetings and writing for Henry Scott Holland's *Commonwealth*. What helped make him one of the best-known journalists of the age, however, was the Saturday column he wrote from 1901 until 1913 for the Liberal *Daily News*. Throughout these years and indeed for the rest of his life he described himself as a Liberal. At the same time he was outspokenly critical of the Liberal party. In 1913, during the Marconi scandal, his discontent with official Liberalism reached a kind of crisis and he began writing a series of articles for the Socialist *Daily Herald* in which he continued his violent criticism of parliamentary government and advocated revolution as a solution to the problems of the age. The pacifist policy of the *Herald* at the beginning of the war and his physical and emotional collapse in November of the same year effectively ended this brief alliance with the Socialists.

The connection with the movement that was later to become Distributism began in 1916 when he took his brother Cecil's place as editor of *The New Witness*. It is true that he had been writing occasional articles and verse for it and its predecessor *The Eye-Witness* from their foundings in 1911 and 1912. And it is also true that many of the ideas about politics from which the Distributist thesis was derived can be found in earlier works, particularly in his writing in Orage's *New Age* and in his 1910 sociological study *What's Wrong With the World*. Thus he had very early enunciated a primary Distributist principle that public life existed for the sake of the private life it was meant to protect and consequently that all political and social efforts must be devoted to securing the good of the family which was the basic unit of society. And in 1911, during a railway strike, he wrote 'The Song of the Wheels'. This poem provides an interesting example of the way in

[1] A. J. Penty, *Distributism: A Manifesto* (London: The Distributist League 1937), p. 7.

which his early political verse anticipates the Distributist protest about the mechanization of life in a Capitalist society:

> Call upon the wheels, master, call upon the wheels;
> We are taking rest, master, finding how it feels,
> Strict the law of thine and mine: theft we ever shun—
> All the wheels are thine, master—tell the wheels to run!
> Yea, the wheels are mighty gods—set them going then!
> We are only men, master, have you heard of men?
>
> King Dives he was walking in his garden in the sun,
> He shook his hand at heaven, and he called the wheels to run,
>
> Sitting in the Gate of Treason, in the gate of broken seals,
> 'Bend and bind them, bend and bind them, bend and bind them into wheels,
> Then once more in all my garden there may swing and sound and sweep—
> The noise of all the sleepless things that sing the soul to sleep'.

The official existence of Distributism as a political movement began only with the publication of *G.K.'s Weekly* in March 1925, or more accurately perhaps with the founding of the Distributist League in September 1926. But Chesterton's own version of the Distributist philosophy had been formulated more or less completely many years before. What seems to have happened in the post First World War years is that the emphasis in his political thought gradually shifted from an attack on what he called the corruptions and hypocrisies of modern political life to an increasingly positive argument in favour of the Distributist programme of land distribution and worker control.

Nevertheless, if there was a classical period of Distributism, it occurred during the years between 1926 and 1936 when he was at once the president of the Distributist League and the editor of *G.K.'s Weekly* which was its political organ. During this last decade of his life, he and his associates produced a considerable body of literature in which they attempted to supply Distributist answers to the political and economic questions of the day. What is most surprising about this large mass of material is how little it tells one about the details of the programme. There seems to have been an extraordinarily wide range of opinion within the movement about its meaning and an extraordinarily limited amount of agreement among its members about their common policy. Chesterton did very little to help establish a party line that would unite his divided followers. He maintained a curious kind of detachment from the movement of which he was the nominal leader. Father Vincent McNabb and to a lesser extent Eric Gill expressed an almost luddite contempt for machinery of every kind. Others wrote approving accounts in *G.K.'s Weekly* of bizarre attempts by League members to set up primitive and self-sufficient rural settlements. The medieval debate continued endlessly. But he refused to take sides in any of the acrimonious arguments that shook the tiny society, and he showed a great reluctance in making decisions that would affect the everyday running of the league and paper. The articles he wrote dealt mainly with Distributist principles, and it was usually possible for any faction however extreme to invoke his authority for the position it held. His view of Distributism as a political theory escapes the summing-up that its simple outline seems to invite. As Ronald Knox remarks, '. . . it is not exactly a doctrine, or a philosophy, it is simply Chesterton's reaction to life'.[1]

One must in fact turn to his fiction for the clearest and most vivid expression of what he meant by Distributism. That the movement was a central concern to him is obvious from the time and energy that he devoted to keeping it in existence. Much has been written about what this effort cost him in terms of money and health, but not enough has been written to explain his obsessive interest in what seems to be an absurd medieval fantasy. A careful reading of the fiction and particularly the novels makes this much easier to understand. The importance of political themes in them is at once obvious. More interesting, however, is the way in which the imaginative statement of his political beliefs presents them in a new light and to a degree alters their meaning. It is as though he could express himself most clearly in a sort of political parable. His imagination was unusually allegorical like that of Blake and Watts about whom he wrote, and like them he was an artist who claimed to see truth as it were in a series of pictures. Any discussion of Chesterton and Distributism, therefore, would be seriously incomplete that failed to take into account the political meaning of his fiction.

What is most interesting about the main body of the fiction is the way in which it questions the Distributist dream of returning to a simpler and more agrarian medieval past. The future crisis of industrialism predicted by Belloc is more or less taken for granted, but there is no idealization of what is supposed to be the medieval age. In fact there is a recurrent anti-medieval theme in the novels, for in them every attempt to create a medieval social order ends in total disaster. In *The Napoleon of Notting Hill*, medievalism is merely the private joke of King Auberon that the humourless Adam Wayne happens to take seriously. The medieval empire they are responsible for establishing

[1] 'G. K. Chesterton: The Man and His Work', *The Listener*, 19 June 1941, p. 880.

eventually becomes so tyrannical that both of them are glad to see it destroyed. The story is supposed to begin in 1984. This may be a coincidence, but it is appropriate that Orwell, who admired the prophetic quality in Chesterton's writing, should choose that date for the title of a novel that draws another horrifying picture of the future. In *The Ball and the Cross*, the medieval theocracy of which the hero dreams turns out to be a nightmare world of terror and injustice.

But the clearest examples of this distrust of medieval politics is found in *The Return of Don Quixote*. In it the bogus League of the Lion at first enjoys a splendid success. But the 'medieval' state which it establishes is finally revealed as a puppet regime of the cynical industrialists who have manipulated it for their own purposes from its creation. It is significant that the romantic idealist who is the hero is enthroned as King in what is called 'the Mussolini manner'.[1] But it is even more significant that he eventually denounces the movement he helped to found and identifies the true medievalist as the syndicalist trade-union official who is bitterly opposed to the medieval league and works instead within a labour union trying to introduce a system of worker control. In the dedication of the novel to the sub-editor of *G.K.'s Weekly*, where it was first serialised, Chesterton calls it 'a parable for social reformers'.[2] The warning to the more enthusiastic medievalists among his Distributist followers ought to have been sufficiently clear, but there is no sign that they understood it. The suggestion, of course, is that they might create a tyranny in their eagerness to escape one. What is more interesting, however, is the implication that the medieval ideal is entirely destructive unless it is seen as a kind of myth providing a perennial social standard by which to judge the modern world.

The criticism of medievalism is also linked to the distrust of political power of any kind that is expressed throughout the novels. Indeed, the central problem in the fiction, which is also the central problem of Distributism, is how to give power to the people without corrupting them by doing so. *Tales of the Long Bow* provides the only detailed description of a successful Distributist revolution, and in it the revolutionary leaders are only interested in talking about their children and their gardens. After the revolutionary war, they simply retire from politics, and there is no reason to believe that the government they have abandoned will not again fall into the wrong hands. In *Manalive*

[1] See Chesterton's note on the novel when the serialization of it ended in his paper. 'An Explanation: You Can End This Story Here', *G.K.'s Weekly*, 30 November 1926, p. 135.
[2] See the dedication to W. R. Titterton in the first edition (London: Chatto and Windus, 1927), p. v.

the private court that is established in the suburban boarding-house spends its entire time examining the political and social implications of the eccentric hero's family-life, but nothing is said about the way in which political power should be exercised. *The Flying Inn* and *The Man Who Knew Too Much* end with victories over the vaguely Eastern armies that are the favourite embodiment of evil in the fiction, but in neither novel is there any hint of the way in which the post-revolutionary society will be governed.

At the end of *The Napoleon of Notting Hill*, Auberon and Adam recognize the failure of their medieval experiment and express the hope that the conflict between irony and idealism which has torn their country apart will be finally resolved when political power is given to the ordinary citizen, 'the equal and eternal human being... [who] sees no antagonism between laughter and respect ... the common man, whom mere geniuses like you and me can only worship like a god'. But the myth of the heroic common man, like the myth of the heroic medieval past, provides no real solution for the practical problems of politics. Indeed, the common man remains an ideal only because he has never had to carry the corrupting burden of actual political power. And the failure to distinguish between what is a mythic and what is a practical reality could result in the same kind of tragedy that destroyed the experiments in medieval politics.

The distrust of power and the inability to come to terms with it help explain the deeply pessimistic tone that is characteristic of Distributism in the fiction. It is true that in late novels such as *Four Faultless Felons* and *The Paradoxes of Mr Pond* the king, who makes a brief appearance, wields power without being destroyed by it. But his success is too easy and complete to be convincing. The way in which he reconciles the bitter divisions between left and right is never really described. It is something that is told rather than shown. In *The Man Who Was Thursday*, the story of the anarchist conspiracy and the philosophical police that combats it is an allegory about the hope that comes through suffering and apparent isolation. But what haunts the imagination of the reader is the hopelessness of the fight against the rich and fanatical 'church' of anarchy. The interesting suggestion that the desire for revolution and radical reform can sometimes mask a longing for annihilation and death is never worked out in any detail and is somewhat weakened by the happy ending. And 'Sunday' is far more convincing as the mad millionaire who directs the international anarchist movement than he is as the good policeman who organizes the group that defeats it.

This pessimism is also implied by the relationship of the Chestertonian hero to the world in which he lives. Not only does the fiction draw a familiar contrast between the corruptions of business-men and politicians and the simple virtues of farmers and workers, but it also emphasizes the isolation and helplessness of those who attempt to challenge the existing social order. The typical hero is always an eccentric whose history creates a further and more tragic contrast between the madness of a world that claims to be sane and the sanity of a character that the world regards as mad. *Manalive* is perhaps the most obvious example of this, since it is supposed to be the record of a trial that is meant to determine the hero's sanity. In *The Napoleon of Notting Hill*, Auberon speaks of himself and Adam as the only people who are sane ('the whole word is mad, but Adam Wayne and me') and at the end of the novel they must both accept responsibility for the mad irresponsibility of their plans. *The Ball and the Cross* presents an even more disturbing situation in its allegory of the world as a universal mad-house in which all the characters are imprisoned and from which they all make an improbable escape that symbolizes the coming social revolution. *The Poet and the Lunatics*, as its title suggests, examines different kinds of madness and in an oblique way attempts to define social sanity. But the hero is finally a fugitive who is regarded as a madman himself and the climax of the novel occurs when he makes his escape from yet another mad-house. The Chestertonian wise fool is in fact always alone and helpless. It is true that, like Chesterton, he claims to be a spokesman for God and the people, but as Charles Williams points out, both these great allies are 'voiceless and unarmed'.[1]

The political pattern of the fiction therefore remains remarkably constant. It might be called with equal accuracy radical tory or moderate anarchist. The anarchist element is provided by the distrust of the modern state and the belief in the value of free and self-governing social units. What the novels celebrate is the love of freedom and the confidence that the weak and dispossessed are able to co-operate to help one another. The feeling that inspires all the fiction is a longing for the destruction of what Chesterton regarded as an unjust and thoroughly discredited Capitalistic system. It is a feeling that is perfectly expressed in the bitter anti-parliamentary song that is sung by the revolutionary army at the conclusion of *The Flying Inn*:

> Men that are men again; who goes home?
> Tocsin and trumpeter! Who goes home?
> For there's blood on the field and blood on the foam
> And blood on the body when Man goes home.

[1] 'Gilbert Keith Chesterton', *Poetry at Present* (Oxford: Clarendon Press, 1930), p. 99.

> And a voice valedictory ... Who is for Victory?
> Who is for Liberty? Who goes home?

But the pattern is also a tory one, since it implies the impossibility of achieving the political success that Chesterton regarded as necessary. The feeling of confidence and trust in human nature that underlies the Distributist call for revolution and the Distributist idealization of the common man is contradicted by the strong undercurrent of pessimism about the corrupting effect of political power. Thus the novels stress repeatedly the dangers of constructing any kind of Utopia, particularly the pseudo-medieval Utopia which held such a fatal and delusive charm for Distributist idealists. What the novels finally express with great force is a tory conviction about the reality of original sin, which Chesterton defined as 'the permanent possibility of selfishness [that] comes from the mere fact of having a self'.[1] And it is this inner weakness that makes the Distributist dream of a free and equal society for all impossible. This contradiction in Chesterton's political thinking and feeling is never resolved. There are indeed occasional hints of what a solution would be in the scattered and moving allusions to the kind of divine grace that would heal and perfect man's wounded nature. But until a cure for man's perennial selfishness is found the Chestertonian common man will continue to be crushed by the hateful system Chesterton so movingly denounced:

> Through the Gate of Treason, through the gate within,
> Cometh fear and greed of fame, cometh deadly sin;
> If a man grow faint, master, take him ere he kneels,
> Take him, break him, mend him, end him, roll him,
>   crush him with the wheels.

---

[1] 'Mr H. G. Wells and the Giants', *Heretics* (London: John Lane, The Bodley Head, 1905), p. 79.

LEO A. HETZLER
Chesterton's Writings in his Teenage Years

FOR a writer with as distinctive a style and as personalized a view of the world as Gilbert Keith Chesterton's, the youthful, formative years are always an interesting period for the critic. In Chesterton's case there is ample and clear evidence of what he was thinking and writing between the ages of 15 and 19, not only in letters and some two dozen notebooks but also in the issues of *The Debater*, a monthly magazine of the Junior Debating Club at St Paul's. This club had been founded by Chesterton on 1 July 1890 when he had just turned sixteen; the early issues of the *Debater* were mimeographed but then professionally printed from March 1891 to June 1893. The magazine contained not only articles and poems by Chesterton, dating from his fifteenth year, but also reports of his remarks in discussions and debates.

In reference to the structure and style of these articles, he had by this time learned to write short but penetrating essays that rapidly surveyed an author or general topic, dipping here and there to praise, chide, or qualify. Typically, he was also attempting to place men and ideas in their larger, historical framework, as he did, for example, with fine, perceptive criticism in this observation on Milton:

> His earliest recollections were of Greene and Marlowe, Fletcher and Shakespeare, and all the crowd of reckless, impoverished, dissolute geniuses who wrote in garrets and taverns the works which are the glory of the Elizabethan age. Their classical learning and their daring imagination he imbibed and carried with him into the narrower and sterner sphere of Puritan piety, and forms, as it were, a link between the two Englands differing in everything but their glory, the England of Elizabeth and the England of Cromwell. (May 1891)

Or again on Gray:

> Between the new natural school of which Cowper and Burns were the forerunners and the old logical school of which Beattie and Collins were the dregs, Gray sits in meditative dignity, mingling the faint prelude of the one with the dying echoes of the other. (July 1891)

Equally significant was his friend Lucian Oldershaw's advice to pay less attention to biographical facts and more to impressionist criticism and personal reflections. Thus in an assessment of Shelley, Chesterton attempted to capture the poet's essential spirit:

He was not a bad man, he was not a good man; he was not an ordinary man; he was a sincere philanthropist and Republican; yet he was often as lonely and ill-tempered as a misanthrope; he had far purer feelings towards women than either Burns or Byron, yet he was a far worse husband than they; he was one of those men whose faults and failures seem due, not to the presence of tempting passions or threatening disasters, so much as to a mysterious inner weakness, a certain helplessness in the hands of circumstances. (May 1892)

In regard to Chesterton's style itself, certain of his later characteristics were already discernible in these *Debater* essays. Even then he had an admirable command of words, images, and rhythms. There is the same outpouring of examples, parallellisms, and webs of connected thoughts as in his later writings. Mention 'dragon', and the Python, the Hydra, the pterodactyla and ichthiosaura immediately marched through his mind. In 'Poetry and Science' (when 17), for example, he illustrated a general statement (that poets, like scientists, closely study nature) with such a torrent of examples:

All the fantastic goblins and chimeras that ever sprawled in the grotesque elfindoms of Callot, or the weird pandemonium of Doré, are only combinations of forms studied from natural life. No one would have imagined a unicorn who had not taken some notice of a horse. No one could have drawn a a griffin who had not first drawn an eagle. Or again, architecture obviously had its origin in a close observation of the natural world. No one could see the costly and intricate tracery of the oriental palaces without feeling that they were conceived under the influence of the varied and glowing vegetation of the country, just as the solemn Gothic cloisters seem to have taken their character from the gloomy, silent pine forests of the north. (Oct. 1891)

Moreover, the endings of his essays had the same polished rounding off that concluded his mature essays—a sense of sweeping or swirling to a completion, like bronze gates swung closed, and when possible an ascent to the mystical. Thus, for example, he wrote this conclusion to 'Dragons: A Sketch' (he had dealt with the dragon as an archetype of evil):

Reader, when you meet him, under whatever disguise, may we face him boldly and perhaps rescue a few captives from his black cavern; may we bear a brave lance and a spotless shield through the crashing mêlée of Life's narrow lists, and may our wearied swords have struck fiercely in the painted crests of Imposture and Injustice, when the dark Herald comes to lead us to the pavilion of the King. (Mar. 1891)

But the device of paradox was used sparingly. Those that do appear are used, as he was to use them generally later, to stress simple, basic truths that needed to be pointed out. For instance, in a school essay when 17, he observed, 'It is precisely the utter spiritualist who is the animalist, for to him all life, physical and mental, is divine, because the centre of the world is divine.'

More significantly, Chesterton's contributions to the *Debater* reveal not only a developing style but also ideas that were not present in his childhood writings, certain intuitions and insights that would become part of Chesterton the man. These are difficult to arrange in any natural order, but one might begin with two principles he invoked in judging literature but that have wider implications. He judged literature to be great if it portrayed human nature in a spirit of 'true' realism; that is, not narrowly as romanticists and naturalists did, but 'in the round'. Thus he disparaged Kipling for straining after a kind of novelty and smartness, and conversely praised Barrie for having a true realism, evoking emotions 'felt by the majority of fellow creatures' (Dec. 1891). Chesterton also found true realism in the harsh, fallen world of George MacDonald's Scotch novels, in which some men strove to resist inner weaknesses and environmental influences—characters such as the alcoholic scholar Cosmo Cupples in *Alec Forbes*.

The second principle was a corollary of the first but applied specifically to the moral dimension: that literature of any worth is rooted in the simple moral values of the common man. Thus in the course of one J.D.C. debate Chesterton condemned the 'new purely aesthetic school of Swinburne and Morris' because their poetry lacked 'the really stirring and popular element in literature', the 'sentiment of moral' (May 1893). By way of contrast, he hailed the old ballads and fairy-tales as the kind of deeper realism he sought (Mar. 1891 and Mar. 1892). Further, in a letter to Edward Clerihew Bentley (1892) he viewed fairy tales as the repository of the ancient, intuitive wisdom of humanity:

> It is all very well, as small children to read pretty stories about Satan and Belial [in *Paradise Lost*], when we have only just mastered our 'Oedipus' and our Herbert Spencer, but when we are men we know that Cinderella is much better than any of those babyish books.

This early interest in the essential humanity of the common man of his folk literature is even more evident in his early views on politics. His tendencies in these years moved strongly towards an egalitarian socialism, a revolutionary movement directed towards a universal reign of perfect justice and freedom. Thus, for example, in his first year at St Paul's (when twelve) he wrote a paper pleading that poor boys be admitted to public schools—not as a matter of charity but of justice. He quoted Kingsley's charge that intermittent acts of philanthropy are not the solution: 'Can your lady patch hearts, that are breaking, with handfuls of coal and of rice?' What Chesterton found lacking among his elders (and even among his companions) was a want of true sympathy and social conscience. Again in 1890 he wrote to Oldershaw (just such a companion) from Italy, 'Italy has not abated my socialism ... because here the peasantry are so supremely superior to the upper classes, morally, physically, and intellectually.' And he continued to be pro-revolutionary: his 'Journal' (Aug. 1890) reports that he is explaining 'Carlyle's Revolution to Innocent Child [his younger brother Cecil]', and that he is asked every day by his schoolmates to 'give a complete history of the life and doings of the Revolutioners'.

Certainly the injustices against the poor were much on his mind. For instance, in his 'Dramatic Journal' (composed during the Christmas Vacation of 1891 when he was 16), he complained that Oldershaw was unmoved by Elizabeth Browning's 'Cry of the Children':

> Your Humble Servant read Oldershaw Elizabeth Browning's 'Cry of the Children,' which the former could scarcely trust himself to read, but which the latter candidly avowed that he did not like. Part and parcel of Oldershaw's optimism is a desire not to believe in pictures of real misery, and a desire to find out compensating pleasures. I think there was a good deal in what he said, but at the same time I think that there is real misery, physical and mental, in the low and criminal classes, and I don't believe in crying peace where there is no peace.

Partly because of his very close friendship with Lawrence and Maurice Solomon and partly because of his strong sense of justice, Chesterton was vehemently pro-Jewish all through his early and late youth. In 1889, for example, in drawing up five propositions setting forth the rights of man in regard to freedom of religion, he quoted favourably (in Proposition IV) from Macaulay's speech for the admission of Jews into Parliament:

> The points on which Jews and Christians differ have a great deal to do with a man's fitness to be a bishop or a Rabbi, but they have no more to do with his fitness to be a Member of Parliament than with his fitness to be a cobbler.

Moreover, there were many references to his anger at the harsh treatment of the Jews in Russia; thus his 'Dramatic Diary' recorded this reaction to the latest dispatch from Russia in the *Review of Reviews*: 'Made me feel strongly inclined to knock somebody down, but refrained' (Dec. 1890). Again, in one of the longest poems he wrote at this time, 'Before a Statue of Cromwell, At the Time of the Persecution of the Jews in Russia', he noted that Cromwell had 'set the Hebrew free', and he chaffed at England's present inaction

'While a brave and tortured people cry the shame of men to God!'

In these years, too, he realized that the struggle to actualize ideals was a near-impossible task. In the *Debater* series of 'Letters of Three Friends', he adopted the role of 'Guy Crawford, an artist of strong socialistic tendencies'. In one incident Guy attempted to defend a young Nihilist revolutionary against Czarist troops in St Petersburg (the youth was significantly named 'Emmanuel', 'God with us'):

> I saw only a poor student dead upon the road, with his white face to the stars, and his blood dark on the stones; unknown, unrecorded, as a champion of justice, like thousands who have fallen for it in the dark records of this dark land. (Nov. 1892)

There was another facet of his mental development at this time, however, that added a further dimension to his political thinking: his discovery that there were some worthwhile ideas inherent in the spirit and tone of medieval society. Yet this new appreciation developed in spite of his strong dislike for the cult of 'Medievalism', inherited from Tennyson, Rossetti, and Morris by contemporary Aesthetes. At the time he entered St Paul's, he used medieval material mainly in rollicking parodies of romances and saints' legends. In point of fact, it seemed to him that the contemporary cult of 'Medievalism' was backward-looking, ineffectually longing for a perfect age of beauty, chivalry, and order. To him this was not only a misreading of history, but such nostalgia diverted energy from what he felt was the task of his generation: that of fulfilling the potentialities of the times in which he lived.

His animosity towards a pseudo-medievalism (an animosity which flared up intermittently in his notebook jottings and is apparent in *The Return of Don Quixote*, 1927) was spurred on by the presence of such a Medievalist among the Junior Debating Club, Robert Vernède. The opposition between him and Chesterton on this point is evident in a draft of 'A History of the J.D.C': Vernède's

> idealism drove him, not forward into the tumult of search and question, but backwards into the weird beauty of old legend and romance, in the vague charm of antique dreams in which unreality he could rest. Rest was the keynote of his poetry, as unrest was that of the Chairman Chesterton ... [Vernède sought] in ancient chivalry and kinghood a refuge from the modern commonplace.

(In an unpublished novel of Chesterton's about 1896, Vernède appears as the character Valentine Amiens.)

However, at the same time Chesterton did discover certain values in the spirit and tone of medieval society that had an effect on his political thinking. These insights were conclusions drawn primarily from his personal contact with medieval art during vacation tours on the continent and his chance reading of a number of medieval texts (edited by Professor York Powell). Such experiences revealed that the Middle Ages had not been, as his history books had taught, an age of utter barbarism and lacking all reason. In the guise of 'Guy Crawford' in the *Debater* he praised the Middle Ages for its sense of the oneness of all humanity, for he had discovered saints, all manner of contemporary townspeople, and angels mixed together in its paintings; he also had found this same sense of oneness in the contemporaneity of the heaven and hell in Dante's *Divina Commedia*, in contrast to the static Renaissance mythology of Milton (July 1892).

From these personal insights into the spirit of the Middle Ages, he attempted to find some satisfying explanations from what he had unexpectedly discerned. Perhaps from his innate religious feelings, he felt that whatever was good in the Middle Ages was a remnant of the spiritual teachings of Christ. In an important *Debater* letter (from Florence, 1892) Chesterton as 'Guy Crawford' spoke of the primitive Church as 'the first true socialists':

> Those early Christian fellows were the first true Socialists. With 'Little Children, love one another', still almost in their ears, they could not help starting socialistically, with equality of rank and community of goods. But in this black, roaring, grinding earthly factory of selfishness, that brave little attempt was soon stifled and swept away, and we have taken 1800 years to come back to the point again. (Mar. 1892)

Perhaps one of the most significant socio-political views the 17-year-old Chesterton held is expressed in this same *Debater* letter of March 1892. He warned that the contemporary socialist movement was in danger of aligning itself, not with timeless, unchanging democratic ideals and desires, but with relativistic, materialistic determinism and Darwinisticism. Now, Chesterton was convinced democracy must rest on unchanging values and ideals arising from the whole man. And he was determined there must be a government-enacted, gradual redistribution of wealth in England. He also made a special plea in this letter that this was not 'impractical' in the sense of being 'impossible':

> Impractical, you say? Yes, and more impractical by every word you say. We do not need to be told the age of unselfish brotherhood is a very long way off. Nothing is more foolish than to connect our socialism with more modern scientific agnosticism. For democracy is an essentially spiritual idea, a

contradiction to the modern materialism which encourages the brute tendency to an aristocracy of the physically 'fittest'. There is no sort of doubt that the eagle is larger and stronger and handsomer than the sparrow. There is no room for equality in the question, except from the spiritual point of view—'For not one sparrow falls to the earth'—you know the rest. No, nor one man either, and I think that he who sees them fall will call humanity to account for them and will not be answered by the survival of the fittest or the law of individual success, or any other paraphrase of the good old formula, 'Am I my brother's keeper?'

Since this letter contains these key ideas, this is one of the most important documents from this period.

Mention of his admiration for the early Christianity community leads to the question of his religious beliefs. During his years at St Paul's he had largely lost interest in Christian doctrines as presented by any of the formal Churches. He considered himself an agnostic—but he was a struggling, thinking agnostic. He had opposed the idea of a Divine inspiration of the Bible when 14 in a poem on that theme; some years later in 1892 he wrote in a notebook, 'The idea of the Bible's literal inspiration was the salt required to keep it, until men could appreciate grand literature for its own sake.' But more significantly, he looked on the Bible and Judaeo-Christian beliefs as too limited and applicable only to part of man's total mystical experience. Yet interestingly enough, he saw the social implications of the biblical tale of Adam's creation: an illustrated poem, 'A Vision of Edens' (1891), portrays five Adams with the physical characteristics of different races with the epigraph, 'I do not believe the Bible story can be true, I believe there were five Adams'. Perhaps they overheard him, because they begin to fight one another, until a red-clay biblical Adam (signifying a common human nature) appears to bring peace.

Like so many others at this time in England, young Chesterton was interested in Oriental religions, and in a notebook (1892) are a list of terms from Buddhism, illustrated by a series of excellent drawings of 'the [Nine] Fetters of the Soul of Man'. Again, in *The Human Club* (about 1891) the choric figure and source of wisdom is Basil Marks, an 'Eastern mystic ... a seeker of totality'. In Chapter 4, 'The Consulting Room of the Philosopher', Basil, in a reception room furnished with things from all over the world (symbolizing universality or totality), leads his patients out of their difficulties with indirect, Buddhist observations.

Generally, he believed that some spiritual Intelligence existed that gave to life and man an eternal significance and value. Indeed, in 'Idolatry' (January 1892) he asserted that modern man cannot so easily eradicate this almost primal conviction that emanates from the very core of human nature:

> Whence came that strange, mystical impulse, with the strength of true sacrifice strong,
> Before symbols of earth and of heaven, before canons of right and wrong?
> Out of the deep, mysterious—we know not whence it began—
> Out of the deep, all-present, from the depths of the Nature of Man.

But he was uncertain whether the Ultimate Vision revealed by that 'poetic' sense would be dark or bright. With overtones of Yeats's later 'The Second Coming', Chesterton in 'Hymn to the Spirit of Religion' viewed this Spirit as a 'doom' and could not say whether it be 'sweet or bitter':

> The shapes and the forms of worship wherein the divine was seen
> Are shattered and cast away on the fields of the things that have been.
> A terrible stir of change and waking through all the land
> Will we know what things to believe, or what knowledge be near at hand?
> Therefore I turn unto thee, the nameless infinite
> Mother of all the creeds that dawn and dwell and are gone
> Voice in the heart of man, imperative, changeless, blind
> That calls to the building of faiths through the ages of all mankind
> Fathomless mystical Impulse, that is and forever has been ...
> Thou art more than Plato or Buddha or Mahomed or Christ the divine
> Thou art more than all the faiths, or false or true that befall
> For thou art the unseen force that is under and shapeth all ...
> But a doom is sweet or bitter has bound us forever to thee.

Although he could find no clear confirmation that the world and man had meaning and although physical and moral evil were all too evident, nevertheless his dominant mood was that of a zestful appreciation of life. Certainly the world contained fearsome adversaries: his six major poems on historical personages (all but one are dramatic monologues) are all idealists who ended as seeming failures: Danton in the tumbrel, William of Orange dying helpless in old age, Simon de Montfort fatally wounded on a lost battlefield, Algernon Sydney on the scaffold, St Francis Xavier dying in the East without success. Yet the overall tone of these poems is that of a self-sacrificial battle endured for the brotherhood of all men with song and joy.

A final instance of an insight from this early period that will later prove important is to be found in an imaginative folk tale, 'The Taming of the Nightmare', written when he was seventeen (in *The Coloured Lands*, 1937). In the story a young hero, setting out to tame the wild Nightmare, passes by a melancholy, pathetic monster, the

Mooncalf, and overhears its poignant soliloquy. Man had once led the Mooncalf away to the sunlit, rationalist world to attempt to put him to some practical use, but there he remained a thing of horror and ridicule; when he escaped back to the land of the moonshine, the Mooncalf sang of the loneliness of his monstrous state:

> I forget all the creatures that taunt and despise,
> When through the dark night-mists my mother doth rise,
> She is tender and kind and she shines the night long
> On her lunatic child as he sings her his song.
> I was dropped on the dim earth to wander alone,
> And save this pale monster no child she has known,
> Without like on the earth, without sister or brother
> I sit here and sing to my mystical mother.

Later the hero does meet the wild Nightmare and wrestles it to submission and to love. 'And Jack took the big, ugly head in his lap and kissed it and guarded it in silence, till at last the Nightmare opened her eyes, now as mild as the Mooncalf's, whinnied sorrowfully and rubbed her head against him.' They talk of the Mooncalf and ride together through the hostile landscape.

'Come,' said the boy dismounting, 'since men will not receive us, we will go on our way together. Perhaps we will visit the Mooncalf again and see your mother and your brothers.' 'My master,' said the Nightmare, sitting down at his feet. 'I have no mother nor brothers. I know no one but you, who does not shrink from me. 'But you are my master and I will go with you whither you will.'

The symbolism of this strange tale cannot be simply equated with abstract terms. Yet one might suggest that the 'Mooncalf' points towards the mystical, intuitive, and imaginative side of man—a valid and necessary part of man that Chesterton's utilitarian age spurned and suppressed. Next, the wild, motherless Nightmare suggests Nature in its lawless, fierce, and terrifying aspects—and man too is part of that Nature. Thus the monstrous horse that runs at night is a multiple symbol for the terror and ugliness of Nature, the near-diabolism of the spirit, and the dark unconsciousness of the mind; it is everything in the world that exists beyond the circle of light cast by human reason and human virtue. In young Chesterton's view terror and darkness were indeed found in man's world; but man can and must conquer the Nightmare. The key phrase is 'We will go on our way together.'

As he faced the future, the various tendencies in his thinking were still a labyrinth without a discernible pattern. When he left St Paul's, he turned to the world that lay before him, part of which would be the daily routine of common life, part the terror of the Nightmare, and part the mystical intuition of the Mooncalf.

R. C. CHURCHILL
The Man Who Was Sunday

'RATIONALLY speaking', Chesterton once remarked, 'there is no more reason for being sad towards the end of a hundred years than towards the end of a hundred fortnights.'

There is no rational reason, either, for celebrating the hundredth anniversary of a writer's birth rather than the ninety-ninth. Nevertheless, we do tend, in our paradoxical way, to celebrate the fact that a famous writer was born a hundred years ago, and Gilbert Keith Chesterton, born in Campden Hill in London on 29 May 1874, must conform willy-nilly to this irrational fashion.

Only once did I see him in the sinful flesh. It was in his last, forgetful years, when he came rolling down Salisbury High Street wearing his famous cloak and a puzzled frown. He was probably due to speak at Birmingham or Beachy Head and had taken the wrong train. A schoolfellow nudged me, amazed, and whispered: 'There's G. K. Chesterton!' and so indeed it was.

Chesterton belonged to the last generation of writers who were also public figures, as beloved of cartoonists as politicians or film-stars. There was no mistaking GKC, as there was no mistaking GBS or Arnold Bennett. Whereas now, if Graham Greene jostled you in some seedy passage, or if Samuel Beckett requested a lucifer to light his stub, you probably wouldn't know them from Adam—or from Godot.

Wells once complained that Shaw had 'invented a most amusing personal appearance', and the same accusation could be levelled against Chesterton. Both GBS and GKC were partly self-creations, figures deliberately built up from promising material, actor-managers who relished their own performance as much as the most delighted of their audience. The more 'GBS' Bernard Shaw became, the less, some people felt, there was left the original Shaw, and Chesterton was, apparently, swallowed up by 'GKC' long before he wrote *The Man Who Was Thursday* in 1908. Like Lucian Gregory in that novel, he was 'helped in some degree by the arresting oddity of his appearance, which he worked, as the phrase goes, for all it was worth'.

But Shaw and Chesterton, like all great humorists, were fundamentally serious writers, writers who used their paradoxical wit for serious purposes, for causes dear to their hearts, as well as for public entertainment. If the writers of their generation could be named, like Chesterton's anarchists, after the days of the week, there is no doubt at all which day Chesterton himself would represent. He had grown up in the 1890s, in the Beardsley period, in a *fin de siècle* which was more than usually 'finished', when literature had fallen into the sere and yellow leaves of *The Yellow Book*, when the aesthetic adventure was at its last gasp. As he said himself, dedicating his 'Nightmare' to his schoolfriend E. C. Bentley:

> The world was very old indeed
> When you and I were young . . .

If the aesthetic survivors from the 1890s could be called Friday—Friday evening, be it understood, and pretty late in the evening—and Shaw and Wells could be called Monday morning, with the fresh light of the twentieth century shining in their eyes, then Chesterton could only be called Sunday, not simply because, like the character in his book, he was 'abnormally tall and quite incredibly fat', but because he championed the traditional Sunday, Christian virtues against the weekday Fabian and secular Humanist virtues of Shaw and Wells. Like Newman, Chesterton had moved from the evangelical Christianity of his forebears (his grandfather had been a Methodist lay-preacher and a leader of the Teetotal Movement) to Anglicanism, and from Anglicanism eventually to Rome, but there was no period in his life, save possibly a few sceptical years in adolescence, when he was not a convinced Christian, though curiously enough he was never, according to his friend Christopher Hollis, a regular churchgoer.

Chesterton it was, aided and abetted by his brother Cecil and their friend Hilaire Belloc, historian and Liberal MP, who borrowed the Shavian and Wellsian technique and turned it against them. Where Shaw and Wells had poured amusing scorn on the traditional Victorian virtues, Chesterton and his allies poured amusing scorn on the 'new' virtues and the 'new' fashions, most of which they found no difficulty in proving were no more novel than Algernon Charles Swinburne or Samuel Butler. At this distance of time, we can acknowledge a certain degree of truth on both sides. Wells, for example, was a truthful prophet on many points: we have, after all, seen 'the first men in the moon' whom he envisaged as long ago as 1901. But Wells grossly underestimated (as he came to realize) the irrational forces that led to Hitlerism, whereas Chesterton and Belloc, whatever the practical deficiencies of their Distributist philosophy, seldom underestimated either the irrational or the sinister aspects of the human soul. They had their own blind spots, particularly their prejudice against the Jews, but Chesterton's *New Jerusalem* (1920) is sympathetic to Zionism, which in general he supported, and he was too humane a man to defend the kind of anti-Semitism that arose in Germany in the 1930s. He was a true prophet when he wrote in *G.K.'s Weekly* in July 1930, that Hindenburg was not so much the Dictator of Germany as 'the man who keeps the seat warm for a Dictator to come'.

We have seen, again, Shavian–Webb social legislation, most of it of considerable benefit to humanity. Yet some of the warnings contained in Belloc's *The Servile State* (1912) and in the journal *The Eye Witness* founded in 1911 by Belloc and Cecil Chesterton—and which became *G.K.'s Weekly* after Cecil's death in the First World War—are by no means out of date.

Chesterton has been called an all-round man, and in his 'rotundity' he was of his Edwardian generation, which seems in this respect, as in most others, very different from our own. Novelist, story-writer, poet, dramatist, critic, essayist, biographer, autobiographer, historian, journalist, editor, theologian, travel-writer ... Chesterton wrote in his lifetime almost a hundred books, besides contributions to journals and newspapers not yet all collected into posthumous volumes. Belloc, Shaw, and Wells were equally prolific, while a 'list of works to date by Arnold Bennett' (dated 1925) numbers over sixty items. How did these writers manage (as Bennett once put it) to 'live on 24 hours a day'? Did they never sleep, eat, or relax? Were they born, and did they die, with a pen in their hand?

Part of the answer lies in their opportunities. The Edwardian age was a very good time indeed for a writer to be alive. There were literally scores of newspapers and magazines, the majority of which have long since disappeared. Debates which now take place on radio or television took place then either on platforms in public halls or on 'platforms' in the press. The public followed the debates between Shaw, Wells, and Robert Blatchford on the one side and Belloc and the Chestertons on the other with an enthusiasm now only faintly inherited by such radio programmes as 'Any Questions'. GKC was always in the thick of it, swopping paradoxes with Shaw or cheerful insults with Wells, nearly as much a favourite with his opponents (most of whom were close friends in private life) as with his supporters. When he summed up the Emancipation of Women with the immortal wisecrack: 'Twenty million young women rose to their feet with the cry *We will not be dictated to*: and proceeded to become stenographers', it was not only conservatives who laughed. And it is not only the politics of the early 1930s which is brought to mind in this typical extract from the later, editorial Chesterton of *G.K.'s Weekly*:

> The Ministerialist demands that strong action should be taken to reduce Unemployment; but the Liberal does not scruple to retort that Unemployment is an evil, against which strong action must be taken. The Liberal thinks that we ought to revive our Trade, thus thwarting and throwing himself across the path of the National Tory, who still insists that our Trade should be revived. Thus the two frowning cohorts confront each other; and I hear the noise of battle even as I write.

With a man of so many aspects, it is best to speak for oneself, for one's own particular interests. As a literary critic, I think more highly of Chesterton's literary criticism than do most critics of my generation. But I distinguish. I believe, for example, that in the field of Dickensian criticism, the thing to read (or to read first) is not his *Charles Dickens* (1906) but his collected Everyman prefaces *Appreciations and Criticisms of the Works of Charles Dickens* (1911), particularly the chapters on *Oliver Twist*, *Bleak House*, and *Hard Times*, where much that today is conceived of as 'post-Chesterton criticism' can be seen in embryo, written by Chesterton himself. This was the book T. S. Eliot quoted in 1927 when he wrote that 'there is no better critic of Dickens living than Mr Chesterton' and I remember owing much to that book when in 1937 I wrote my own first, undergraduate paper on Dickens, later expanded into my *Scrutiny* essay of 1942. I was similarly obliged to the brilliant first chapter of Chesterton's *Robert Browning* (1903) when in 1946 I was preparing a radio feature for the BBC on the centenary of the Browning marriage.

It cannot be denied, of course, that Chesterton was extremely careless, often as 'unbuttoned' in his criticism as in his dress. The Browning volume is full of misquotations, the Dickens volume of 1906 combines critical genius with many extremely questionable assertions. These are not all so forgivable, or so amusing, as Chesterton's statement that every postcard Dickens wrote was a work of art. Dickens wrote no postcards, but he might well have done had postcards been issued in his time. Chesterton seldom corrected his mistakes and he was the first to admit that he was atrociously careless about mere accuracy.

I have never considered him a great novelist. But he was surely one of the supreme story-tellers of the early twentieth century, particularly good in the short story or the *nouvelle*. He confessed, I believe, a certain debt, when he started, to Stevenson's *New Arabian Nights*, one of the favourite books of his boyhood. But Chesterton's 'New Christian

Nights', as they might be called—I refer to such masterpieces of their kind as *The Man Who Was Thursday*, *The Club of Queer Trades*, and the best (which are usually the earliest) of the Father Brown stories—such things are completely original, of the finest Chestertonian, imaginative essence. No one else could have written them, as no one else could have written the songs in *The Flying Inn*.

His biographies differ from those of his friend and colleague Belloc in that we are more conscious in reading them of the personality of the author. There is little of Belloc in *Danton*; there is a good deal of Chesterton in *William Cobbett* and *Chaucer*. Only when he could submerge his GKC-persona entirely in his subject—as in his late study of St Thomas Aquinas (1933)—was he a completely reliable biographer. For much the same reason, his *Orthodoxy* (1908) is a more serious work than his superficially more attractive *Heretics* (1905) or *George Bernard Shaw* (1909), though only to his fellow-Anglicans, as later his *Everlasting Man* (1925) to his fellow-Catholics, can his arguments seem wholly convincing.

Chesterton illustrated more of Belloc's books than he did his own, a modesty all the more annoying when we consider what a fine job he could have made of illustrating as well as writing *The Man Who Was Thursday* or *The Napoleon of Notting Hill*. We should never forget that he was a trained artist, had studied at the Slade, and owed something of his feeling for the Middle Ages to those Pre-Raphaelite artists and poets whom he criticized on other grounds. His stories are full of dawns, sunsets, storms, wild skies, observed with the eye of a Turner. It is curious that, among his ninety volumes, he wrote only one work of art criticism, the early study of G. F. Watts (1904), though his *William Blake* (1910) had the advantage of being written by a literary critic who was also a draughtsman.

We have not yet completely circumnavigated his 'rotundity'. There is as much of Chesterton in literary history as there was in life. Some of his prolific work will remain on the shelf, merely historical items which have had their day; some will continue to be read, among them most of those mentioned in this article. After a period of comparative neglect, such as usually overtakes a writer esteemed by our fathers or grandfathers, he is coming into favour again, as witness essays in recent years by such poets and critics as W. H. Auden, Anthony Burgess, and Kingsley Amis. There will inevitably be a certain feeling of nostalgia for a vanished age in some of our rereading, but this in itself is no paradoxical tribute to a writer who in his books and Distributist propaganda alike insisted so much on the traditional virtues and on the perils of mere novelty.

KATHERINE WHITEHORN
The Return of G. K. Chesterton

G. K. CHESTERTON would have been a hundred next Wednesday, which means that he's still considered 'modern' in schools, and plenty of people can remember this strange, shambling bear of a man.

They remember him as boisterous and huge: an American coming to London in the 1930s remarked on his vast figure, despairingly draped by his women-folk in a cloak and a wide hat, leaning his pad against a wall to write an article and reading it aloud as he went; 'The delightful thing was,' he said, 'that no one took the slightest notice.' They recall his cracks: his dislike of jelly ('I can't stand food that's afraid of me') or his reply to the lady who asked why he wasn't out at the Front: 'If you go round to the side, madam, you'll see that I am.' And there's a stream of stories about absent-mindedness; not just the telegram AM IN MARKET HARBOROUGH WHERE OUGHT I TO BE which might happen to anyone; but the splashing, followed by a thud, heard outside his bathroom door, followed by an even louder splash and the groan 'Dammit, I've been in here already.'

Poetry? Mostly the rollicking stuff: 'Lepanto', account of a battle that's like a drinking song sung at a gallop; the rolling English drunkard and the rolling English road, or Higgins, who was heathen and to lecture rooms is forced, where his aunts, who are not married, demand to be divorced; or things like his Answers to the Poets, such as the one where Lovelace's girl explains she much prefers him as a soldier: 'Yet this inconstancy forgive Though gold lace I adore I could not love the lace so much Loved I not Lovelace more.' Not his drawing, though they're reissuing *Greybeards at Play* for this centenary— usually the joke was verbal anyway, like the picture of Shaw, vegetarian socialist, 'Refusing after the Revolution to Drink the Blood of Aristocrats'. Father Brown waggles his umbrella through the anthologies still, though this first of psychological detectives seems to be working on a somewhat dated base: who now would conclude that no anarchist in his wildest dreams would feel his cause furthered by slaughtering any one bloated capitalist?

The titles of his fantastic novels—*Manalive, The Man Who Was Thursday, The Napoleon of Notting Hill*—are known to people who've never read them. And as for Distributism, the social theory embodied in a leaky little paper called *G.K.'s Weekly*, into which he and his friends poured money for over ten years, you'd be hard put to it to find one man in the street who even knew what it was.

You could almost say that the idea died with him, in 1936; in that stark landscape, swept by the great winds of socialism, darkly overhung by Fascist stormclouds, a policy of sharing out property into little units, insistence on the rights of small farmers and small shops, seemed to make no sort of sense; 'three acres and a cow', their ideal for Everyman, sounded, as GK regretfully admitted, more like the name of a pub than the name of a philosophy.

But times change. I grew up so intoxicated with the way he put things that I was prepared to believe almost anything he said, but was soon forced to concede that most of his ideas wouldn't wash; what's remarkable now is how many of them seem to be coming round again. GKC said: 'It was my instinct to defend liberty in small nations and poor families; that is, to defend the rights of man as including the rights of property; especially the property of the poor.' It hardly seems dated now. Imperialism, which Chesterton loathed, is dead and discredited. Prohibition, which Shaw his sparring partner said would shortly be universal, has hardly prospered; bigness is regarded with a dark suspicion it wouldn't have attracted even ten years ago. Social workers meet in droves to wring their hands over the decline of the family and ask, as Chesterton asked, what on earth one can possibly put in its place.

He believed in nationalism, when the word equated either with the Empire, for the Right, or with sentimentalism, for the Left. 'The first duty of an English nationalist is to uphold other people's nationalism', he said. And look, indeed, what's happened. All the African federations came unstitched as soon as the Colonial Office packed their needles and string and left. Scotland and Wales are more fiercely independent than ever; America, supposed like a boa to absorb and eliminate all petty national prejudices, is gripped nigh unto death with ethnic indigestion. And whatever the Common Market's shown it isn't that people are willing to sink their national feelings—witness, if you can bear to, the French. We are all little Englanders now—even if it's like being a non-profit making company in that notice people have on their desks: 'We didn't plan it that way, but that's the way it is.' And now we know why.

Territory is *in* again—and one notes in passing that although Robert Ardrey's remarks about it were hissed and spitted upon when they appeared ten years ago, the *idea* of a territorial imperative is now the common coin of conversation, even among people who would never give Ardrey the time of day. Chesterton's best on the subject was *The*

Napoleon of Notting Hill, the Napoleon being a humourless madman who thought the people who lived in a place ought to decide what was done with it: a view shared by the Defenders of Cublington (a good many of whom really were prepared to go to the barricades about it); by every tenants' association or furious local fanatic; it's no surprise to know that Michael Collins, Ireland's great independent, carried a copy of *Napoleon* in his pocket, even when he was on the run.

In GKC's day, socialists hadn't come up against anything so inconvenient as Cuban or Syrian nationalism; they were against such (as they saw it) anachronistic hangovers. Chesterton started out, well before the First World War, as more or less in sympathy with both Liberals and Socialists. He left the Liberals 'before the next great move for democratic reform removes the right to strike'; and he was clear enough where he stood about Socialism: 'Roughly speaking: (1) I praise [the Socialists] to infinity because they want to smash modern society, (2) I blame them to infinity because of what they want to put in its place. As the smashing must, I suppose, come first, my practical sympathies are mainly with them.' Where they parted company was on the simple issue of property. As he saw it, socialism said that because some people were obviously hogging an unfair share of property, nobody should have any; GK thought this left out what ordinary people actually *want*. They want a place of their own, a family of their own; some one spot on earth where they can do as they please. Before the days when you could make the middle classes cry (like R. L. Stevenson's harlots) merely by naming their state, Chesterton had noted that the house, garden, and slammed front door, which is seen as a middle class privilege, is exactly what any working class man would like—if only anyone took the trouble to ask him.

'A property-owning democracy' is a dirty phrase because it was simply a euphemism for saying no one need bother about the housing needs of anyone who can't afford a mortgage. But in GK's opinion property was a word already defiled by the corruption of the great capitalists—and they were *not* on the side of property. If they wanted it all in one lump, for themselves, that was the opposite of wanting some of it, enough, for everybody: 'It is the negation of property that the Duke of Sutherland should have all the farms on one estate, just as it would be the negation of marriage if he had all the wives in one harem.'

Again, Chesterton was considered very unreconstructed, for his loathing of what he saw as interference, others saw as philanthropy; the do-gooding lack of respect for the rights of anyone who was unlucky enough to be poor. (He regarded it as the extreme of saintliness 'to love the man I saw yestereen who knocked not when he came with alms'.) In this respect approved thinking has caught up with him: look at all the elaborate social work paraphernalia of 'non-judgemental' help for the 'client', the insistence that people must do things for themselves. Mind you, I think he overdid it. He had no understanding, for example, of the extent to which mothers can abuse their own children or husbands their wives; and I don't think he, any more than any other Edwardian, had the faintest idea of just how unspeakable the conditions of living and labour were in London at the turn of the century; maybe the drawing-room reformers hadn't either.

And I wouldn't agree with him about schools—but there are plenty nowadays who do: who think that what's taught there is a strange substance invented by, and only of use to, educationists. Whether he'd actually have been in sympathy with the current deschooling movement I'm not sure. He would certainly have found it funny that the name for it was coined from a simple mis-hearing—'schools are there to screw you' being heard as 'schools are there to school you'—an equally vile activity.

I wouldn't try to maintain that every one of his ideas is now spankingly up to date—if only because he was highly scornful of using the word 'modern' as a compliment, just as he thought attacking a thing because it was old was like being regarded as brave for knocking down your grandmother. He had his unlikable tastes: he was often accused of antisemitism—though what he was really against was international money, meaning (on the whole) as little enmity to real Jews as Harold Wilson meant to real gnomes, whether resident in Zurich or not (and GKC was a convinced Zionist).

He was, or so I have always thought, wildly wrong about women; or at least he'd got it right about one sort of woman, but didn't seem to realize that there were any other varieties. Nor *all* secretaries, after all, bring that 'sacred stubbornness' meant to defend home and children to the task of protecting their employer.

But what remains to me irresistible is his ability to see some hackneyed thing completely fresh—he is his own Manalive, who was perpetually breaking into his own house and eloping with his own wife to penetrate 'the grey disguise of sleep and custom in between'. Phrases, too: everyone says 'You can't put the clock back'—he says 'You can. A clock, being a piece of human construction, can be restored by the human finger to any figure or hour.'

A hundred people reading F. E. Smith saying that the Welsh Disestablishment Bill had shocked the conscience of every Christian country in Europe would have read on regardless: he visualized what it *would* have meant if it had.

> In the mountain hamlets clothing
> Peaks beyond Caucasian pales,
> Where establishment means nothing
> And they never heard of Wales,
> Do they read it all in Hansard
> With a crib to read it with—
> 'Welsh Tithes: Dr Clifford
> Answered.' Really, Smith?

And there still seem to be half a dozen of his statements that seem so unaswerably true that I would like to have them read again once a year, like the Collect for the Day or the 'Yes, Virginia, there is a Santa Claus' article reprinted each Christmas in the *Journal American*. Any time anyone attacks juries, I want them to reprint his essay, 'The Twelve Men', in which he sees 'what a jury really is and why we must never let it go': a check on the trend towards specialism and professionalism; the real case for juries being, not that they are better at sifting evidence than trained men, but that they see exactly what the trained men who are used to it cannot: that it is a terrible thing to mark a man out for the vengeance of men.

When anyone applauds business or vocational education, I go panting back to his essay where he points out that to train a citizen is to train a critic: which is exactly what you can't do if you train a man *in* a system:

And when anyone asks me what is the use of telling my son all about ancient Athens and China and the medieval monasteries when he is to be a superior scientific plumber in Pimlico . . . I say that it will not only prevent him from supposing that Pimlico covers the whole of the planet but enable him to realize that even Pimlico . . . may conceal, here or there, a defect.

Order? 'What my anarchist friends do not seem to realize is that if they do not have rules, they will have a ruler.' Animals? 'I am very fond of dog, so long as he is not spelt backwards.' Tennyson? 'He could not quite think up to the height of his own towering style.'

Even Chesterton's remarks about his friends seem to apply even more to my friends, as when he says to H. G. Wells: 'Don't you sometimes find it convenient, even in my case, that your friends are less touchy than you are?'

Chesterton always prided himself on being a journalist; and Kipling's epitaph for the journalists was 'they served their day'. I reckon his day is not done, and he could well serve ours as well.

RICHARD INGRAMS
The Mystic beneath the Sombrero

IN the years before the Great War, a huge shambling man, six feet four inches tall, with long chestnut-coloured hair, and weighing about twenty stone, could often be seen wandering down Fleet Street. Dressed in a black cloak and sombrero hat, with pince-nez perched incongruously on his nose, he carried a swordstick and his pockets bulged with 'penny dreadfuls'. From time to time he stopped to read a book or wrote something down, or paused in the middle of the road, while the traffic whirled about him, apparently struck by a sudden and important thought. Most of the time he seemed to be chuckling over some secret jokes of his own.

This was G. K. Chesterton. Poet, prophet, artist and clown he was the best known Fleet Street character since Dr Johnson. Bernard Shaw described him as 'a man of colossal genius'. Everybody had a story about him. The biographer Hesketh Pearson told how a journalist spotted him one day turning the corner of the street when a gust of wind blew his hat off. The journalist ran after it, followed by GKC. Eventually, they retrieved it from the jaws of an oncoming bus. Mopping his brow, Chesterton panted his thanks: 'But you shouldn't have taken the trouble,' he said. 'My wife has bought me a new hat, and she will be most disappointed—*most* disappointed—when she hears that the old one has only just been saved from well-merited destruction.' In that case, why had he bothered to run after it? the indignant journalist queried. 'It's an old friend,' Chesterton replied, 'and I wanted to be with it at the end.'

On another occasion, when he was a young man, he was sitting arguing with a friend in a cafe. Chesterton ordered two poached eggs. These were eventually brought, but Chesterton, intent on the debate, took no notice of them. Then, as he lifted his hand to emphasize a point he swept the eggs into his lap. Even there, they went unnoticed. When the waitress came, Chesterton said: 'Will you bring two more poached eggs. I seem to have lost the others.'

Like any other mystic, Chesterton was 'not all there'. He found it hard to remember where he was, or where he was supposed to be. This proved to be a hazard, particularly when he took to lecturing up and down the country. An audience which had assembled to hear him talk about Dickens was treated to a lecture on Browning. A Women's

Rights organization was told about coalfields and the necessity of working them. And then there was the celebrated occasion when he sent a telegram to his wife which read: 'Am in Market Harborough. Where ought I to be?' No wonder his long-suffering wife confided to a friend: 'I always give money to pavement artists as I am quite sure that is how Gilbert will end.'

Chesterton, in fact, would have made a very good pavement artist. He began to draw at the age of five, or earlier. His father Edward Chesterton, the estate agent, whose firm still flourishes in Kensington, was himself a keen amateur artist who dabbled in water colours, stained glass, magic lanterns and medieval illumination. By the age of eleven, Chesterton's artistic style was quite mature and his drawings to illustrate E. C. Bentley's famous book of clerihews were done when both were still pupils at St Paul's. Chesterton spent much of his time at school drawing all over his text books. He drew goblins, grotesques and countless pompous-looking men some of whom resembled Gladstone, Balfour and Joe Chamberlain, 'Chesterton, Chesterton, have you *no* care for books?' said one master, seeing his Greek primer covered with the usual frieze of gargoyles.

In common with most great writers, Chesterton never went to university. Instead he became an art student at the Slade School, but the head of the school, Professor Tonks, told his parents that 'they could not teach him anything without spoiling his originality'. He then left and became a writer. But he never gave up drawing and later in life always had a pocket full of chalks to entertain himself and the many children who doted on him. Nicolas Bentley, himself a brilliant illustrator, whose father, E. C. Bentley, was Chesterton's greatest friend from the time they first met at the age of twelve, says: 'If any one person more than another disposed me towards the profession of artist, it was Gilbert. To watch him at work, drawing forth handfuls of luminous powdery chalks from his pockets and sketching embattled Greeks or Persians, or perhaps it was Cyrano or the Cid or Mephistopheles or Mr Balfour—watching him put them down on brown paper, with clear, deliberate strokes, suggested that it was the easiest thing in the world.' Had he not become a writer, Bentley thinks Chesterton would certainly have achieved an equal fame as an artist. 'There is no doubt that purely and simply as a decorative draughtsman he would have had very few equals.'

In 1909, Chesterton, at his wife's insistence, left London to live in Beaconsfield. Some of his friends resented his departure from Fleet Street, where most of his newspaper articles appeared in the *Daily News* and later in the *Daily Herald*. But others, who knew what the effects of heavy drinking and late hours could be, sympathized with his wife. Not that Chesterton was ever an alcoholic. 'Drink because you are happy, never because you are miserable' was his recipe. But he was a heavy drinker, largely because he drank absentmindedly and took no account of the consequences. Although a convinced Christian all his life—he became a Roman Catholic in 1922—he never had any time for narrow puritanism. 'There is more simplicity in the man who eats caviar on impulse,' he said, 'than in the man who eats Grape-Nuts on principle.'

At Beaconsfield, Chesterton played host to a score of visitors, especially children. He had a collection of toy theatres and devised countless pantomimes, writing the play and painting the figures on cardboard. The stories are now forgotten, but the paintings survive . . . —giants, dragons, heralds, trumpeters, beadles, knights. One of the plays was about Beatrice and Sidney Webb—whom Chesterton was continually making fun of, as he did of all planners and 'do-gooders'. A cardboard angel from this entertainment bears a *Daily Mail* placard proclaiming 'GOD SENDS FIERY CAR FOR MRS WEBB but—she will not desert SIDNEY'. Being himself childlike, Chesterton struck up an immediate rapport with children. A favourite amusement of his was walking round his garden, shooting arrows into the air, or, if he was working upstairs, throwing birds cut out of brown paper down into the garden where the children were playing.

Dictating in his study, he paced up and down smoking cigars and stabbing at cushions with his sword stock, or sat chipping bits out of his desk with a large African knife. Once he shot an arrow out of the window and accidentally hit a dog passing down the road. Such behaviour naturally endeared him to children. One small boy, asked how he had enjoyed taking tea with the famous Mr Chesterton, replied: 'He taught me how to throw buns in the air and catch them in my mouf.'

It was perhaps hard for people to accept that this apparent lunatic was, in fact, a profound and serious figure. It was the kind of paradox he himself delighted in. But it was true. Despite scribbling over his text books at school, he nevertheless acquired, somehow, an encyclopaedic knowledge of English history and literature which helped him to write scores of books almost out of his head. His memory was phenomenal and he could recite whole chapters of Dickens by heart. He never went near a library, let alone the British Museum, and he refused to verify any of his many quotations. He wrote a book about Thomas Aquinas and was half way through it before he thought it might be advisable to refer to the saint's writings.

His mind was such, that he was capable of simultaneously writing one article by hand and dictating a completely different one to his secretary. He had an intuitive insight into everything, of which the superficial observer, seeing only the absent-minded buffoon, was unaware. Hesketh Pearson, who had talked once to Chesterton for only a few minutes and met him again after an interval of eight years, was surprised to be told: 'Yes, I remember you well. In fact you appear in one of my Father Brown stories.'

Unlike some of his fellow Roman Catholics Chesterton was not only charitable, he was a democrat. He never felt the lure of titles and country seats. When once he did stay in a 'grand house' he was horrified to discover that the valet had unpacked his pockets. 'There laid out on the quilted silk eiderdown were several stubs of pencils, a paper-backed murder story, some coloured chalks and a small cigar or two. By the expressionless face of the valet I could see he was thinking, "How different, how *very* different are these from the jewelled trifles that I am accustomed to removing from the tidy pockets of the well-to-do."'

'If a man is genuinely superior to his fellows,' Chesterton said, 'the first thing he believes in is the equality of man.' His views have a very contemporary appeal. 'Small is beautiful' is only one modern maxim which would have appealed to him. 'He who lives in a small community,' he said, 'lives in a much larger world. He knows much more of the fierce varieties and compromising divergences of men. The reason is obvious. In a large community we can choose our companions. In a small community our companions are chosen for us.'

Chesterton was the natural champion of the patriotism of small groups. It is a curious fact that his best known book, *The Napoleon of Notting Hill*, a fantasy set in the future, begins in the year 1984. And it tells, among other things, how the citizens of Notting Hill go to war against the rest of London in order to save a small street of shops, which is being knocked down in order to build a motorway. The idea has often been dismissed as ridiculous but, as a prophecy, written in 1904, it stands up rather better than George Orwell's. The villagers of Cublington and the shop keepers of Covent Garden could all bear witness to that.

BERNARD LEVIN
The Case for Chesterton

WELL, did Chesterton matter?

Ours is not a propitious time in which to be asking such questions. Whatever else Chesterton was, he was a most extravagant romantic, and we live in a decidedly unromantic age. His worst work (and among his gargantuan output a lot was inevitably bad) is almost all bad in the same way; empty bombast, strained paradox, gritty alliteration, a hollow booming which suggests that the vessel of inspiration has temporarily run dry. But these faults are always hovering around even much of his best work: even *The Man Who Was Thursday*, his masterpiece, is littered with literary and dramatic effects that seem strained, and the highly regarded Father Brown stories seem to me so full of such artificial devices as to be extremely tiresome if read in any but the smallest quantities.

His philosophy, too, in so far as he had one, which was not very far, wearies us with his constant bawling of his religion, a sadly unsophisticated Catholicism that he seemed at times to be unable to distinguish from drinking. (Belloc, though I remain convinced he was the finer artist, had an appalling influence on his hero.)

There is also his anti-semitism, which matters where that of, say, Dostoievsky or Wagner does not, being extraneous to their work: in Chesterton it had 'got into the wine' with a vengeance, and since Mr G. C. Heseltine, in his essay ('G. K. Chesterton, Journalist') even has the gall to claim that Chesterton was not anti-semitic, I had better stop for a moment and demonstrate that he was. In the Acknowledgments for *Greybeards at Play*, the editor (Mr Sullivan) says demurely of 'The Logical Vegetarian' that 'the third stanza is omitted', an omission which I shall now proceed to repair. It runs like this:

> Oh, I knew a Doctor Gluck
> And his nose it had a hook,
> And his attitudes were anything but Aryan;
> So I gave him all the pork
> That I had, upon a fork
> Because I am myself a Vegetarian.

Mind you, the second stanza of 'The Song of Quoodle', which Mr Sullivan *does* include, rivals that:

> They haven't got no noses,
> They cannot even tell
> When door and darkness closes
> The park a Jew encloses.
> Where even the law of Moses
> Will let you steal a smell.

And ripest of all was the verse (not included anywhere, I think, except in Maisie Ward's standard biography) which he wrote when asked by a doctor to scribble something when he had hurt his arm. This is what leaped instantly from his subconscious:

> I am fond of Jews
> Jews are fond of money
> Never mind of whose.
> I am fond of Jews
> Oh, but when they lose
> Damn it all, it's funny.

The best thing one can say of Chesterton's anti-semitism is that it was not as vile as Belloc's; let us leave it at that.

So much for the prosecution; now for the defence. I would not be surprised if Chesterton were to come back into fashion. His medievalism, his hatred of the industrial society, his horrified and prophetic vision of the stifling of human personality in the mass state—these are themes that we hear now too and the tradition that he thus represented never entirely died, so perhaps the ecological argument owes something to him. But it is in those categories that his hope of survival must lie. For all its faults, his was a splendid vision, and it stemmed from his genuine love of his fellow-men, his genuine compassion for their suffering. Such a poem as 'The Song of the Wheels' makes up for a good deal of nonsense and nastiness, in its great cry of vicarious pain for all those whom soulless labour turns into something akin to the machines they tend. And this in turn comes from his fierce conviction that there is such a thing as the soul, that it is within every human being, and that the greatest sin is to drive it out.

That was the spark he lived to fan, and one can hear the bellows on almost every page of his work. *The Napoleon of Notting Hill* is absurd, but gorgeously and bravely absurd, and though it is absurd because men would not in fact be willing to die for Kensington, it is gorgeous and brave because they have throughout history been willing, over and over again, to die for *something* outside themselves. Such an essay as 'The Twelve Men' (still, in its tiny compass, the finest defence of the jury system yet written, a sword directed with absolute precision to the heart of the argument) exhibits, in its very different form, the same quality, the need to keep in view the reality of the individual in the dock, which the professionals in the court cannot see because 'they can only see the usual man in the usual place'. The theme is also stated clearly in the pendant to *The Hammer of God*, in which Father Brown will not deliver the murderer to human justice, but will not let him commit suicide either; the priest is not interested in the sinner's body, but is quick in defence of his soul.

Viewed in this light, many of his weaknesses become more easily understandable, more readily forgivable. The naïvety of his Catholicism is balanced by the force it acquired from his belief in it as, most literally, a religion of love (how he would have enjoyed Pope John); his romanticism springs from his advocacy of all those things that give the individual his individuality; even his mad anti-semitism is not *just* mad, but based on his essentially medieval conviction that the usurer Jew is behind the impersonal economic organization of the world, with its assault on the human personality. The valuable in him must have outweighed the dross, for he passes with ease that searching test: he was loved by too many good men for it to be an accident. Walter de la Mare's famous quatrain said everything that can be said or need be said on his side:

> Knight of the Holy Ghost, he goes his way,
> Wisdom his motley, Truth his loving jest;
> The mills of Satan keep his lance in play,
> Pity and Innocence his heart at rest.

Did Chesterton matter? If no, very nearly. But I think we can be a shade more charitable: Only just, perhaps, but yes.

BENNY GREEN
Father of Father Brown

MANY of Chesterton's hobby-horses are gathering dust: his militant medievalism, his duo-decimo Catholicism, the lame duck of his Distributism, the cheerful callousness of his armchair army games. Even his literary exercises are frowned upon today by a foolishly functional world.

The essay form, in which he could excel, is derided, usually by people incapable of practising it; his polemics, warped by the unfortunate influence of his brother Cecil and of Belloc, pay lip-service to the wrong deities, while his novels appear suspect because of their apparent lack of any discernible link with reality, which is probably why they have remained readable for so long.

There remains, however, a book like *The Man Who Was Thursday*, a most extraordinary work which creates its own category, and which raises the most baffling question; is the byzantine lunacy of the counter-counter-espionage in that tale an example of its author's prescience or has the madness of our modern world simply overtaken his benign Edwardian joke?

Repeatedly the excesses of our own epoch find echoes in the whimsical insanities of his fiction. And it is remarkable, for instance, how *Manalive* whose hero, Innocent Smith, puts the romance back into his marriage by wooing his own wife and who represents his creator's lifelong preoccupation with the interaction between the strange and the familiar, is a kind of chaste anticipation of Pinter's *The Lovers*.

W. H. AUDEN
The Gift of Wonder

ALL lovers of poetry, I imagine, would rather quote from poems they admire than talk about them. Once read or listened to, their merits should be immediately apparent. One feels this all the more strongly in the case of a poet whom, like Chesterton, one suspects to be out of fashion and little read. Aside from a few stock anthology pieces, like 'The Donkey' and 'Lepanto', how many of his poems are known to the contemporary reading public? Not very many, I fear. Editorial footnoting, as distinct from aesthetic judgement can, however, sometimes be essential. *The Divine Comedy*, for example, is full of references to the names of persons and places with which the contemporary reader, even if he is Italian, is unacquainted, and he needs to be informed of the facts. When Chesterton is obscure, this is usually because the stimulus to his poem came from some public event with which he assumed his readers were familiar, but which has now been forgotten. 'A Song of Swords' is prefaced from a newspaper cutting: 'A drove of cattle came into a village called Swords, and was stopped by the rioters.' But, since I know nothing about this incident in Irish history, I cannot make head or tail of the poem. With the help of a footnote, perhaps I could. I know a little more about the history of the First World War, but the meaning of 'The Battle of the Stories' (1915) eludes me and I would welcome editorial assistance. In only one of his poems, 'The Lamp Post', do I feel that my lack of comprehension is Chesterton's fault. It seems to be based upon some private mythology of his own, which neither I nor any other reader can be expected to decipher.

Consciously or unconsciously, every poet takes one or more of his predecessors as models. Usually, his instinct leads him to make the right choice among these, but not always. In Chesterton's case, for example, I think that Swinburne was a disastrous influence. That he should ever have allowed himself to be influenced by Swinburne seems to me very odd, when one thinks how utterly different their respective views about Life, Religion, and Art were, but he was and always to his harm. It is due to Swinburne that, all too often in his verses, alliteration becomes an obsessive tic. In Anglo-Saxon and Icelandic poetry where the metrical structure is based upon alliteration, its essential function is obvious. In modern verse, based upon regular feet and rhyme, alliteration can be used for onomatopoeic effects, but only sparingly: in excess it becomes maddeningly irritating. The other vice Chesterton acquired from Swinburne was prolixity. Too often one feels that a poem would have been better if it had been half the length. 'Lepanto', it seems to me, exhibits both faults.

> He sees as in a mirror on the monstrous twilight sea
> The crescent of his cruel ships whose name is mystery;
> They fling great shadows foe-wards, making Cross and Castle dark,
> They veil the plumèd lions on the galleys of St Mark;
> And above the ships are palaces of brown, black-bearded chiefs,
> And below the ships are prisons, where with multitudinous griefs,
> Christian captives, sick and sunless, all a labouring race repines
> Like a race in sunken cities, like a nation in the mines.
> They are lost like slaves that swat, and in the skies of morning hung
> The stairways of the tallest gods when tyranny was young.
> They are countless, voiceless, hopeless as those fallen or fleeing on
> Before the high Kings' horses in the granite of Babylon.

In the case of his longest and perhaps, greatest 'serious' poem, 'The Ballad of the White Horse', I do not, however, I am happy to say, find the length excessive. When, for example, Elf the Minstrel, Earl Ogier and Guthrum express in turns their conceptions of the Human Condition, what they sing could not be further condensed without loss. Here Guthrum.

> 'It is good to sit where the good tales go,
>   To sit as our fathers sat;
> But the hour shall come after his youth,
> When a man shall know not tales but truth,
>   And his heart shall fail thereat.
>
> 'When he shall read what is written
>   So plain in clouds and clods,
> When he shall hunger without hope
>   Even for evil gods.
>
> 'For this is a heavy matter,
>   And the truth is cold to tell;
> Do we not know, have we not heard,
> The soul is like a lost bird,
>   The body a broken shell.
>
> 'And a man hopes, being ignorant,
>   Till in white woods apart
> He finds at last the lost bird dead;
> And a man may still lift up his head
>   But never more his heart.

'There comes no noise but weeping
   Out of the ancient sky,
And a tear is in the tiniest flower
   Because the gods must die.

'The little brooks are very sweet,
   Like a girl's ribbons curled,
But the great sea is bitter
   That washes all the world.

'Strong are the Roman roses,
   Or the free flowers of the heath,
But every flower, like a flower of the sea,
   Smelleth with the salt of death.

'And the heart of the locked battle
   Is the happiest place for men;
When shrieking souls as shafts go by
And many have died and all may die;
Though this word be a mystery,
   Death is most distant then.

'Death blazes bright above the cup,
   And clear above the crown;
But in that dream of battle
   We seem to tread it down.

'Wherefore I am a great king,
   And waste the world in vain,
Because man hath not other power,
Save that in dealing death for dower,
He may forget it for an hour
   To remember it again.'

Guthrum's pessimistic conclusions about the nature of the internal and invisible life are based, it should be noticed, upon his observations of objects in the external and visible world, clouds and clods, flowers, a dead bird, etc. Chesterton, however different his conclusions, does the same. Both in his prose and in his verse, he sees, as few writers have, the world about him as full of sacramental signs or symbols.

Wherein God's ponderous mercy hangs
On all my sins and me,
Because He does not take away
The terror from the tree
And stones still shine along the road
That are and cannot be.

> Men grow too old for love, my love,
> Men grow too old for wine,
> But I shall not grow too old to see
> Unearthly daylight shine,
> Changing my chamber's dust to snow,
> Till I doubt if it be mine.

I would not call him a mystic like Blake, who could say: 'Some see the sun as a golden disk the size of a guinea, but I see a heavenly host singing Holy, Holy, Holy.' Chesterton never disregards the actual visible appearance of things. Then, unlike Wordsworth, his imagination is stirred to wonder, not only by natural objects, but by human artifacts as well.

> Men grow too old to woo, my love,
> Men grow too old to wed:
> But I shall not grow too old to see
> Hung crazily overhead
> Incredible rafters when I wake
> And find I am not dead.

Probably most young children possess this imaginative gift, but most of us lose it when we grow up as a consequence, Chesterton would say, of the Fall.

> They haven't got no noses,
> The fallen sons of Eve;
> Even the smell of roses
> Is not what they supposes;
> But more than mind discloses
> And more than men believe . . .
>
> The brilliant smell of water,
> The brave smell of a stone,
> The smell of dew and thunder,
> The old bones buried under,
> And things in which they blunder
> And err, if left alone . . .
>
> And Quoodle here discloses
> All things that Quoodle can,
> They haven't got no noses,
> They haven't got no noses,
> And goodness only knowses
> The Noselessness of Man.

In verses such as these, there is little, if any, trace of Swinburnian influence. Behind them one detects the whole tradition of English

Comic Verse, of Samuel Butler, Prior, Praed, Edward Lear, Lewis Carroll and, above all, W. S. Gilbert. It was from such writers, I believe, that Chesterton, both in his verse and his prose, learned the art of making terse aphoristic statements which, once read or heard, remain unforgettably in one's mind. For example:

> Bad men who had no right to their right reason,
> Good men who had good reason to be wrong . . .

> God is more good to the gods that mocked Him
> Than men are good to the gods they made . . .

> And that is the Blue Devil that once was the Blue Bird;
> For the Devil is a gentleman, and doesn't keep his word.

> But Higgins is a Heathen,
>   And to lecture rooms is forced,
> Where his aunts, who are not married,
>   Demand to be divorced . . .

> For mother is dancing up forty-eight floors,
> For love of the Leeds International Stores,
> And the flame of that faith might perhaps have grown cold,
> With the care of a baby of seven weeks old.

I cannot think of a single comic poem by Chesterton that is not a triumphant success. It is tempting to quote several, but I must restrain myself. Instead, I recommend any reader unacquainted with them to open *The Collected Poems* (Methuen) and sample 'The Shakespeare Memorial' (p. 156), 'Ballade d'une Grande Dame' (p. 190), and 'A Ballade of Suicide' (p. 193). His parodies of other poets are equally good, especially those of Browning and Kipling.

I shall, however, now quote from a volume called *Greybeards At Play* originally published in 1900, reprinted in 1930, not included in the Collected Poems and now, I believe, out of print. Until it was sent me by John Sullivan, Chesterton's bibliographer, I had never heard of its existence. I have no hesitation in saying that it contains some of the best pure nonsense verse in English, and the author's illustrations are equally good.

> The million forests of the Earth
>   Come trooping in to tea.
> The great Niagara waterfall
>   Is never shy with me . . .

> Into my ear the blushing Whale
>   Stammers his love. I know
> Why the Rhinoceros is sad,
>   —Ah, child! 'twas long ago . . .

Come fog! Exultant mystery—
  Where, in strange darkness rolled,
The end of my own nose becomes
  A lovely legend old.

Come snow, and hail, and thunderbolts
  Sleet, fire, and general fuss;
Come to my arms, come all at once—
  Oh photograph me thus! . . .

The Shopmen, when their souls were still,
  Declined to open shops—
And Cooks recorded frames of mind
  In sad and subtle chops . . .

The stars were weary of routine:
  The trees in the plantation
Were growing every fruit at once,
  In search of a sensation.

The moon went for a moonlight stroll,
  And tried to be a bard,
And gazed enraptured at itself;
  I left it trying hard.

Surely, it is high time such enchanting pieces should be made readily available.

By natural gift, Chesterton, was, I think, essentially a comic poet. Very few of his 'serious' poems are as good as these. (His one translation from Du Bellay makes one wish he had done many more.) But here is a poem of his which any poet would be proud to have written.

The Sword of Surprise

Sunder me from my bones, O sword of God,
Till they stand stark and strange as do the trees;
That I whose heart goes up with the soaring woods
May marvel as much at these.

Sunder me from my blood that in the dark
I hear that red ancestral river run,
Like branching buried floods that find the sea
But never see the sun.

Give me miraculous eyes to see my eyes,
Those rolling mirrors made alive in me,
Terrible crystal more incredible
Than all the things they see.

Sunder me from my soul, that I may see
The sins like streaming wounds, the life's brave beat:
Till I shall save myself as I would save
A stranger in the street.

V. S. PRITCHETT
Secret Terrors

How touching and hapless one's elders looked when one spotted them on their own: lonely Shaw, looking into the window of the gun shop in the Strand, Wells with a fly button undone in a club, Yeats asleep with a detective novel fallen on his chest, Chesterton at the window of a pub restaurant off Leicester Square, also asleep, with his head on the marble-top table—the clown in all the passive dignity of private life. Edwardians all of them—but what was an Edwardian? P. N. Furbank defines a general quality of these writers:

> the imagination of the 'Edwardian' writers is nurtured not so much by the sap of nature as by printer's ink. It is an ad hoc imagination, a product of the will and opinion.

Chesterton was franker than the rest in admitting this. Weary he must have become of the inexhaustible fizzing of topsyturvy argumentation that seemed to come from an intellectual vat. We now know—through Maisie Ward's memoir—that his astonishing size had a peculiar origin. Owing to some glandular disturbance, he did not reach puberty until he was 18 or 19 and had become, as Furbank suggests,

> a victim of phantasmagorical sexual obsessions which convinced him of the close and immediate presence of the devil. And, more to the point, it gave to the intellectual movements of the day, in which he had already taken a cheerful debating society interest, a sulphurous flavour of diabolism in him. His reaction was to combat them with the weapons of his childhood—with a toy theatre medievalism, pasteboard swords and debating society high jinks. It was a reaction of genius, restoring his moral balance and leaving his intellect untrammelled and free.

It led also to aberrations: he saw something sinister in Impressionism; and, in a schoolboy way, something diabolical in the Jews. It is a mystery that a man so loved for his simple innocent goodness should have nurtured this latter gross obsession, though possibly the innocent have more need of a devil than the virtuous.

Furbank is excellent on one aspect of Edwardian life: the appetite among the self-educated for debate on 'ideas' of every description—the universities were closed to the mass public—and the prophets had to be entertaining. One feels, he says, a nostalgia for a period so spirited and witty, but points out that this stirred the feeling, in the next generation, that all 'ideas' were dangerous to art. That is indeed why I was to lose, very early, my adolescent delight in Chesterton.

For all the stir in the mind, the inky debate had little of the sap of life in it. Chesterton's 'little man' might enjoy the fantasies of self-inflation but he knew life was either graver or meaner than these mock-heroic dreams. He died in hospital, not in some knightly fight, and when real combat did come he was slaughtered like cattle. Still, the debate was so free that Chesterton had one or two successes. Mr Furbank quotes at length from a splendid piece of polemic on Indian nationalism which made such a deep impression on Gandhi that the Indian leader immediately wrote in *Hind Swaraj* his first statement on the need for the 'Indian' solution of his country's problems.

Even so, the Chesterton that survives from his excess of facility—there are a hundred books and piles of journalism—belongs to the first ten years of his life, when the influence of Wilde's paradoxes and Swinburne's alliterations had not become a dreadful mechanical habit; when the world was not so continuously viewed as an opportunity for side-splitting puns and aphoristic somersaults. At his best, Chesterton is a travelling writer, seeing new things strangely on small journeys. His hatred of Impressionism seems really to spring from a regard for the freshness of the exact: he had a boy's eye. Writing on the verse ... Auden notes his fascination with objects. He sees 'as few writers have, the world about him as full of sacramental signs and symbols'. In his comic verse he owes most to W. S. Gilbert. But the verses and drawings in *Greybeards at Play* are only passable jokes. The edge has gone. In the Father Brown stories which Chesterton himself affected to despise, everything depends on the seemingly casual but illuminating observation of ordinary things and ordinary language:

If you walk down the street to the nearest hatter's shop, you will see that there is, in common speech, a difference between a man's hat and the hats that are his.

Father Brown is not a symbolical figure; though he is a priest it is not for Romish reasons; his innocence—like Chesterton's—has given him the child's secret terror of a complicity within himself. Father Brown has no special spiritual powers. He is not thaumaturgic; some appalled sense—which may be linked with sexual or sadistic horror—enables him to identify himself with the murderer. 'Are you a devil?' the exposed criminal asks. 'I am a man,' replies Father Brown, 'and therefore have all devils in my heart.' There is a good deal of the aesthetic influence of the decadents in the Father Brown stories, and that is where a reader like myself finds the modest priest somewhat of a flirt with evil.

There is general agreement that the pugnacious and rhetorical Belloc was a bad influence. In many ways, Chesterton was made helpless by his wit. It is conventional to say that Chesterton's book on *Dickens* is the best thing he ever wrote. I read it again this week. It is not merely good; it is a masterpiece and contains, among other things, the most enlightening portrait of Dickens himself that I have ever read. Once more it is the decent, sensible liveliness of Chesterton's detail that spreads, like a kind of wisdom, into the portrait. The book is an early one, written before facility swept him out into the sea of print.

DAVID LODGE
Dual Vision: Chesterton as a Novelist

'HOW many Roman Catholics have been good novelists?' George Orwell once rhetorically asked. 'Even the handful one could name have usually been bad Catholics. The novel is practically a Protestant form of art.' Orwell was writing in 1940, before Evelyn Waugh and Mr Graham Greene had produced the full range of their mature work (though for that matter both these writers have been accused of being 'bad' Catholics in their time), and was more likely to have been thinking of writers like James Joyce and Ford Madox Ford. Certainly Chesterton's literary career would have provided evidence for, rather than against, Orwell's argument that Catholicism was not a spiritual climate nourishing to the fictional imagination. Chesterton produced only one novel, not his best—*The Return of Don Quixote* (1927)—after his reception into the Church in 1922; and even if we take into account the fact that his 'conversion' was unduly delayed out of tenderness for his Anglican wife's feelings, this does not greatly affect the issue. The earliest evidence we have of Chesterton's intention of becoming a

Catholic is his tentative conversation with Father O'Connor in 1912[1] by which time he had written his best novels: *The Napoleon of Notting Hill* (1904), *The Man Who Was Thursday* (1908), *The Ball and the Cross* (1909), and *Manalive* (1912). Only *The Flying Inn* (1914) and—after a very long interval—*The Return of Don Quixote*, came later. It is true that Chesterton went on writing many short stories, notably about Father Brown, but even the best of these belong to his 'pre-Catholic' period, e.g. *The Innocence of Father Brown* (1911).

There are of course other possible explanations for the decline of Chesterton's fictional imagination. As he grew older his prolific creative energy began to fade under the colossal demands he made upon it—in 1927 he published seven other titles as well as *The Return of Don Quixote*. But he did produce many fine books after his conversion—*The Everlasting Man* (1925), for instance, *Chaucer* (1932) and *St Thomas Aquinas* (1933), though they are not fictional works. I suggest, therefore, that Chesterton's acceptance of the most precise and systematic of all Christian creeds imposed upon his thought a certain rigidity which did not combine easily with the form of fantastic novel he had developed. The novels expressed the most unorthodox aspects of his orthodox mind; they expressed his sense of mystery and confusion, of the ambivalence of all opinions and precepts. After his conversion Chesterton was spokesman for one philosophy; but in his best novels he had been (dialectically) spokesman for two—often with equal persuasiveness.

In the period when his most important novels were written, Chesterton was obsessed with the idea of metaphysical duality: a tension at the heart of things which allowed almost any idea to be turned inside out and remain equally, and sometimes more significant. This expresses itself in the conflict of antithetical characters: Quin and Wayne in *The Napoleon of Notting Hill*, MacIan and Turnbull in *The Ball and the Cross*, and Dalroy and Ivywood in *The Flying Inn* for instance. 'The story is told', runs Chesterton's definition of the novel-form in *The Victorian Age In Literature* (1912), 'for the sake of some study of the difference between human beings.' As Chesterton's orthodoxy crystallized, his sympathies became more marked. In the earlier novels there is a suggestion that all oppositions are ultimately reconciled. 'When dark and dreary days come, you and I are necessary, the pure fanatic, the pure satirist' says Wayne to Quin at the end of *The Napoleon of Notting Hill*. This is consistent with the idea expressed in *Orthodoxy* (1908), that in the whole man apparent opposites, such as

---

[1] J. O'Connor, *Father Brown On Chesterton* (1937), p. 85.

Love and Hate, Optimism and Pessimism, would coexist in a kind of tension:

... we want not an amalgam or compromise, but both things at the top of their energy; love and wrath both burning.

Can [man] in short, be at once not only a pessimist and an optimist, but a fanatical pessimist and a fanatical optimist? Is he enough of a pagan to die for the world and enough of a Christian to die to it?

In *The Ball and the Cross* the atheist and the Catholic are equally matched in debate, and the latter admits at the end of the novel, 'There must be some round earth to plant the cross upon.' The denouement, however, in which the hermit Michael walks unscathed through the flames, justifies MacIan against Turnbull. In *The Flying Inn* we approach the traditional hero-villain opposition of romance, but Ivywood is by no means to be despised. He has his own kind of fanatical courage and idealism, and the heroine of the novel admires him for it.

In Chesterton's novels we find a reflection, often grotesquely distorted by his sense of humour, of that characteristic paradox of modern Christianity, expressed by T. S. Eliot in his essay on Baudelaire, that it is better to be spiritually alive, even spiritually in danger, than spiritually dead: 'it is better, in a paradoxical way, to do evil, than to do nothing: at least we exist.' For this reason Chesterton never ceases to respect characters who stand heroically for principles which he could not approve—the atheist Turnbull or the teetotaller Ivywood, for example. The fact that a man would fight for his beliefs made him significant, and his beliefs significant. 'I say ... there were never any necessary wars but the religious wars. There were never any just wars but the religious wars' (*Orthodoxy*, p. 167).

War, i.e. violent physical conflict, runs through all Chesterton's early fiction: the great battles around Notting Hill; the brawl in the last chapter of *The Club of Queer Trades* (1908); the theme of pursuit in *The Man Who Was Thursday*; the constantly interrupted duel of *The Ball and the Cross*; Innocent Smith's 'life-dealing' revolver in *Manalive* (1912); the skirmishes and final battle in *The Flying Inn*. Chesterton's battles are exciting but curiously painless. The First World War (in which Chesterton's brother Cecil was killed), with its shocking evidence of the unheroic squalor of war, may have had an inhibiting effect on his imagination in this respect, for there is less violence in *The Return of Don Quixote* than in any of the earlier novels.

The sword is a recurring motif in Chesterton's novels. The duel in *The Ball and the Cross* is fought with two seventeenth-century swords (a deliberate anachronism). Captain Dalroy in *The Flying Inn* has his 'straight naval sword'. In his story *The Sword of Wood* (1928) Chesterton plays with the idea that an inverted sword makes a cross; and one recalls that at the end of *The Ball and the Cross* the two duelling swords are found in the ashes of the razed asylum 'in the pattern of a cross'. The sword is the symbol of Wayne's fanaticism in *The Napoleon of Notting Hill*. Early in the novel Wayne, as a young boy, challenges King Auberon with a wooden sword. Whimsically Quin commends him for being 'so stalwart a defender of your old inviolate Notting Hill'. From this incident stems Auberon's fantastic scheme of 'a revival of the old medieval cities applied to our glorious suburbs', and Wayne's fanatical local patriotism. The adult Wayne later says to Quin:

> 'I know of a magic wand, but it is a wand that only one or two may rightly use, and only seldom. It is a fairy wand of great fear, stronger than those who use it—often frightful, often wicked to use. But whatever is touched with it is never again wholly common ... If I touch, with this fairy wand, the railways, the roads of Notting Hill, men will love them, and be afraid of them for ever' ...
> 'What is your Wand?' cried the king, impatiently.
> 'There it is,' said Wayne; and pointed to the floor, where his sword lay flat and shining.

And yet Chesterton was the very opposite of a militarist, which in late Victorian, and Edwardian England, meant being an imperialist. *The Napoleon of Notting Hill* was, in one of its aspects, a pro-Boer pamphlet:[1] it was a defence of the rights of small nations.

Another aspect of *The Napoleon of Notting Hill* was its attempt to answer 'the problem of how men could be made to realize the wonder and splendour of being alive, in environments which their own daily criticisms treated as dead-alive, and which their imaginations had left for dead' (*ibid.*, p. 134). This is perhaps the most personal and the most permanent of Chesterton's contributions to literature. He formulated the problem in a characteristic antithesis, 'How can we contrive to be at once astonished at the world and yet at home in it?' (*Orthodoxy*, p. 13), and offered a solution in his novels, which he called 'Romances' because they comprised 'that mixture of the familiar and the unfamiliar which Christendom has rightly named romance' (*ibid.*).

The most elaborate working-out of this idea, that we must look at commonplace reality with a fresh vision, is *Manalive* (1912). In the organization of the book a characteristic duality is to be observed. The first part is called 'The Enigmas of Innocent Smith', and the second

---

[1] See Chesterton's *Autobiography* (1936), p. 114.

part 'The Explanations of Innocent Smith'. The form of the book is thus an extended paradox. Smith is accused of being a dangerous lunatic. He is alleged to have threatened an Oxford don with a revolver, to have broken into a house, and to have deserted his wife, subsequently abducting several other women. His defence, however, reveals that he threatened the don with death in order to shock him out of an affected pessimism, and to restore his zest for life; that he broke into his own house to invest domesticity with a sense of adventure; that he left his wife and went round the world for the pleasure of coming home, and that he eloped with his own wife several times for the sheer delight of falling in love with her again. Smith 'the Allegorical Practical Joker' is a mystic in the Chestertonian sense of the word: 'a mystic is one who holds that two worlds are better than one.' He comes like an incarnation of the great West Wind which blows so magnificently through the first pages of the novel, sweeping life into the moribund bourgeois household of 'Beacon House'. Michael Moon finds himself drinking poor wine on the rooftop with Smith, and enjoying himself for the first time in nine years. It seems to him that 'he had come out into the light of that lucid and radiant ignorance in which all beliefs had begun'.

The theme of *Manalive*—the rediscovery of life by investing it with danger—is a valuable clue to the interpretation of an earlier novel, *The Man Who Was Thursday*. We recall that when Syme was fighting his duel with the Count, and thought that his last moment had come:

> He felt a strange and vivid value in all the earth around him, in the grass under his feet; he felt the love of life in all living things. He could almost fancy that he heard the grass growing; . . .

This mystical awareness fades when the Count proves to be, not an incarnation of the devil, but only another policeman in disguise:

> That tragic self-confidence which he had felt when he believed that the Marquis was a devil had strangely disappeared now that he knew that the Marquis was a friend.

The ultimate word of the novel is, of course, Sunday's staggering reply to Syme's question, 'have you ever suffered?'—'Can ye drink of the cup that I drink of?' which, as Chesterton admitted with a revealing hesitation, 'seems to mean that Sunday is God'.[1] *The Man Who Was Thursday* is one of the most obscure novels in modern literature, and the responsibility for this obscurity rests with Chesterton himself. We accept, of course, that the novel is subtitled 'A

---

[1] Maisie Ward, *Gilbert Keith Chesterton* (1944), p. 169.

Nightmare', and that probability is not to be looked for. But one suspects that when Chesterton began *The Man Who Was Thursday* he had no clear conception of how he would end it. This can be demonstrated by comparing Syme's retrospective description of his first encounter with Sunday with the original account of the meeting. The two accounts conflict. In the second Syme says that only the *back* of Sunday had impressed him with a sense of evil (thus suggesting the duality of his nature). But in the first account:

> this sense became overpowering as he drew nearer to the great President.
> The form it took was a childish and yet hateful fancy. As he walked across the inner room towards the balcony, the large face of Sunday grew larger and larger; and Syme was gripped with a fear that when he was quite close the face would be too big to be possible, and that he would scream aloud.

I have not yet mentioned the satirical aspect of Chesterton's fiction, which links him more closely than the element of fantasy with Hilaire Belloc. His satire, however, lacks the bitter ironic edge of Belloc's; as the latter observed in his book on Chesterton, he lacked the capacity for hate.[2] *The Ball and The Cross* and *The Flying Inn* offer good examples of Chesterton's satire. The basic theme of *The Ball and the Cross* is that there are only two men in England who care about the existence of God sufficiently to fight about it—one because he believes in God and the other because he doesn't—and all the machinery of the Law is organized to stop them. The comedy begins when MacIan, the wild Catholic Highlander, assaults the atheist Turnbull, and breaks his office window, for publishing an attack on the Virgin Birth of Christ. He is brought before a tolerant, liberal-minded magistrate Mr Cumberland Vane, who fancies himself a wit:

> 'Come, Mr MacIan, come!' he said, leaning back in his chair, 'do you generally enter your friends' houses by walking through the glass?' (Laughter).
> 'He is not my friend', said Evan, with the stolidity of a dull child.
> 'Not your friend, eh?' said the magistrate, sparkling. 'Is he your brother-in-law?' (Loud and prolonged laughter.)
> 'He is my enemy,' said Evan, simply; 'He is the enemy of God.'
> Mr Vane shifted sharply in his seat, dropping the eyeglass out of his eye in a momentary and not unmanly embarrassment.
> 'You mustn't talk like that here,' he said roughly, and in a kind of hurry, 'that has nothing to do with us.'

This is typical of Chesterton's critical attitude to secular institutions, which rises to a tone of high indignation in the description of a lunatic asylum where MacIan and Turnbull are detained, hygienic and

[2] *On The Place of Gilbert Chesterton in English Letters* (1940), pp. 78-9.

pitiless, where 'the unpiercable walls were washed every morning by an automatic sluice. There was no natural corruption and no merciful decay by which a living thing could enter in.' The certification of eccentric but sane men seems to have been a permanent preoccupation of Chesterton's. It recurs in *Manalive* and *The Return of Don Quixote*.

The Ball and the Cross suggests that Chesterton's religious beliefs had not completely crystallized by 1910, the date of its publication. There is still an aura of mystery about the great metaphysical debate between good and evil. MacIan believes Turnbull's atheistic opinions to be evil, but he cannot help liking the man.

> 'Then you are quite certain that it would be wrong to like me?' asked Turnbull, with a light smile. 'No', said Evan, thoughtfully, 'I do not say that. It may not be the devil, it may be some part of God I am not meant to know.'

In the course of the novel the two characters have nightmares in which they are tempted with a diabolic embodiment of their respective idealisms: Turnbull by 'the Revolution', MacIan by an ordered, ritualistic despotism. Both reject the temptation offered to them by 'Professor Lucifer': Turnbull when the latter asks for his approval of the maxim 'No man should be unemployed. Employ the employables. Destroy the unemployables', MacIan when he is asked to approve the maxim 'Discipline for the whole society is surely more important than justice to an individual'. In other words, each man rejects the rigid and heartless extension of his respective belief. Perhaps MacIan's nightmare was a projection of Chesterton's own hesitation to commit himself to the hierarchic and uncompromising discipline of the Catholic Church.

The Flying Inn has most of the characteristics of the kind of Catholicism generally associated with Chesterton and Belloc—robust, alcoholic, anti-progressive and mildly Rabelaisian. It chronicles the resistance of an Irish naval captain, Dalroy, and his publican friend Humphrey Pump, against the attempts of the fanatical Lord Ivywood and his associates to impose Prohibition and other fads, such as Vegetarianism, on England. The cant of Progress is guyed in the person of Lord Ivywood, and it is linked with the fantasy of a Mohammedan mission in England. The satire on Mohammedanism seems rather eccentric until we remember that Belloc's and Chesterton's xenophobia was directed with peculiar virulence against the Levantine races, and that their interpretation of European history insistently emphasized the seriousness of the Mohammedan threat to Christianity.

The Return of Don Quixote (1927) is the only full-length novel which Chesterton produced after his conversion, and it was planned and partly written long before that event. Although there are splendid things in the book, it is the least successful of his full-length novels because it is the least unified. It is interesting from a technical point of view, however, because of the evidence that Chesterton was tending towards a more conventional style of fictional writing. His comedy in places relies more on observation than, as in the earlier novels, on invention. The treatment of the cultural receptions at Seawood Abbey, and of the clientele of the working-class pub (particularly the character of George, the contented butt) are good examples; but it is more convenient to quote the description of the sales assistant in a big store whom Murrel asks for some rather rare Illumination Paints:

> How she said they had got illumination colours and produced water colours in a shilling box. How she then said they had not got illumination colours and implied that there were no such things in the world; that they were a fevered dream of the customer's fancy. How she pressed pastels upon him, assuring him that they were just the same. How she said, in a disinterested manner, that certain brands of green and purple ink were being sold very much just now. How she asked abruptly if it was for children, and made a faint effort to pass him on to the Toy Department. How she finally relapsed into an acid agnosticism, even assuming a certain dignity, which had the curious effect of appearing to give her a cold in the head, and causing her to answer all further remarks by saying: 'Dote know, I'be sure.'

Such excellent observation is absent from the earlier novels. But it would have been out of place there, and it does in fact conflict with the fantastic element which emerges unexpectedly exactly halfway through *The Return of Don Quixote*. It is only then that the main theme of the book becomes clear. Michael Herne, the eccentric and retiring librarian of Seawood Abbey is persuaded to take part in a medieval costume-play. He becomes so obsessed with medievalism as a result that he refuses to take off his costume. He organizes the League of The Lion, composed of enthusiastic young aristocrats who dress up as medieval knights and follow the fanatical Herne for the fun of the thing. The shrewd Prime Minister Eden encourages and exploits the new movement which he sees as the Right's heaven-sent answer to the threat of the brilliant young Socialist Braintree, who is organizing strike action in the local industry. It is safe to assume that this was the basic scheme of the novel as Chesterton had first conceived it.

In dramatic terms Chesterton proves his point—that feudalism was more genuinely democratic than capitalism—brilliantly, if somewhat speciously. Braintree is brought to trial before Herne, accused of attempting to compel Sir Howard Pryce, Lord Seawood, and the Earl of Eden, to hand over the local Paints and Dye industry, which they control, to the workers. To observers, the result of the trial seems a foregone conclusion: it seems inevitable that Herne, with his fanatical enthusiasm for the hierarchic medieval society, will find against the socialist. Moreover, Herne is in love with Lord Seawood's daughter Rosamond. But Herne startles everybody by applying to the case the medieval requirement of a Masterpiece from Masters of Crafts. He says of the three capitalists, 'I have no note of the date or occasion of their presenting Masterpieces in the manufacture of dyes or pigments'. He proceeds to expose the underhand methods by which the three capitalists acquired their present properties and titles. He concludes:

> These three men have claimed the mastery of a craft and the obedience of all their workmen; and their cause is judged. They make the claim of mastery and they are not masters. They make the claim of property and they are not proprietors. They make the claim of nobility and they are not nobles. The three pleas are disallowed.

Consequently Herne is repudiated by the men who had exploited his fanatical idealism for their own ends. With Murrel, his Sancho Panza, he sets out astride an old cabhorse, a new Don Quixote, to seek adventures and glory on the highways of England.

In the original scheme of the novel, the story may have ended here. But Chesterton appended a kind of footnote or epilogue in which his Catholicism becomes explicit. One day Herne and Murrel return to the neighbourhood of Seawood Abbey. Murrel goes off to obtain news of the Seawoods, and returns stunned:

> '... a most stunning and crashing catastrophe has fallen on that Abbey.'
> 'What do you mean? What has happened to the Abbey?'
> 'It has become an Abbey,' said Murrel gravely.

This development has been foreshadowed by a conversation between the two heroines of the novel, Rosamund and Olive (Braintree's beloved). Rosamund, torn between her love for Herne and pity for her humiliated father, says: 'It's as if there were a curse on this place.' Olive replies; 'There's a curse because there is a blessing ... The curse is in the name of this house ... This place doesn't belong to the old families any more than the new families. It belongs to God.' We learn later that after Lord Seawood's death Rosamund becomes a Catholic and restores her home to the monks. Herne finds her working as a nurse in a Catholic slum settlement, but to his relief she has not become a nun. He learns that Olive has also become a Catholic, 'and the odd thing is that John Braintree doesn't seem to mind a bit'. Then Herne says:

'I suppose I am a heretic.'
'We will see about all that,' she said with serene magnificence.

The effect of this passage is to make the previous action seem irrelevant. All conflicts, all enthusiasms are reconciled and absorbed in the Catholic Church. Michael Herne, the last of Chesterton's fanatics, is also the first to admit that he is wrong. It seems not unreasonable to suppose that the acceptance of Catholicism (or rather, the particular form of Catholicism which he embraced) had, in Chesterton's case, the effect of subduing that keen appreciation of conflict between different idealisms which had been the source of his best imaginative writing.

GARRY WILLS
The Man Who Was Thursday

> Chesterton restrained himself from being Edgar Allan Poe or Franz Kafka, but something in the makeup of his personality leaned toward the nightmarish, something secret, and blind and central.
>
> Borges, *Other Inquisitions*

THIS 1908 novel has long enjoyed a kind of underground cult among those with a special interest in fantasy. It is the story of a conspiratorial council of seven anarchists, each one named for a day of the week, with the mysterious Sunday as their president. Admirers of the tale have included J. R. R. Tolkien, C. S. Lewis, W. H. Auden, Jorge Luis Borges, and T. S. Eliot. Kingsley Amis has frequently written about it. Yet the wider reading public remains largely unaware of it.

No wonder. It is a detective story that seems to solve itself too easily, and lose its mystery. But those who stay with it, even after they think they have seen through it, are teased back and back by its ultimately unresolved nature, all the puzzles that remain after the last pages are read. It does not give up its secrets at a glance. Even Mr Amis, despite his enthusiasm for the tale, seems to misunderstand it—as when he writes: 'What I find indigestible in the closing scenes is ... the person of the fleeing Sunday, who at one point makes off mounted on a Zoo elephant and who bombards the pursuit with messages of elephantine facetiousness.'[1] He is attacking the finest clue of all. But, more than

---

[1] *Encounter*, October 1973. Cf. *New York Times Book Review*, 13 October 1968, and Introduction to Kingsley Amis, ed., *G. K. Chesterton, Selected Stories* (Faber and Faber, 1972), pp. 15–16.

that, he lapses into the condescending attitude he came to criticize—the view that Chesterton cannot resist buffoonery, even when he is onto something bigger and more startling than a good joke (or a bad one).

But Sunday's riddles go beyond joking, good or bad; they show a cruelty in humour like the cruelty of nature itself—they are taunts thrown back at men who have been tortured. The best parts of this racy entertainment, as Borges understood, are moments of weird near-break-down:

> As he now went up the weary and perpetual steps, he was daunted and bewildered by their almost infinite series. But it was not the hot horror of a dream or of anything that might be exaggeration or delusion. Their infinity was more like the infinity of arithmetic, something unthinkable, yet necessary to thought. Or it was like the stunning statements of astronomy about the distance of the fixed stars. He was ascending the tower of reason, a thing more hideous than unreason itself.

The book is all a chase, an evasion, and a dream; a benign nightmare prolonged, page by page, beyond our waking. It has the compelling inconsequence of nightmare, its tangle of mutually chasing loves and hates, where the impossible becomes inevitable and each wish comes partnered with its own frustration. Nightmare is described in the book itself as a world of 'tyrannic accidents'. Auden and others have noticed Chesterton's power to evoke the despotic mood of dreams. Borges compared him to Poe in this aspect, and C. S. Lewis to Kafka. The reason we go on reading Chesterton's tale—after we have cracked its first secret (that all the conspirators are also, unbeknownst to each other, anti-conspirators)—is that a dream mood leads us on, linking all its incidents. It aims at an effect that intrigued Chesterton in his own disturbing dreams, one achieved in some of his favourite works of literature.

> Here is the pursuit of the man we cannot catch, the flight from the man we cannot see; here is the perpetual returning to the same place, here is the crazy alteration in the very objects of our desire, the substitution of one face for another face, the putting of the wrong souls in the wrong bodies, the fantastic disloyalties of the night. . .[2]

So, even after we know that the anarchists are also cops, the dream-suspension of things in air continues—the flight from Age, through a crippling ache of snow; a slow climb up the mad tower of pure reason;

---

[2] Essay on 'A Midsummer Night's Dream', now available in W. H. Auden, *G. K. Chesterton: A Selection* (1970), p. 95.

the duel with a phantom who comes apart like meat being carved but will not bleed; the endless chase by anonymous somebodies who gradually become Everybody, embodying paranoia's logic. Then, after running as the quarry, the book's accumulating heroes turn and reach new stages of bewilderment as the pursuers. They knew more as the hunted than as hunters. Desperation gave them solidarity; but at a hint of victory they come apart again, each teasing at the private riddles addressed to him by Sunday. But this dominance of a nightmare mood should not blind us to the riddles addressed to us as readers. These are nicely differentiated, and cluster around two questions. Who are the conspirators? And: Who is Sunday?

THE CONSPIRATORS

> Then thou scarest me with dreams
> and terrifiest me through visions.
> Job 7.14

The first set of clues is almost too obvious—which makes men overlook further hints, to which the first set was only our introduction. The point is not only that everyone is in disguise, but that his disguise is revealing. Each man's secret is unwittingly worn like a shield instead of an emblem. The biggest clue can be overlooked because Chesterton has placed it so prominently in the title. A man can be Thursday only because other men are already Friday, Monday, etc. Granted, the Council of Days is a device that readers quickly penetrate; and most of them focus thenceforth on the identity of Sunday. But the riddle of Monday is not disposed of simply by knowing that he is Sunday's Secretary and also the hidden Detective's right-hand man. Chesterton tries to keep reminding us of this; but readers, so far as I can tell, still keep forgetting. When Dr Bull says, toward the end, 'We are six men going to ask one man what he means', Syme replies: 'I think it is six men going to ask one man what they mean.'

What does the Secretary, the first and most persistent of the Council, mean—with the cruel tilt of laughter as a doubt across his face? At the final banquet he will wear robes that make him more real—a pitch black garment with the struggle of first light down its expanse. He is Monday, light out of darkness, the first unstoppable questioning that is man's last boast—'And God said: Let there be light.' He comes after his fellow-conspirators in the long dream-scene of chase with a black mask on, his face a pattern of light and dark echoed in all his followers. He dwells in darkness, only to fight it, and is described from the outset as tortured with thought in its most naked form. Syme wonders why, when the Secretary gets tossed from the hood of the car, darkness comes on so soon—a minor riddle, but part of a large pattern. Monday, with his complex mind, is the simplest and truest of them all in his quest for truth. He will not stop asking impertinent questions even in the unknowable Emperor's palace.

Gogol, shaggy under his load of wild tresses, but transparent and easily found out, is as simple as the waters of the Second Day, Wednesday is the Marquis, whose absinthe philosophy brightens to the green clothing of earth. Thursday is Syme, a poet, a divider of planet from planet on a plan—as Michelangelo's sculptor-God on the ceiling shoulders moons off from the sun. Friday is the Professor, who has a nihilist's ethic of bestiality, but a deeper kinship, also, with the innocence of animals. Saturday is the last day, Man, a thing almost too open and childish to wear a disguise, an optimist of reason, the tale's French revolutionary, declaring the patent rights of man as king of the creation—each man a king.

All six of the men are puzzles, but elemental puzzles, the kind that one cannot really 'solve'. They represent man's status as a partner in his own creation—the question of man's questioning; the open energy of Gogol; the dim recesses of the Marquis; Syme's swagger; Friday's depths of despair; and Saturday's insaner hope. When Syme grieves that the conspirators have looked only on the fleeing back of the universe, we think his talk deals only with Sunday, since he is often glimpsed from behind in the story. But later, in the garden, all things—in dancing—turn a sudden face on the Council, each tree and lamppost. Everything has a story untold, an episode wandered into, a history only half-understood. And what is true of the clues is true of the detectives, who are themselves the main clues they must read. Each of them deceived the others because he was seen from behind or partially, at an odd angle. The 'back' of intellect is doubt; of subtlety, deviousness; of energy, rage. Everything in the tale, as in the world, needs deciphering, nothing more than oneself. We are all walking signs, signaling urgently to one another in a code no one has cracked. If anyone could understand himself, he would understand everything. So the last person to guess what the man called Monday means will be Monday. Sunday is not a greater mystery than the other Days, except in one respect—he is not only a clue, and a reader of clues; he also plants the clues. He may have cracked the code. That is why they go in search of him.

The tale is not an idle play with symbols. It gets its urgency and compression from the fact that it is the most successful embodiment of the seminal experience in Chesterton's life, his young mystical brush with insanity. In that sense, it is full of clues to his own mental crisis—his depression and near-suicide as an art student in the decadent 1890s. At the centre of Chesterton's best fiction there is always a moment of aporia, the dark seed of all his gaudy blossomings. In *Thursday* that moment comes when the chase is urged on by the masked Secretary:

> The sun on the grass was dry and hot. So in plunging into the wood they had a cool shock of shadow, as of divers who plunge into a dim pool. The inside of the wood was full of shattered sunlight and shaken shadows. They made a sort of shuddering veil, almost recalling the dizziness of a cinematograph. Even the solid figures walking with him Syme could hardly see for the patterns of sun and shade that danced upon them. Now a man's head was lit as with a light of Rembrandt, leaving all else obliterated; now again he had strong and staring white hands with the face of a negro. The ex-Marquis had pulled the old straw hat over his eyes, and the black shade of the brim cut his face so squarely in two that it seemed to be wearing one of the black half-masks of their pursuers. The fancy tinted Syme's overwhelming sense of wonder. Was he wearing a mask? Was anyone wearing a mask? Was anyone anything? This wood of witchery, in which men's faces turned black and white by turns, in which their figures first swelled into sunlight and then faded into formless night, this mere chaos of chiaroscuro (after the clear daylight outside), seemed to Syme a perfect symbol of the world in which he had been moving for three days, this world where men took off their beards and their spectacles and their noses, and turned into other people. That tragic self-confidence which he had felt when he believed that the Marquis was a devil had strangely disappeared now that he knew that the Marquis was a friend. He felt almost inclined to ask after all these bewilderments what was a friend and what an enemy. Was there anything that was apart from what it seemed? The Marquis had taken off his nose and turned out to be a detective. Might he not just as well take off his head and turn out to be a hobgoblin? Was not everything, after all, like this bewildering woodland, this dance of dark and light? Everything only a glimpse, the glimpse always unforeseen, and always forgotten. For Gabriel Syme had found in the heart of that sun-splashed wood what many modern painters had found there. He had found the thing which the modern people call Impressionism, which is another name for that final scepticism which can find no floor to the universe.

The dragging in of impressionism here makes no sense except by its connection with the morbid experiences of Chesterton at the Slade School during the years 1892 through 1895, when a fashionable pessimism was cultivated by the same people who were taken with fashionable 'impressionism'.

Much of the material for *Thursday* comes directly out of the notebooks and poems of those art-school years, almost a decade and a half behind him when he wrote the novel. An early poem on suicide lies behind Chapter 10. The account of an art-school conversation is drawn on for the lantern episode in Chapter 12. The emergence from solipsism into fellowship, described in Chapter 8, lies behind much of his poetry from this period—like 'The Mirror of Madmen', from which I quote just the opening and closing stanzas:

> I dreamed a dream of heaven, white as frost,
> The splendid stillness of a living host;
> Vast choirs of upturned faces, line o'er line.
> Then my blood froze for every face was mine.
>
> Then my dream snapped and with a heart that leapt
> I saw, across the tavern where I slept,
> The sight of all my life most full of grace,
> A gin-damned drunkard's wan half-witted face.

The same experience lies behind the novel's dedicatory poem, with its tribute to the two men who meant so much to him in his personal ordeal—Stevenson of Tusitala, who also rebelled against the aesthetes of *his* art school in Paris; and Whitman of Paumanok, who praised the mere existence of multiple things in a democracy of existence. Indeed, the first sketch of what would become *Thursday* was written as an exercise in Whitman pantheism. It appears in an unpublished Chesterton notebook from the early nineties:

> The week is a gigantic symbol, the symbol of the
>   creation of the world:
> Monday is the day of Lent. [Light? Ed.]
> Tuesday the day of waters.
> Wednesday the day of the Earth.
> Thursday: the day of stars.
> Friday: the day of birds.
> Saturday: the day of beasts.
> Sunday: the day of peace: the day for saying
>   that it is good
> Perhaps the true religion is this
> that the creator is not ended yet.
> And that what we move towards
> Is blinding, colossal, calm
> The rest of God.

Chesterton opposed the chaos in himself and the life around him by considering each man's life a re-enactment, day by day, of the first verses of Genesis. One of his student letters has this passage: 'Today is Sunday, and Ida's birthday. Thus it commemorates two things, the creation of Ida and the creation of the world. . . . Nineteen years ago the Cosmic Factory was at work; the vast wheel of stars revolved, the archangels had a conference, and the result was another person. . . . I should imagine that sun, wind, colours, chopsticks, circulating library books, ribbons, caricatures and the grace of God were used.' Chesterton took as the ground of his hope that very sense of dissolution that threatened his sanity. By the energy of existence things keep re-emerging from dissolution. Creation uses chaos as its working material—just as the spirit, freed in dreams, uses the world as a set of signs, shifting their meaning in ways that terrify man while making him the master of 'unsignified' matter:

> If we wish to experience pure and naked feeling we can never experience it so really as in that unreal land. There the passions seem to live an outlawed and abstract existence, unconnected with any facts or persons. In dreams we have revenge without any injury, remorse without any sin, memory without any recollection, hope without any prospect. Love, indeed, almost proves itself a divine thing by the logic of dreams; for in a dream every material circumstance may alter, spectacles may grow on a baby, and moustaches on a maiden aunt, and yet the great sway of one tyrannical tenderness may never cease. Our dream may begin with the end of the world, and end with a picnic at Hampton Court, but the same rich and nameless mood will be expressed by the falling stars and by the crumbling sandwiches. In a dream daisies may glare at us like the eyes of demons. In a dream lightning and conflagration may warm and soothe us like our own fireside. In this subconscious world, in short, existence betrays itself; it shows that it is full of spiritual forces which disguise themselves as lions and lamp-posts, which can as easily disguise themselves as butterflies and Babylonian temples. . . . Life dwells alone in our very heart of hearts, life is one and virgin and unconjured, and sometimes in the watches of the night speaks in its own terrible harmony.[3]

Chesterton was drawn back, constantly, to the Book of Genesis because of its beginning in chaos. Once one has experienced that nothingness, the emergence of any one thing into form and meaning is a triumph, the foundation for a 'mystical minimum' of aesthetic thankfulness. Then, as Blake saw, each sunrise becomes a fiery chariot's approach.

When we say that a poet praises the whole creation, we commonly mean only that he praises the whole cosmos. But this sort of poet does really praise creation, in the sense of the act of creating. He praises the passage or transition from nonentity to entity. . . . He not only appreciates everything but the

---

[3] From an early (1901) essay on 'Dreams', in *The Coloured Lands* (Sheed and Ward), p. 83. Cf. *Daily News*, 9 July 1904: 'A world in which donkeys come in two is clearly very near to the wild ultimate world where donkeys are made.'

nothing of which everything was made. In a fashion he endures and answers even the earthquake irony of the Book of Job; in some sense he is there when the foundations of the world are laid, with the morning stars singing together and the sons of God shouting for joy.[4]

The Council of Days not only praises this transition, but effects it—as God creates through his six days. Creation is not only *the* beginning, but is *always* beginning—with the Council of Days in on the battle against chaos from the outset. They overthrow their own darker side, their evil brother, as God had to wrestle the sea-god into bonds in the Book of Job. When the six Days gather in Sunday's garden, they have gone back beyond their childhood 'where a tree is a tree at last'—to the primordial self they could only accomplish by a struggle that, illogically, *forms* that self. Their end is to arrive at their own beginning, in a puzzle Chesterton often returned to:

It is at the *beginning* that things are good, and not (as the more pallid progressives say) only at the end. The primordial things—existence, energy, fruition—are good so far as they go. You cannot have evil life, though you can have notorious evil livers. Manhood and womanhood are good things, though men and women are often perfectly pestilent. You can use poppies to drug people, or birch trees to beat them, or stones to make an idol, or corn to make a corner, but it remains true that, in the abstract, before you have done anything, each of these four things is in strict truth a glory, a beneficent specialty and variety. We do praise the Lord that there are birch trees growing amongst the rocks and poppies amongst the corn; we do praise the Lord, even if we do not believe in Him. We do admire and applaud the *project* of a world, just as if we had been called to council in the primal darkness and seen the first starry plan of the skies. We are, as a matter of fact, far more certain that this life of ours is a magnificent and amazing enterprise than we are that it will succeed.[5]

[4] *St Francis of Assisi* (1928. Image edition, 1957), p. 77.
[5] *T.P.'s Weekly*, 1910.

BENNY GREEN
Defender of the Faith

STYLISTICALLY Chesterton could be a tiresome trickster, politically a pantaloon and, as for his religious philosophy, I always found it surprising that there was anyone to take it seriously. What is revealing about [a] collection of reviews of Chesterton's work is that all these reservations were commonplace at a very early stage of his development and that Chesterton ignored the strictures most unwisely.

An unsigned notice in *The Saturday Review* of 1905 points out that 'paradox is fatal as a foundation of style', a sentiment we heartily applaud, pausing only to add that unsigned reviews are fatal as a foundation of intellectual honesty. A year earlier *The Times*, 'facile extravagance of style'; *The Manchester Guardian*, 1906, 'his habit of antithesis becomes wearisome'; *The Times* 1910, 'antitheses . . . jaded and superfluous'. None of this scrupulously honest comment seems to have had the slightest effect on Chesterton's style, with the result that, to this very day, the attempt to read one of his books from cover to cover soon induces a kind of exhaustion; in the end his relentless exuberance is indigestible, in a way that say, Beerbohm's in *Zuleika Dobson* is not. The quality which Chesterton seemed to lack utterly was stylistic restraint.

Similarly, there appear to have been literary complaints about his relentless Christian optimism almost from the beginning. In 1902, Charles Masterman, reviewing the long since forgotten *The Defendant*, actually referred to this blatant optimism as 'blasphemous', and found Chesterton an insensitive philosopher because he was so smugly reconciled to badness. To be fair, Chesterton clearly baffled Masterman as he baffled so many other sober reviewers. By 1903 Masterman has decided that 'it is as a poet that Chesterton will make his contribution to literature'; a year later he changes his tune once more with the sudden spectacular appearance of that extraordinary piece of virtuosity *The Napoleon of Notting Hill*. But one senses that neither Masterman nor any of the other critics willing to grant Chesterton the accolade of serious review ever quite forgave him his antithetical cartwheels or his persistent jolly laughter in the Vale of Tears. And in *The Westminster Gazette* in 1908 someone points out the deadly fallacy behind the argument that Religion is the direct opposite of Reason.

Having rejected Chesterton the religious theorist and Chesterton the versifier, some of the reviewers were tempted to clasp to their bosoms Chesterton the literary critic. The trouble was that every book Chesterton ever wrote was about himself. So far from being a fault, it says much for his unquenchable originality as a man. It says very little, however, for his chances as a critic. I will never forget my disappointment at first reading Chesterton on Dickens, perhaps the most overpraised work of literary criticism of its generation. Having been assured by Shaw and all the rest of them that this book was wonderful, I tried it and found the intermittent crassness too blatant to compensate for the insights. A writer who can tell us that the trouble with Dickens was that he was 'a little too happy' is not the man to whom to entrust our literary sensibilities: predictably Chesterton, who remained emotionally a child to the end, denies Dickens the right to leave childhood behind, insisting that it is the young Boz of the coaching inns and the cherubic cheeks that is the real stuff, rather than the miserable later books. Apart from the fact that the choice is as phoney as the one between Religion and Reason, there is something to be said for the idea that the reader who rejects Frederick Dorrit for Ralph Nickleby is one who has allowed his ideas of social virtue to cloud his aesthetic judgements.

For this reason the most revealing of all the criticisms of Chesterton's thought are those contributed by another creative writer with Catholic sympathies, T. S. Eliot. Eliot finds the style 'offensive to my vanity' and then adopts the true Chestertonian paradoxical stance by ending his review of Chesterton on Stevenson with the observation that Stevenson is 'well enough established to survive Mr Chesterton's approval'. The most imperceptive remark in over five hundred pages is E. C. Bentley's stupefying 'to him all men were brothers', the most revealing Rebecca West's shrewd diagnosis of our hero as a man whose 'worst blasphemy lies in trying to preach the Gospel when heaven has sent him down a comic song'. Let that be Chesterton's epitaph; when all the evangelizing and the theorizing fall away, we are left with a small but priceless residue of far-fetched fables like *The Napoleon of Notting Hill* and *The Man Who Was Thursday*.

JOHN CAREY
For Beer and Liberty

CHESTERTON had a body like a slag heap, but a mind like the dawn sky. He saw the world new, as if he'd just landed from another planet. Mankind's very existence struck him as wildly improbable. How could anyone take seriously, he demanded, an intelligent being which sustained itself by stuffing alien substances through a hole in its head? His amazement at normality unlocked for him the poetry of Edwardian London, which becomes in his books a magical stage-set, alive with barrel-organ music and fiery orchards of gas-lamps. But Chesterton's innocent eye rested with most incredulity on the poor. Like the fair-minded cannibals who visit Europe in Montaigne's essay, he couldn't understand why the poor didn't simply fall upon the rich and cut their throats. Wealth filled him with almost physical loathing: he dreamed of bloody revolutions which would smash the 'fat white houses' in Park Lane.

This hatred came partly from his Christianity, partly from his conviction that the rich were, in any nation, the scum of the earth. But it was fuelled, too, by his phobia about international finance. When he first saw the neon signs in New York, trumpeting the monopolies of the rich, he remarked what a glorious sight they'd be if only you couldn't read. He believed there was a worldwide conspiracy of millionaires intent on reducing man to docile uniformity and stamping out patriotic urges. The poor, not the rich, were necessarily the real patriots, he argued: the rich could always scuttle off to the Caribbean in their yachts if times got bad. Great Jewish families like the Rothschilds, with ties that spanned national frontiers, appalled him. They were treason personified: all Jews, he suggested, should be made to wear Eastern dress, so that they couldn't pass as loyal natives. His support for Mussolini can likewise, if we switch off hindsight, look quite healthy. It stemmed solely from sympathy with the poor. Mussolini, Chesterton reported, had proved more independent of the rich and done more to coerce employers than any English government. He admired Fascism's contempt for parliamentary rule: parliament, he'd always thought, was a charade run by a governing clique who were in league, as like as not, with the dark plutocrats of his nightmares. When his brother Cecil unearthed some crooked share dealings involving government ministers, it confirmed his worst fears.

In fact he probably saw through his own obsessions half the time: in his novel *The Man Who Was Thursday*, the creepy conspirators are all, it turns out, honest policemen in disguise. What remained stubbornly sane about him was the treatment of the poor, though his sanity took the form of battling against all current proposals for their relief. Socialism was no good, he decided, because it aimed to take capital out of the hands of the few and put it into the hands of fewer—the politicians. The abolition of private property was a mad idea: the poor existed precisely because someone had abolished theirs. Everyone needed something to express himself through, if it was only a roll of wallpaper or a cabbage patch; property was the art of a democracy. As for the reforms being introduced by the Liberal party, they were just high-minded elitism. They compelled the poor to be thrifty, hygienic, and sober, exactly as potential employers would wish. Chesterton stuck out for beer and liberty. The philanthropist, he warned, is not a brother but a supercilious aunt.

He was right, too. The unemployed, it was proposed, should be segregated from their families to stop them breeding and confined to remote labour colonies where idlers would be firmly disciplined. Such schemes were favoured by Socialists like Beatrice Webb, as well as by Liberals. The same tyrannical paternalism inspired the movement for the prohibition of the sale of alcohol, and measures like Lloyd George's National Insurance Act, which involved compulsory deductions from wages but denied unemployment relief to anyone who'd been discharged for 'insolence' or left his job 'without just cause'. Old age pensions, introduced in 1908, were to be withheld from those who had been in prison or 'persistently failed to work'. Under the Mental Deficiency Act, a prime Chestertonian bugbear, 'defective' children of the poor could be forcibly removed and subjected to remedies like craniectomy—an operation which killed a quarter of them and left the rest no brainier than before.

The assumption behind all these charitable enterprises was that the poor needed reformative treatment. Chesterton thought they needed a house and a bit of garden like everyone else. This humdrum, sensible opinion reflects his lifelong respect for the common man, as against intellectuals and other cranks. He helped to found the Distributist League, which campaigned to create a prosperous English peasantry by settling the destitute on smallholdings. A beautiful idea, and a sure flop politically.

Sadly Chesterton watched his League dwindle into a talking-shop for simple-lifers and vegetarians. But then, he'd never reckoned himself a practical politician. Asked what he'd do if he were made

Prime Minister, he said he'd resign at once. He remains a vital force because he held aloft great flaring half-truths about poverty and wealth without which justice dies, and which shuffling practicality prefers to ignore.

JOHN ATKINS
Styles in Treachery

SOME of the old pros say that the best spy book ever written was Chesterton's *The Man Who Was Thursday*. It seems at first sight a send-up of the average spy novel but for the professional agent it contains a deep-lying truth. It is that if you say quite openly that you are a spy or an anarchist, no one will believe you and you will be able to do pretty much as you please. Attempts to hide your identity lead to suspicion, and then everyone will know you are a spy (or an anarchist). There have been some attempts, especially since the Bond era, to satirize the spy novel but most of them have been heavy-handed compared with Chesterton's little masterpiece. McCormick actually claims that *Thursday* still provides a model for aspiring spy writers, to prevent them from 'keeling over too far in the direction of fantasy'. This is in itself a Chesterton-type paradox which means it must be treated seriously.

P. J. KAVANAGH
Chesterton Reappraised

G. K. CHESTERTON began as a Party Liberal. In fact he began as an art student, at the Slade, but as soon as he started to write as a journalist it was in Liberal newspapers and in what he considered to be the Liberal interest. He early became disillusioned with the Party, and then with the Party system, because he did not believe it represented people's real needs. To him it was more like a share-out of jobs and privileges between several, often related, families on both sides of the House, with newcomers like Lloyd George heartily joining in.

For a man of his views, who wanted radical reform, the obvious course was to become a Socialist, like Bernard Shaw (who revered

Chesterton), or, in some degree, like H. G. Wells. But Chesterton could not do this because, detesting the abuse of privilege, he also feared too great an increase in the power of the State and, even more important, he was haunted from the beginning by a sense of a broken historical continuity, which he wanted to help mend. This remained one of his central themes. Programmes of Socialist reform concerned themselves with social needs: Chesterton wanted something that satisfied the whole of a man, his spiritual needs as well as his material ones, and came to believe this balance had once been achieved in the past, and then was destroyed by the greed of a rising middle class. This led him eventually to be considered the advocate of a return to a sentimentalized 'Merrie England', which in a sense he was, but Chesterton was more intelligent than that and can be allowed to explain himself. (A chapter of *A Short History of England* is called 'The Meaning of Merry England'. He always delighted in meeting his critics head-on.)

This standing apart from his time, both from received opinion and from progressive ideas, had the advantage of keeping him clear of movements that were merely fashionable—which he called 'fads'—but it also kept him out of the general intellectual drift of his generation, which is something more powerful and significant than fashion, though of course it can be wrong. He became the odd man out among his peers (Belloc hardly counts in this respect because he was a professional odd man out, which Chesterton was not). This was dangerous for Chesterton, and still is, because he wanted his ideas listened to by the unconverted. There are two things to note about this isolation: one is that it never made him sound strident or angry, the other is that he was so popular and widely read it is extraordinary, looking back, to see how intellectually alone he was.

You would never guess it from his tone, which is confident and genially magisterial. He seems always to be addressing a large and friendly audience, past the heads of the 'moderns', the 'professors', the 'scientists' (or, worst of all, the 'modern scientific professors') and telling a receptive gathering what it half-knew already and would now, after his explanation, agree to be obviously true. Possibly these are 'the people of England, that never have spoken yet' and he is speaking for them, as well as to them. Very possibly: 'modernist' Ezra Pound was heard to sigh, despairingly, 'Chesterton *is* the mob!'

He is as much a political writer as George Orwell. Politics, in the narrow, as well as the widest, sense, informs and inspires his writing, which is why he remained all his life a journalist, to the despair of those who recognized the enormous scale of his gifts. But that is what

Chesterton wanted to be, a 'jolly journalist' (as he unfortunately described himself, the phrase has affected his reputation) because he thought his message important and from Fleet Street it could reach the people who need it most, and who perhaps had the best chance of understanding it.

He would have appreciated Pound's remark about the mob because for him the mob—unintellectual, fundamentally decent, duped—was not a mindless thing, if it was given a chance. He uses the word in his peroration to a chapter ('The Return of the Barbarian') in *A Short History of England*, published in 1917; he is referring to the Great War: 'The English poor, broken in every revolt, bullied by every fashion, long despoiled of property, entered history with a noise of trumpets, and turned themselves in two years into one of the iron armies of the world. And when the critic of politics and literature, feeling that this war is after all heroic, looks around him to find the hero, he can point to nothing but a mob.'

It is possible that even now Chesterton's use of the word 'mob' could be misunderstood. He was frequently and astonishingly misunderstood by his opponents in his own day. He meant of course that the 'mob', the ordinary people, was the best thing about the nation. This was a belief he could not always find among Socialists.

He was involved as a writer in almost every day-to-day controversy of his time and, among his contemporaries, though sometimes they privately cheered him on, was usually in a minority of one. There is insufficient space in a collection such as this to give many examples, though the subjects remain topical: birth control, genetic engineering, the position of women, the secularization of Christianity, and so on. But it is odd that so dedicated a battler should be remembered by many as the author of the 'Father Brown' stories and of charming trifles with titles like 'A Piece of Chalk' or 'What I Found in my Pockets'. There are reasons for this—you don't poke intelligent fun at your generation and then expect it to praise your judgement—and his own refusal to be pompous is one of them; but, as he is one of the great intellects England has produced, it is a pity.

His biographer, Maisie Ward, quotes from one of the many lengthy political exchanges in which Chesterton exulted, and it is worth reproducing here because, apart from the continual pleasure to be derived from Chesterton's *tone*, it gives proof of the extraordinary difficulty some minds experienced, despite his patient clarity, in grasping what he was driving at.

He had written to the editor of the Liberal *Nation* pointing out, among other things, the illiberality of the new licensing laws. The editor, publishing the letter, had been unwise enough to add a brief comment of his own. Here is Chesterton's reply:

Jan 26 1911,
Sir,
In a note to my last week's letter you remark, 'We must be stupid; but we have no idea what Mr Chesterton means.' As an old friend I can assure you that you are by no means stupid; some other explanation of this unnatural darkness must be found; and I find it in the effect of that official party phraseology which I attack, and which I am by no means alone in attacking. If I had talked about 'true Imperialism', or 'our loyalty to our gallant leader', you might have thought you knew what I meant: because I meant nothing. But I do mean something; and I do want you to understand what I mean. I will, therefore, state it with total dullness, in separate paragraphs; and I will number them.

(1) I say a democracy means a State where the citizens first desire something and then get it. That is surely simple.

(2) I say that where this is deflected by the disadvantage of representation, it means that the citizens desire a thing and tell the representatives to get it. I trust I make myself clear.

(3) The representatives, in order to get it at all, must have some control over detail; but the design must come from the popular desire. Have you got that down?

(4) You, I understand, hold that English M.P.s today do thus obey the public in design, varying only in detail. That is a quite clear contention.

(5) I say they don't. Tell me if I am being too abstruse.

(6) I say our representatives accept designs and desires almost entirely from the Cabinet class above them; and practically not at all from the constituents below them. I say the people does not wield a Parliament which wields a Cabinet. I say the Cabinet bullies a timid Parliament which bullies a bewildered people. Is that plain?

(7) If you ask why the people endure and play this game, I say they play it as they would play the official games of any despotism or aristocracy. The average Englishman puts his cross on a ballot-paper as he takes off his hat to the King—and would take it off if there were no ballot papers. There is no democracy in the business. Is that definite?

(8) If you ask why we have thus lost democracy, I say from two causes; (a) The omnipotence of an unelected body, the Cabinet; (b) the Party system, which turns all politics into a game like the Boat Race. Is that all right?

(9) If you want examples I could give you scores. I say the people did not cry out that all children whose parents lunch on cheese and beer in an inn should be left out in the rain... etc.

It is hard to see that much of what Chesterton complained of in our method of government has changed. This is perhaps unsurprising if the opacity of the editor is anything to go by. He still could not see what was worrying Chesterton. He now complained that Chesterton's criticisms only concerned 'small' things. Considering this was a large part of his meaning Chesterton is remarkably patient in his reply: 'What', he asks with deceptive mildness, 'can be more fundamental than food and drink and children?'

On the particular point, anyone who has travelled about Britain with small children has had cause to curse the Children's Act of 1910. Only now is some small compromise being attempted, with gloomy 'Children's Rooms' in some public houses and hotels. The larger point, that it is precisely the 'small' and everyday that matters and to this our rulers pay too little attention, is Chesterton's constant theme. Indeed, again and again, with a thousand illustrations, he points out that we non-rulers cannot see the infinite significance of small things either, that we can barely see at all. This could be described, fundamentally, as the religious point of view; it is certainly that of a poet. Chesterton's poems are usually thoughts expressed in verse, or emotion expressed thoughtfully. He is as much a poet in his prose as in his verse, and this can be a useful way of approaching his work. It is certainly as a poet that he dislikes the way we prefer the general to the particular, because it is easier. He puts the point in verse (albeit light verse, but effectively) in a book—*New Poems*, 1932—published twenty years after the letter just quoted:

THE WORLD STATE

'Oh, how I love Humanity,
   With love so pure and pringlish,
And how I hate the horrid French,
   Who never will be English!

The International Idea,
   The largest and the clearest,
Is welding all the nations now,
   Except the one that's nearest.

This compromise has long been known,
   This scheme of partial pardons,
In ethical societies
   And small surburban gardens—

The villas and the chapels where
   I learned with little labour
The way to love my fellow-man
   And hate my next-door neighbour.'

It is a simple criticism that Chesterton never tires of making but, if we look around fifty years later, there is little sign that it is too simple to be worth repeating.

Thus, among other things, it was the generalized nature of Socialist plans for amelioration that Chesterton did not like, though he was eager for change and, if necessary, revolution; although it is never quite clear how he envisaged such a thing. It seldom is, even in more practically minded reformers. His political thinking took the form, roughly, of somehow giving back to the people what had been taken from them and this later came to be called 'Distributism', with *G.K.'s Weekly* as its mouthpiece. The idea attracted a following and, having established the idea, the way Chesterton interpreted his political role—apart from being badly out of pocket and dangerously overburdened with work because of it—is somewhere between the lines of this gentle, rather sad, rebuke to the more zealous of his followers, in 1929. There had been quarrelling among them:

... I could only manage to keep this paper in existence at all, by earning money in the open market; and more especially in that busy and happy market where corpses are sold in batches: I mean the mart of Murder and Mystery, the booth of the Detective Story. Many a squire has died in a dank garden arbour, transfixed by a mysterious dagger, many a millionaire has perished silently though surrounded by a ring of private secretaries, in order that Mr Belloc may have a paper ... Many an imperial jewel has vanished from its golden setting, many a detective crawled about on the carpet for clues, before some of those little printers' bills could be settled which enabled the most distinguished and intelligent of Distributists to denounce each other as Capitalists and Communists, in the columns of the 'Cockpit'and elsewhere. This being my humble and even highly irrelevant contribution to the common team-work, it is obvious it could not be done at the same time as a close following of the various shades of thought in the Distributist debates. And, this ignorance of mine, though naturally very irritating to people better informed, has at least the advantage of giving some genuineness of my impartiality. I have never belonged distinctively to any of the different Distributist groups. I have never had time. (*Gilbert Keith Chesterton*, Maisie Ward).

Nevertheless, to have an alternative programme to hand which you consider practical is a great assistance towards the clarification of ideas and the identification of what is wrong. Even more important, his overwork and lack of political quietism—so deplored by friends who wanted him to give up journalism, in order to write his 'great' book—his journalistic and editorial immersion in affairs and his attempts at the proffering of a solution, led him to a detestation of the group-soul and a veneration of the individual one; led him, in other words, to the religious, more specifically, Christian and, latterly, Roman Catholic position. This of course separated him even further from his time; something he bore with his accustomed cheerfulness.

It is important to recognize that it was his practical political concerns that led him to Roman Catholicism; it was not his increasingly dogmatic (in the true sense) Christianity that originally inspired his politics. In a way he *discovered* Christianity, found that for him it contained the solutions he sought, and he makes sure we can track him in this slow process from *Orthodoxy* in 1908, when he was hardly a Christian at all, but an enquirer, to *The Everlasting Man* in 1925, by which time he had become a Christian apologist. He was not received into the Roman Catholic Church until 1922, when he was nearly 50.

He is always clear about the stages in his thought and then clear about his position; too clear for some. But the clarity is undeniable, though many still refused to believe that he meant what he said. In 1922 in *What I Saw in America* he puts it in a sentence: 'There is no basis for democracy except in a dogma about the divine origin of man.'

It is a challenge thrown down. It contains two words guaranteed to raise nearly every British hackle, 'divine' and 'dogma'. If the first doesn't do it the second certainly will. But Chesterton believed that dogma, willingly assented to, was the basis of intellectual and political liberty, because it was a shared premise. Perhaps he knew too little of what it was like to live in a Church-dominated country. Shaw tried to warn him, citing the example of Ireland. But that is perhaps not so relevant after all: he was writing in England. He would have had different targets in Ireland, or Spain—or for that matter in the Soviet Union. For him dogma was the frame within which man could move freely, with tolerance. Also, of course, he came to believe in the divine.

But he does not just leave it there, for the reader to agree or disagree. Even when apparently talking about something else, Chesterton is concerned to explain why he believes as he does, step by step, inviting the reader to argue with him. Any reader would have to be more in love with the secularized corporate state than he is likely to be, for him to turn away in disgust at the invitation, even if the writing itself were not so entertaining. However, it is possible that the entertainment Chesterton so prodigally provides can prevent an unwary reader (like the editor of the *Nation*) from hearing what he says. Chesterton disliked the way the world was going and tried to divert that progress: that he failed has caused him to be neglected, sometimes derided, but that does not prove him wrong.

He is, for example, frequently accused of being wilfully paradoxical, of standing ideas on their heads for fun. In fact he rarely does so. His gift is for brilliant analogies, often absurd ones. Belloc called it 'his genius for illustration by parallel . . . I can speak here with experience, for in these conversations with him or listening to his conversation with others I was always astonished at an ability in illustration which I not only have never seen equalled but cannot remember to have seen attempted. He never sought such things; they poured from him as easily as though they were not the hard forged products of intense vision, but spontaneous remarks' (Maisie Ward, op. cit.). Belloc also says it is the way Chesterton '*taught*' (he italicizes the word) and Chesterton certainly was a teacher, wanting to help us see, with our eyes as well as, figuratively, with our minds. For so speculative a man he is surprisingly visual. One remembers that he began as an art student, and drew all his life (frequently demons: behind his geniality, or rather at the root of it and of all his work, is a lively sense of the power of evil). But, although his mind was illustrative, he was myopic, and this may have affected his method; he preferred the broad sweep, both pictorially and intellectually, to the focused detail. Not that this led to any vagueness or evasiveness in argument, or blurring of outline; on the contrary, it added vigour to the forward march of his prose, as though it refused to be distracted by brief roadside glimpses. Nevertheless, though he seldom numbers the petals on a rose, his description of landscape (and townscape) and climatic *effects* in the Father Brown stories, for example, are often the best things in them; and they came from a man who was not only short-sighted but who could seldom bring himself even to go for a walk.

His work is filled with pictures, usually absurd ones; here are a couple (from *A Short History of England*):

It is almost necessary to say nowadays that a saint means a very good man. The notion of eminence merely moral, consistent with complete stupidity or unsuccess, is a revolutionary image grown unfamiliar by its very familiarity, and needing, as do so many things of this older society, some almost preposterous modern parallel to give its original freshness and point. If we entered a foreign town and found a pillar like the Nelson Column, we should be surprised to learn that the hero on top of it had been famous for his politeness and hilarity during a chronic toothache. If a procession came down the street with a brass band and a hero on a white horse, we should think it odd to be told that he had been very patient with a half-witted maiden aunt. Yet some such pantomime impossibility is the only measure of the Christian idea of a popular and recognized saint. It must especially be realized that while this kind of glory was the highest, it was also in a sense the lowest. The materials of it were almost the same as those of labour and domesticity: it did not need the sword or sceptre, but rather the staff or spade. It was the ambition of poverty.

It is the 'preposterous' nature of these analogies that reinforce our surprise at being reminded of something we already knew: that great saints were sometimes spectacularly unsuccessful men, and yet were revered. It is in fact a paradox. But the analogies are also right in another way; it is of course the 'small' things which all of us know, which are without attendant drama—toothache, elderly relatives—that test our heroism. These are the same small things the importance of which was beyond the comprehension of the unfortunate editor of the *Nation*. Chesterton's work is all of a piece throughout. And the 'preposterous' can lead without strain to a noble and subtle sentence: 'It was the ambition of poverty.' This is only an apparent, not a genuine, paradox. The ambition to be poor we can understand, but now we understand afresh the spiritual poverty of our usual definition of ambition.

Chesterton's paradoxes are therefore not a form of play, they are literary devices, analogues to the paradoxical nature of history and truth. Even the Gospels are paradoxical—'the meek shall inherit the earth', and so on. The meek, in the form of monks (with the ambition of poverty), did indeed inherit the earth because by working it and putting it in good order their monasteries became rich. (One of the best chapters of *Orthodoxy* is called 'The Paradoxes of Christianity'.)

Chesterton's use of analogy and visual illustration is demonstrated in the opening pages of the early novel *The Napoleon of Notting Hill* (1904), a book so important to Chesterton (not for its quality— he never thought of his work in that way—but for what he hoped it contained) that he said if he had not written it he could not have gone on writing. Auberon Quin, a Government official, is walking behind two colleagues, dressed in the style of their class in 1904, which is to say in frock-coats.

> So the short Government official looked at the coat-tails of the tall Government officials, and through street after street, and round corner after corner, saw only coat-tails, coat-tails, and again coat-tails—when, he did not in the least know why, something happened to his eyes.
> Two black dragons were walking backwards in front of him. Two black dragons were looking at him with evil eyes. The dragons were walking backwards, it is true, but they kept their eyes fixed on him none the less. The eyes which he saw were, in truth, only the two buttons at the back of a frock-coat; perhaps some traditional memory of their meaningless character gave this half-witted prominence to their gaze. The slit between the tails was the noseline of the monster: whenever the tails flapped in the winter wind the dragons licked their lips. It was only a momentary fancy, but the small clerk found it embedded in his soul ever afterwards.

It is a commonplace experience sometimes to see in this way but here it has a purpose: Quin is not seeing dragons; he is seeing frock-coats, as though for the first time, and seeing their absurdity. He sees that they represent all that is pompous and solemn and self-satisfied in the official, middle-class world—and they have evil eyes.

Born in 1874, Chesterton was for roughly half his life a Victorian, a period during which the gulf between the official and middle classes and the working class was probably wider than it has ever been, before or since. He was born into the comfortable middle class himself (the estate agents, Chesterton, still exist in Kensington), fortunate in his settled and interesting (and agnostic) parents, but for him the unhealthiness of the class division was self-evident. He remarked that people laughed at medieval barons who put their vassals below the salt who themselves put their servants below the stairs. He therefore detested the uniform—the frock-coats—that was used to emphasize this separation and, although never self-consciously a bohemian, would not, or could not, wear it. After he was married in 1901, his wife, despairing of keeping him tidy, invented with a stroke of genius the Chesterton style: a cloak, which he did not have to try to keep buttoned, a slouch hat, which he did not have to keep brushed. But Chesterton is not only satirizing, in some sense illustrating, the cruel pretension of his class, he is saying in the novel that the dragon exists and St George can slay it, that pageantry and chivalry and local loyalty can reveal below the unpromising appearance of Notting Hill a place that is, or could be, magical, fantastical, alive with romance and with romantics. He is talking of the wonder-filled landscape of his childhood, he was born near Notting Hill, and is saying, '*Look*! Can't you see?'

It is not surprising, with this perception of the drama of the ordinary and gift for truth-bearing exaggeration, that he should feel kinship with Robert Browning and Charles Dickens. About each of them he wrote a book at the outset of his career, and on his chosen subjects he became, at once, an outstanding literary critic before there was such a thing as 'literary studies' and before Dickens, certainly, was fashionable.

In 1903, *Robert Browning* marked his commencement as a writer of books, as opposed to short pieces and poems. He was 29, and the book bursts with pent-up energy and ideas. He said of it later that it was really filled with themes of his own, 'a book in which the name of Browning was introduced from time to time, I might almost say with considerable art ...'. There ought to have been a law to prevent Chesterton saying things like that about his work because there were too many prepared to take him at his word. It is a brilliant book, illuminating about Chesterton (as it ought to be) and illuminating about Browning because Chesterton understood him.

But Chesterton always refused to speak seriously of himself and therefore there are those, bewilderingly, who cannot regard the work as serious. Because his touch is light they think his head is, whereas it would not be difficult to make a case that there is no more serious writer. Hardly a line of the enormous number he piled together is not concerned with the sort of questions liable to empty a room, such as 'the meaning of life', how we can best understand it, and live it. His touch was light because he was more interested in winning an audience than in status. He continued to work through Fleet Street because, doubtless, he enjoyed the fray and the companionship, but also because he cared more for his message than he cared for himself. Few men of genius have been so little interested in posthumous fame.

He probably did not care for himself enough. The size of his output was certainly bad for his health. Apart from commissioned books he wrote a long column in the *Illustrated London News* every week for thirty years (refusing to ask for an increased fee when he could have asked for anything he liked, because they had helped him with regular work when he needed it most). There was also, for years, the column in the *Daily News*, a multitude of regular or semi-regular contributions elsewhere, lectures all over the country, 'Father Brown' stories and novels and poems and long public controversies like the one quoted from the *Nation*; there were also public debates with people like Shaw. It was too much. In 1909 his wife managed to move them from London to Beaconsfield (which, contrary to the murmurs of disgruntled friends, who missed him, Chesterton makes clear he enjoyed) but the output continued as great. Between 1909 and 1915, apart from a quantity of journalism that would break any normal man, twenty-one books were published. Some of these were collections of newspaper pieces and stories, but the list includes three novels (*The Ball and the Cross*, *Manalive*, *The Flying Inn*), studies of Bernard Shaw and William Blake, *The Victorian Age in Literature*, a full-length epic poem *The Ballad of the White Horse* and a play, *Magic*. His absence of mind became legendary, but it was inevitable; he must have been composing sentences in his head, when he was not actually writing them, most of his waking hours. The jolly, bibulous journalist that Chesterton was happy to be considered had become almost pure mind.

But of course he kept that a secret, until in 1915 he collapsed and lay in a coma, or near-coma, for several months, his life despaired of.

After his recovery (when he immediately asked for newspapers; back numbers so that he could trace in detail the course of the war), Shaw wrote to tell him, 'You have carried out a theory of mine that every man of genius has a critical illness at 40, Nature's object being to make him go to bed for several months. Sometimes Nature overdoes it: Schiller and Mozart died.'

There is a lessening of *élan* after the recovery. There were many reasons for this: the war, of course; the misery of the Marconi Affair (in which his brother Cecil more or less took on the Government, accusing people connected with it of corruption). The malign atmosphere of this episode, in which he supported his brother, deeply distressed Chesterton; it seems he lost the last of his trust in public men. Then in 1918 Cecil died and, out of love for him, Chesterton continued his brother's political crusades; these involved him in burdensome editorships, culminating in *G.K.'s Weekly*, which were not only a drain on his pocket but also on his health.

But apart from these griefs and burdens that weighed down the wonderful springiness of his early style, there was also the natural disappointment, bound to come as the years pass, of a champion who sees that his causes will not triumph in his day, that for all his efforts the world goes on as before, or goes on worse. But Chesterton never became exasperated, as most men do. Although it is true that after the death of his brother he was guilty for the first and last time of violent language in public, it was to do with one of his dead brother's bitter battles, the Marconi affair.

Yet, if one suspects a final diminution of energy, there comes *St Thomas Aquinas* (1933), a brilliant book on so tricky a subject that the Thomist philosopher Etienne Gilson threw up his hands in admiration and despair: 'The few readers who have spent twenty or thirty years studying St Thomas Aquinas and who, perhaps, have themselves published two or three volumes on the subject, cannot fail to perceive that the so-called "wit" of Chesterton has put their scholarship to shame.' Before that, in 1925, there had been *The Everlasting Man*, a massively ambitious work in which Chesterton attempts to pull together and in some sense systematize his whole view of the world. Its aim is to adjust an anti-Christendom imbalance in H. G. Wells's enormously successful *Outline of History* (which Chesterton admired) and so, despite its length, it can be considered yet another salvo in the continuous battle Chesterton conducted, more or less single-handed, with his time. Some have considered it his most important book. And there is the posthumously published *Autobiography* (1936) which is among his best books.

When he was received into the Church, thirteen years before he died, it was natural that the Church should use, and over-use, so notable a convert. His explicitly Roman Catholic work often suffers both from a too great sense of responsibility towards his subject, which is a form of humility, and a new sense of exclusiveness; for the first time he can sound too aggressively flippant to his opponents. The faltering in tone is only occasional but in a man whose tone is normally so exceptionally agreeable, it is audible. Also, his explicit Christianity, his view of 'Christendom' as a continuing imaginative and spiritual entity, could lead him not so much into mistakes as into mistaken emphases. He undervalued Oriental mysticism, for example. He was much less than fair to the influence of Islam and, notoriously, was led to make some (but not many) regrettable remarks about the Jews.

The violent language mentioned earlier, after the death of his brother, was addressed to Jewish politicians and financiers such as Godfrey Isaacs, whom his brother had attacked. In an Open Letter to Sir Rufus Isaacs (by this time Lord Reading) he talks of the Isaacs 'tribe' and says: 'You are far more unhappy than I; for your brother is still alive.' Grief had unbalanced him, but he detested Imperialism and Capitalism anyway, and suspected some highly placed Jews (it is important to remember that his criticisms were aimed at men of great power) of manipulating both of these to their own, supra-national ends. He did not believe they could be loyal to his concept of Christendom because they did not belong to that tradition. He was not a racialist, he disbelieved in the existence of continuing racial characteristics in the displaced and frequently says so, scornful of the 'Teutonism' that was becoming fashionable in pre-War England. He did not object to these men on grounds of race but because of what he suspected about their personal behaviour, which he thought derived from their religion, whether they were religious or not. He was a Zionist because he believed a man could only be loyal to his own tradition. That Jews could become loyal Englishmen he appears not to have believed. If he did not do so he was wrong, but so would we be wrong, with our special and terrible hindsight, if we condemn him too much. Such views were a distasteful component of the atmosphere of his time and are perhaps the only example of Chesterton, consciously or unconsciously, breathing that atmosphere without caution. There are those who blame this on loyalty to his brother, and on the influence of his friend Belloc. But Chesterton can bear the responsibility himself. He early detected the threat of Hitler, and warned against him when hardly anyone else had noticed. He continued to do so when the Government began to parley with the Nazis. To say that some of Chesterton's best friends were Jews will only raise a sad smile, but it was true; however, financiers, and above all non-Christian financiers, clearly worried him to distraction.

But what is most remarkable about Chesterton, considering what a battler he was, is not his enmity but his lack of it. It is astonishing how few enemies he made, if he made any. It was always to be seen that it was the idea he hated, not the man. The man he enjoyed; and, if he could, liked. The more one reads Chesterton, and thinks about him, the more it becomes clear, quite apart from his views, how extraordinarily 'Christian' he was, in the everyday use of that word. He approaches most people's idea of a good man. His brother made many enemies, so did Belloc, who gloried in doing so: 'There is something sundering about Hilary's quarrels,' Chesterton was heard to murmur, sadly. His were not like that. Indeed, Belloc regarded this as a weakness in him, and was wrong, because Chesterton's lack of interest in drawing an individual's blood on particular issues keeps the issues clearer, makes them remain of interest to us.

Those he argued with remained his friends. H. G. Wells, hearing after his death of some secret kindness Chesterton had done him, said he was not surprised: 'I have never known a man so steadily true to form as GKC.' People who worked for him loved him, and stayed. His final secretary, Dorothy Collins, became almost an adopted daughter (the Chestertons, who loved children, were able to have none of their own) and she has devoted the rest of her life to making well-chosen collections of his scattered pieces, which have been of great use to this selection.

There remains the popular notion that Chesterton was never quite as good as he could have been, 'a genius who wrote no masterpiece'. It is impossible to argue with such a view, whether true or not. If it is indeed a 'popular notion', Chesterton would say its popularity guaranteed that it contained a truth, as it might, for all he cared. 'I have no feeling for immortality,' he said, as a young man. 'I don't care for anything except to be in the present stress of life as it is. I would rather live now and die, from an artistic point of view, than keep aloof and write things that will remain in the world hundreds of years after my death. What I say is subject to some modification. It so happens that I couldn't be immortal: but if I could, I shouldn't want to be. What I value in my own work is what I may succeed in striking out of others.'

It is true he never wrote an unflawed novel, if such a thing exists. *The Man Who Was Thursday* comes nearest to being a conventional story, a sequence of events leading to a predetermined end. But the sustained extravagance of *The Napoleon of Notting Hill*, the exuberance (tinged with sadness) of *The Flying Inn*, the clowning of *Manalive*, the ingenuity of *The Ball and the Cross* all seem parts of a larger whole. They are not masterpieces but pieces of a master.

Likewise, it is true that great books like *Orthodoxy, A Short History of England, St Thomas Aquinas,* can give the impression of being collections of essays attached to a theme; closely attached, but not an obviously developing sequence. Yet it has been found, when trying to separate one chapter from another for [a] selection, that it is difficult to do so without loss; there is an inner connection.

This sense of the indivisibility of Chesterton grows. Thus, though it might be argued that the emphasis is on Chesterton the essayist and controversialist, rather than the inventor of fictions, it can also be suggested that his novels were as much a part of the continuing discussion he conducted with the world as his more explicitly argued pieces. It is remembered that he said if he had not written *The Napoleon of Notting Hill* he could not have gone on writing. He had something to say in it, a point to make; he dressed this up in a fabulous narrative but the narrative was not the reason for the writing. In fact it is the narrative of his novels which usually lets them down. The end of that novel, which is a battle with real swords in which nobody gets really hurt, is the only silly thing in it, and dates it badly. Such an account of a battle could not have been written after 1918.

But always, in poems and novels and everywhere else, it is Chesterton the teacher, the helper, at work. His attitude to writing was the same as Dr Johnson's, and as serious: he wrote in order to be useful, and did so from the beginning. In this he never changed . . . For Chesterton did not 'develop', in the ordinary sense. He sprang into the arena with enough in his head to last a lifetime, and all he saw confirmed his early insight. In fact he was always saying the same thing, or things. Most writers do this, but in Chesterton the surprise is in the continually fresh way he manages to say them, and this has to do with his sense of the inexhaustible importance of what he has to say.

As a young man he exploded into the literary life like a display of fireworks, and if there is a 'development' (though he could often erupt again) it is in the direction of self-limitation, as though tired of pyrotechnics, aware they have dazzled too much and too many, aware now that everything must be spelled out again, without impatience but more slowly, more plainly, as though to the lazy ones at the back of the class—for they too must be helped to pass their Final Examination. He dims his lamp to stop it shining in their eyes. Evelyn Waugh is said to have wished he could re-write *The Everlasting Man,* because its content was so good, in order to take out the exasperating Chesterton mannerisms. But it is clear in that comparatively stately book that Chesterton was trying to do so himself. This is perhaps why many like it best of all, it is so obviously serious. But, though now attempting to bring everything together, he is saying in it few things he has not said elsewhere quicker and more brilliantly.

As examples of this flattening and slowing-up of the style (which must have been conscious because, as has been said, he continued to be able to speed up and dazzle when he wanted to) it might be instructive to put two passages, chosen more or less at random, side by side; one from an early book, *Orthodoxy* (1908), and one from *The Everlasting Man* (1925).

Imagination does not breed insanity. Exactly what does breed insanity is reason. Poets do not go mad; but chess-players do. Mathematicians go mad, and cashiers; but creative artists very seldom. I am not, as will be seen, in any sense attacking logic: I only say that this danger does lie in logic, not in imagination. Artistic paternity is as wholesome as physical paternity. Moreover it is worthy of remark that when a poet really was morbid it was commonly because he had some weak spot of rationality on his brain. Poe, for instance, really was morbid: not because he was poetical, but because he was specially analytical. Even chess was too poetical for him; he disliked chess because it was full of knights and castles, like a poem. He avowedly preferred the black discs of draughts, because they were more like the black dots on a diagram. Perhaps the strongest case of all is this: that only one great English poet went mad, Cowper. And he was definitely driven mad by logic, by the ugly and alien logic of predestination. Poetry was not the disease, but the medicine; poetry partly kept him in health. He could sometimes forget the red and thirsty hell to which his hideous necessitarianism dragged him, among the wide waters and the white flat lilies of the Ouse. He was damned by John Calvin; he was almost saved by John Gilpin.

That is the earlier passage. Here is the later one:

In considering the elements of pagan humanity, we must begin by an attempt to describe the indescribable. Many get over the difficulty of describing it by the expedient of denying it, or at least ignoring it; but the whole point of it is that it was something that was never quite eliminated even when it was ignored. They are obsessed by their evolutionary monomania that every great thing grows from a seed, or something smaller than itself. They seem to forget that every seed comes from a tree, or from something larger than itself. Now there is very good ground for guessing that religion did not originally come from some detail that was forgotten, because it was too small to be traced. Much more probably it was an idea that was abandoned because it was too large to be managed.

Both passages are recognizably by the same writer. The second is more obviously reflective, whereas the first grabs hold of the reader as he hurries past. The later passage is grave with responsibility, the subject is of consequence and the reader is presumed ready to listen.

In the earlier, Chesterton is aware he has to startle in order to be heard at all. As a result the antitheses come fast on each other, and there is clear delight in the play of his own mind, which leaps from juxtaposition to juxtaposition, with dabs of colour on the way ('the white flat lilies of the Ouse'), until arriving, with apparent delight, at the brilliant and wholly successful pairing of John Calvin and John Gilpin, an insight of such appositeness it would keep most men aglow with self-satisfaction for a lifetime. The slow, pondering, abstract movement of the later passage makes impossible such sudden leaps of the imagination. Chesterton, as has been said, holds himself in check.

Of course this is unfair, the passages are from different books with somewhat different intentions. But enough will have been said to explain why the present [writer] is more excited (though there are exceptions) by the earlier Chesterton.

Nevertheless, the life's work seems more of a whole than is the case with most writers. If Chesterton was 'a genius who never wrote a masterpiece' a sense of masterpiece persists, though it is difficult to point at a book and confidently say, 'There, throughout, he is at his best.' Early in his career he was accused by outraged publishers of villainously misquoting Browning. He replied, rather grandly, 'I quote from memory both by temper and on principle. That is what literature is for; it ought to be a part of the man.' Perhaps that is where the impression of a masterpiece lies, in the part of the man that is his work. As his 'critic of politics and literature', looking for a hero of the Great War, could only point to the mob, so perhaps, looking for a masterpiece, we could point to the mob Ezra Pound called Chesterton.

A. N. WILSON
Glimpses of Chesterton

G. K. CHESTERTON had a multiplicity of talents; not least among them were facile draughtsmanship, and the almost invariable capacity, when throwing buns in the air, to catch them in his mouth. His drawings, like his handwriting, are playful, idiosyncratic and very slightly mad.

Chesterton began his grown-up life as an art student, vaguely attracted to the Swinburnian decadence of the 1890s (he wrote the best parody of Swinburne—it isn't in this book). He also liked the occult rubbish which appealed to the young Yeats and Arthur Machen.

When he emerged from that phase, he took up with the no less fantastical pursuit of politics and journalism, and this period, when he wore strange knickerbockers, large cloaks and wide-awake hats, saw no diminution in his love of the grotesque.

A taste for GKC must involve at least a tolerance of this side of his humour. I love the man who could say that 'lying in bed would be an altogether perfect and supreme experience if only one had a coloured pencil long enough to draw on the ceiling'. I love his essay on the advantages of having one leg:

> The poetry of art is in beholding the single tower; the poetry of nature is in seeing a single tree; the poetry of love in following the single woman; the poetry of religion in worshipping the single star. And so, in the same pensive lucidity I find the poetry of all human anatomy in standing on a single leg. To express complete and perfect leggishness the leg must stand in sublime isolation like the tower in the wilderness.

Another side of Chesterton which I find attractive is his poetic lyricism, best represented in the narrative poems 'Lepanto' and 'The Ballad of the White Horse'. And a third thing is the short stories, particularly those which feature Father Brown.

When Chesterton accepts the restraints of storytelling, he plays down his tendency to be a windbag, paradoxically playing with his famous ideas. 'I will not say that this story is true; because, as you will soon see, it is all truth and no story,' as one of his essays ('The Secret of a Train') begins.

I am not suggesting that Chesterton was an unwise man; nor am I denying that he was an *anima naturaliter Christiana*, who frequently wrote devastingly true things. I would vehemently deny Mr Kavanagh's surprising assertion that the author of *Orthodoxy* was 'hardly a Christian at all.' But at the same time I would say that we do Chesterton an injustice if we take him too seriously. He was too good a journalist (and I would say too good a Christian) to allow himself to be turned into a sage; and by treating him as a sage, we lose a lot of his essentially playful flavour.

Still, *chacun à son Chesterton*. Any excuse to re-read Chesterton, even to re-read the duller bits, is welcome. Chesterton being dull is livelier than most men's best.

ALLAN MASSIE
The Master Writer beneath the Card

CHESTERTON has suffered from the scrappiness of publication (all those little books of essays and cheap editions); from the English distrust of intellect, and tendency to prefer Teutonic muddiness to Latin lucidity.

He resembles Johnson in more ways than the obvious damage done to his reputation as a writer by his standing as a 'character'. Both were professional writers who knew that they must please the reader if they were to influence him (only solemn and tenured blockheads can afford not to do so). Both were moralists. Chesterton was proud to be a journalist, if only because he knew that more people read newspapers than books. His preferred form was the essay because it is by nature delightful and didactic. He used paradox as Johnson employed antithesis.

Both men were exponents of uncommon sense and exposers of common nonsense. Both distrusted novelty, because both respected the past. There is, for instance, an essay in which Chesterton considers Nietzsche's theory of the Superman. It is sometimes supposed, he says, that the great writers of the past did not think of such ideas. The notion is false. They thought of them; only they did not think much of them. Shakespeare knew all about the Superman; he placed Nietzsche's arguments in the mouth of Richard III; enough said.

Like Johnson Chesterton is simultaneously empirical and dogmatic. He is ready to trust to his experience of the world and learn from his observation of mankind; but he tests his experience against the given body of dogma that is the Christian faith as represented by the Church.

He was a radical in politics, but his radicalism was based on a profound conservatism. He did not want to uproot society, he sought to prune it of excrescent briars and restore it to a more seemly shape. He distrusted modern individualism. He would have agreed with Burke in being 'afraid to put men to live and trade each on his private stock of reason'. He equally distrusted socialism, with its elevation of the abstract and irresponsible State.

His life was dominated by two beliefs; the rightness of the Christian faith and the wrongness of modern society. His optimism was based on his certainty not that things would come right in the end, but that they had been right in the beginning, and that life was therefore fundamentally good, to be enjoyed and praised as the work of a benevolent Creator. He had no time for the woolly thinking which claims that all religions are essentially the same.

Here too he is at odds with modernists, as in his view of history. That was founded on the evidence before him of the selfishness of capitalism and irresponsibility of socialism. His *Short History of England* may not give us the Middle Ages as we get them in Stubbs, Tout or Maitland; but there is something truthful there which is usually kept out of history books. There is an understanding of the medieval mind and what it aspired to.

ROY HATTERSLEY
Dragon-maker

GILBERT KEITH CHESTERTON was an addicted essayist. Of course, he was also a poet, a polemicist, a biographer, a historian, a perpetual chauvinist, an occasional antisemite, a professional journalist and an amateur theologian. But, his poetry aside, whatever he wrote turned out to be an essay or series of essays—narrative essays called short stories, biographical essays put together and called biographies, historical essays bound in the same volume and called histories. Each chapter of his longer books is complete, well rounded, discursive, didactic. Often he allows the essayist's personality to intrude in places where the historian or literary critic would never permit it to appear. Most significantly of all, he understands the importance of carrying his readers along on the strength of his language. Devotees of Father Brown will insist that the ingenuity of the plots and the subtlety of the characterizations hold them entranced. But when, in the mystery of *The Queer Feet*, we read that 'In the heart of a plutocracy, tradesmen become cunning enough to be more fastidious than their customers', we recognize an author who struck a proper balance between style and substance—the essential mark of the true essayist.

Chesterton was also a man of generous disposition, and usually his natural happiness was reflected in what he wrote—another essential attribute of the successful essayist. 'He seems', writes P. J. Kavanagh in his introduction to *The Bodley Head G. K. Chesterton*, 'always to be addressing a large and friendly audience.' That is something of an overstatement. But we ought not to complain when a Chesterton devotee exaggerates his idol's good nature. For when we read Chesterton's judgements on his own literary heroes, we applaud his affectionate reluctance to find fault. There is, in his assessment 'Browning as a Literary Artist', a passage in which the subject of the biographical essays is compared to Wordsworth and Shelley:

> The *Ode on the Intimations of Immortality* is a perfectly normal and traditional ode, and *Prometheus Unbound* is a perfectly genuine and traditional Greek lyrical drama. But if we study Browning honestly, nothing will strike us more than that he really created a large number of quite novel and quite admirable artistic forms.

Rivals for places in the pantheon of English poetry having been thus admonished for their comparative lack of originality, the panegyric sweeps on to applaud the personal as well as the poetic qualities of both the Brownings. I hope that it is more than my own admiration for that extraordinary couple which attracts me to Chesterton's account of their marriage. Even readers who are moved by neither the mystic wonder of Robert's poetry nor the strength of Elizabeth's character and the quality of her sonnets must be recruited to their cause by the succinct brilliance with which Chesterton described the liberation of the sick daughter from her father's domination. After a slightly self-conscious sub-Freudian analysis of their relationship he moves on to the stronger ground: 'she took a much more cheerful view of death than her father did of life'.

Not that Chesterton is invariably, and therefore tediously, in favour of everything and everybody. He asserts that 'what attracts Mr Kipling to militarism is not the idea of courage but the idea of discipline', and he clearly disapproves of such an attraction. He classifies the aphorisms of Oscar Wilde with ruthless accuracy into the work of 'the true humorist. . . . the charlatan. . . . the fine philosopher and . . . the tired quack'. The notion that 'Good intentions are invariably ungrammatical' he dismisses as 'tame trash'. Chesterton is instinctively against the popular and the fashionable, because he is on the side of the common man. Caught by 'two excited policemen' while 'throwing a big Swedish knife at a tree' he is gratified when 'the leading constable became so genial and complimentary that he ended up by representing himself as a reader of my work'. But he wonders 'how he would have got on if he had not been the guest at a big house'. It is the possession of such instincts which makes it easy to forget his excesses.

G. K. Chesterton overdid it—especially in support of favoured causes. In the excellent introduction we are reminded of the assaults on the Liberal Party establishment which he made on behalf of his dead brother, Cecil, who had sought to expose the corruption of the Marconi affair. In the text we are given an example of absurd determination to defend the period of medieval history when the Catholic Church occupied the position in society to which he would have liked it restored. In his *Short History of England*—written as the kindly light was leading him towards Rome—he observed in apparent seriousness that

> torture so far from being peculiarly medieval was copied from pagan Rome and its most natural political science. Its application to others besides slaves was really part of the slow medieval extinction of slavery. Torture indeed is a logical thing common to states innocent of fanaticism.

From his more sensible writing we know that Chesterton could not have tortured a worm—indeed, rather, if its shape, size, and religion had attracted him, he might have announced that in truth it was a dragon. Overstatement was one of Chesterton's greatest pleasures.

ACKNOWLEDGEMENTS

I wish to thank the staffs of the Library of the Universitaire Faculteiten Sint-Ignatius te Antwerpen (Universiteit Antwerpen) and of the Kent County Library branch in Canterbury for their assistance.

The editor and publishers wish to thank the following for permission to reproduce copyright material:

Kingsley Amis: 'An Unreal Policeman' from *What Became of Jane Austen* (Jonathan Cape, 1970), reprinted by permission of Intercontinental Literary Agency; 'The Poet and the Lunatics' reprinted from the *New Statesman*, 26 February 1971, by permission; 'Four Fluent Fellows' from *G. K. Chesterton. A Centenary Appraisal*, reprinted by permission of Intercontinental Literary Agency and the estate of the late Mr J. J. Sullivan.

Anonymous: from the *Times Literary Supplement*, 16 June 1950. Reprinted by permission of Times Newspapers Ltd.

Michael Asquith: Reprinted from the *Listener*, 6 March 1952, by permission of the editor and Lady Oxford and Asquith.

John Atkins: from *The British Spy Novel* (1984). Reprinted by permission of John Calder Ltd.

W. H. Auden: 'Chesterton's Non-Fictional Prose' from *G. K. Chesterton. A Selection from his Non-Fictional Prose*, reprinted by permission of Faber & Faber Ltd.; 'The Gift of Wonder' from *G. K. Chesterton. A Centenary Appraisal*, reprinted by permission of the estate of the late Mr J. J. Sullivan.

Hilaire Belloc: from *The Place of Gilbert Chesterton in English Letters*. Reprinted by permission of A. P. Watt Ltd.

Bernard Bergonzi: from *Critical Quarterly*, Spring 1959. Reprinted by permission of the editors and the author.

Ian Boyd: from *New Blackfriars*, June 1975. Reprinted by permission of the editor and the author.

Neville Braybrooke: from *John O'London's Weekly*, 8 February 1962. Reprinted by permission of the author.

Patrick Braybrooke: from *The Dickensian*, March 1945. Reprinted by permission.

Denis Brogan: from the *Spectator*, 25 April 1969. Reprinted by permission.

Ivor Brown: from the *Observer*, 16 April 1944. Reprinted by permission.

Anthony Burgess: from *G. K. Chesterton: Autobiography*. Reprinted by permission of the Gabriale Pantucci agency and of the author.

John Carey: from the *Sunday Times*. Reprinted by permission of Times Newspapers Ltd.

R. C. Churchill: from the *Contemporary Review*. Reprinted by permission of the editor and the author.

Herbert Evans: from *The Le Bas Prize Essay* 1938. Reprinted by permission of Cambridge University Press and the author.

Benny Green: 'Father of Father Brown' from the *Sunday Telegraph*, reprinted by permission; 'Defender of the Faith' from the *Spectator*, 9 April 1977. Reprinted by permission of the editor.

ACKNOWLEDGEMENTS

Graham Greene: from the *Spectator*, 21 April 1944. Reprinted by permission of the editor and the author.

John Gross: from *The Rise and Fall of the Man of Letters*. Reprinted by permission of Weidenfeld (Publishers) Ltd.

Kenneth M. Hamilton: from *The Dalhousie Review*, Autumn 1951. Reprinted by permission of the editor and the author.

Robert Hamilton: from the *Quarterly Review*, October 1967. Reprinted by permission of John Murray (Publishers) Ltd. and the author.

Roy Hattersley: from the *Times Literary Supplement*, 12 July 1985. Reprinted by permission of Times Newspapers Ltd., and the author.

Leo A. Hetzler: from *The Chesterton Review*. Reprinted by permission of the editor and the author.

Christopher Hollis: 'Chesterton's Paradoxes' from *Time and Tide*, 25-31 July 1963, reprinted by permission of the author; 'G. K. Chesterton' from *New Knowledge*, Vol. 8, No. 12, reprinted by permission of the editor and the author.

Richard Ingrams: from the *Daily Telegraph*, 24 May 1974. Reprinted by permission.

P. J. Kavanagh: from *The Bodley Head G. K. Chesterton*. Reprinted by permission of the Bodley Head.

Hugh Kenner: from *Paradox in Chesterton*. Reprinted by permission of Sheed and Ward Inc. and the author.

Hugh Kingsmill: from the *New English Review Magazine* (1948) and *The Best of Hugh Kingsmill* (Gollancz). Reprinted by permission of the executor of the estate of Hugh Kingsmill Lunn.

Ronald Knox: 'G. K. Chesterton: The Man and his Work' reprinted by permission of the *Listener* and the executor of the estate of R. A. Knox; 'Chesterton's Father Brown' from *Father Brown. Selected Stories*, reprinted by permission of Oxford University Press and the executor of the estate of R. A. Knox.

Bernard Levin: 'Pantomime Horse' from the *Spectator*, 5 December 1958. Reprinted by permission; 'The Case for Chesterton' from the *Observer*, 26 May 1974, reprinted by permission.

C. S. Lewis: from *Time and Tide*, 9 November 1946. Reprinted by permission.

David Lodge: from *The Month*, May 1974, reprinted by permission of the editor and the author.

H. Marshall McLuhan: 'G. K. Chesterton: A Practical Mystic' reprinted by permission of *The Dalhousie Review*, and the author; 'Where Chesterton Comes In' from H. Kenner, *Paradox in Chesterton*, reprinted by permission of Sheed and Ward Inc. and the author.

Michael Mason: Reprinted by permission of *Twentieth Century*, 1968, and the author.

Allan Massie: from *The Times*, 20 June 1985. Reprinted by permission of Times Newspapers Ltd.

Theodore Maynard: from *The Commonweal*, 15 October 1943. Reprinted by permission.

Malcolm Muggeridge: from the *New Statesman*, 23 August 1963, reprinted by permission.

Alfred Noyes: from the *Quarterly Review*, January 1953. Reprinted by permission of John Murray (Publishers) Ltd.

George Orwell: from *Collected Essays, Journalism, and Letters* (Secker, 1968). Reprinted by permission of A. M. Heath and the executors of the Orwell estate.

ACKNOWLEDGEMENTS

Herbert Palmer: from *Post-Victorian Poetry*. Reprinted by permission of J. M. Dent & Sons Ltd.

Hesketh Pearson: 'G. K. Chesterton' from the *Listener*, 28 June 1956, reprinted by permission of the editor and Mrs J. Pearson; 'Gilbert Keith Chesterton' from *Lives of the Wits*, reprinted by permission of William Heinemann Ltd. and Mrs J. Pearson.

R. G. G. Price: from *Punch*, 23 July 1958, reprinted with permission.

V. S. Pritchett: from the *New Statesman*, 7 June 1974, reprinted with permission.

John Raymond: from the *New Statesman and Nation*, 23 March 1957. Reprinted with permission.

Dorothy L. Sayers: from *G. K. Chesterton's 'The Surprise'*. Reprinted by permission of the executor of the estate of Miss D. L. Sayers.

Wilfrid Sheed: from *G. K. Chesterton: Essays and Poems*. Reprinted by permission of Penguin Books Ltd.

Lance Sieveking: from the *Listener*, 3 January 1957, reprinted by permission of the editor and Mrs Maisie Sieveking.

James Stephens: from the *Listener*, 17 October 1946. Reprinted by permission of the editor.

L. A. G. Strong: from *G. K. Chesterton: Wine, Water and Song*. Reprinted by permission of A. D. Peters & Co. Ltd.

Frank Swinnerton: from *The Georgian Literary Scene*. Reprinted by permission of J. M. Dent & Sons Ltd.

John Wain: from *Punch*, 4 April 1962, reprinted with permission.

Evelyn Waugh: from *The Commonweal*, 21 March 1947, reprinted with permission.

Katherine Whitehorn: from the *Observer*, 26 May 1974, reprinted with permission.

Garry Wills: 'Rhyme and Reason' from *Chesterton: Man and Mask*, and '*The Man Who Was Thursday*' from G. K. Chesterton's *The Man Who Was Thursday*, reprinted by permission of Sheed and Ward Inc.

A. N. Wilson: from the *Sunday Telegraph*, 16 June 1985, reprinted with permission.

D. B. Wyndham Lewis: from *G. K. Chesterton: An Anthology*. Reprinted by permission of Oxford University Press and A. D. Peters & Co. Ltd.

While every effort has been made to secure permission, we may have failed in a few cases to trace the copyright holder. We apologize for any apparent negligence.